T3-BNX-970

BIOLOGICAL AND MEDICAL PHYSICS,
BIOMEDICAL ENGINEERING

For further volumes:
http://www.springer.com/series/3740

BIOLOGICAL AND MEDICAL PHYSICS, BIOMEDICAL ENGINEERING

The fields of biological and medical physics and biomedical engineering are broad, multidisciplinary and dynamic. They lie at the crossroads of frontier research in physics, biology, chemistry, and medicine. The Biological and Medical Physics, Biomedical Engineering Series is intended to be comprehensive, covering a broad range of topics important to the study of the physical, chemical and biological sciences. Its goal is to provide scientists and engineers with textbooks, monographs, and reference works to address the growing need for information.

Books in the series emphasize established and emergent areas of science including molecular, membrane, and mathematical biophysics; photosynthetic energy harvesting and conversion; information processing; physical principles of genetics; sensory communications; automata networks, neural networks, and cellular automata. Equally important will be coverage of applied aspects of biological and medical physics and biomedical engineering such as molecular electronic components and devices, biosensors, medicine, imaging, physical principles of renewable energy production, advanced prostheses, and environmental control and engineering.

Rui Bernardes
José Cunha-Vaz
Editors

Optical Coherence Tomography

A Clinical and Technical Update

With 124 Figures

Editors
Rui Bernardes
José Cunha-Vaz
IBILI, Faculty of Medicine, University of Coimbra and AIBILI
Azinhaga Sta. Comba, Celas, 3000-548 Coimbra, Portugal

Biological and Medical Physics, Biomedical Engineering ISSN 1618-7210
ISBN 978-3-642-27409-1 ISBN 978-3-642-27410-7 (eBook)
DOI 10.1007/978-3-642-27410-7
Springer Heidelberg New York Dordrecht London

Library of Congress Control Number: 2012938448

Printed on acid-free paper

Springer is part of Springer Science+Business Media (www.springer.com)

Preface

This book has been written to provide readers with a set of complementary viewpoints and perspectives on what is probably the most impressive imaging technique currently available, optical coherence tomography (OCT). The speed of data acquisition, its noninvasive imaging, and its excellent image resolution make this technique distinct in any field of medicine.

Two different communities contributed to the contents of the book. Part I provides viewpoints on the impact that OCT has made on imaging the human retina in vivo from reputed ophthalmologists, its users. In Part II, leading experts in the field discuss the technique itself and describe developments of further potential. In addition, new applications are discussed that can open OCT to an even broader use in ophthalmology from the traditional structural application.

We hope that readers interested in this methodology find that this book combines both the medical and technical information required to understand the principles and the application of this notable technique that has completely modified the way information on the living human retina can be obtained.

We are very grateful to all of our colleagues who have contributed in making this book possible.

Coimbra, Portugal — Rui Bernardes
Coimbra, Portugal — José Cunha-Vaz

Contents

1 Diabetic Macular Edema 1
Conceição Lobo, Isabel Pires, and José Cunha-Vaz
1.1 Introduction 1
1.2 Pathophysiology of Retinal Edema 2
1.3 Extracellular Edema 2
1.4 Starling's Law 3
1.5 Incidence and Prevalence of DME 3
1.6 Clinical Evaluation of Macular Edema 4
1.7 Diagnosis and Classification 4
References 18

2 Ischemia 23
Suk Ho Byeon, Min Kim, and Oh Woong Kwon
2.1 Retinal Ischemia 23
2.2 Brief Anatomy of Retinal Circulation 24
2.3 Fluorescein Angiography (FA) and Optical Coherence Tomography (OCT): Anatomic Correlations 24
2.4 "External Fovea" in OCT 26
2.5 Detection of Ischemic Retinal Area: Comparisons Between Fluorescein Angiography and OCT 27
2.6 Pathologic Changes and Relevant OCT Findings in Ischemic Tissues 28
2.7 Prominent Middle Limiting Membrane 28
2.8 Retinal Artery Occlusion 29
2.9 Central Retinal Artery Occlusion 29
2.10 CRAO with Cilioretinal Artery Sparing 32
2.11 Cilioretinal Artery Occlusion 33
2.12 Branched Retinal Artery Occlusion 33
2.13 Ophthalmic Artery Occlusion 33
2.14 Cotton Wool Spots 35
2.15 Combined CRAO and CRVO 36

2.16 Ocular Ischemic Syndrome 37
2.17 Evaluation of Retinal Perfusion State on OCT 37
2.18 Central retinal vein occlusion (CRVO) 37
2.19 Central retinal vein occlusion 40
2.20 Diabetic Retinopathy 41
2.21 Diabetic Ischemic Maculopathy 41
2.22 Choroidal Ischemia 43
References 45

3 Optical Coherence Tomography and Visual Acuity: Photoreceptor Loss 51
Adzura Salam, Ute Ellen Kathrin Wolf-Schnurrbusch, and Sebastian Wolf
3.1 Introduction 52
3.2 Principle of SD-OCT 52
3.3 SD-OCT in Normal Eyes 53
3.4 SD-OCT in Retinal Diseases 53
3.5 Correlation Between Visual Acuity and Integrity of Photoreceptor in Retinal Diseases 54
3.5.1 Photoreceptor 54
3.5.2 Diabetic Macular Edema 54
3.5.3 Retinal Vein Occlusion 59
3.5.4 Retinal Artery Occlusion 62
3.5.5 Age-Related Macular Degeneration 65
3.5.6 Central Serous Chorioretinopathy 71
3.5.7 Retinitis Pigmentosa 71
3.5.8 Foveal Microstructure and Visual Acuity in Selected Surgical Retina Cases 75
3.6 Conclusion 81
3.7 Summary for Doctor 81
References 81

4 Optical Coherence Tomography Updates on Clinical and Technical Developments. Age-Related Macular Degeneration: Drusen and Geographic Atrophy 87
Monika Fleckenstein, Steffen Schmitz-Valckenberg, and Frank G. Holz
4.1 Introduction 87
4.2 Early AMD 88
4.2.1 In Vivo Visualization of Drusen and Pigmentary Changes by Spectral-Domain Optical Coherence Tomography 88
4.2.2 Quantification of Drusen Volume by Spectral-Domain Optical Coherence Tomography 91

4.3 Geographic Atrophy 93
4.3.1 Phenotypic Features of Atrophy and Identification of Predictive Markers by Spectral-Domain Optical Coherence Tomography 93
4.3.2 Correlation with Other Imaging Modalities 94
4.3.3 Quantification of Atrophy by Spectral-Domain Optical Coherence Tomography Imaging 100
4.4 Future Perspectives 102
References 103

5 Optical Coherence Tomography in Glaucoma 109
Fatmire Berisha, Esther M. Hoffmann, and Norbert Pfeiffer
5.1 Introduction 109
5.2 Imaging in Glaucoma 110
5.3 Time-Domain Optical Coherence Tomography in Glaucoma 111
5.4 Spectral Domain Optical Coherence Tomography in Glaucoma 112
5.5 Diagnostic Ability of Spectral Domain OCT in Glaucoma 117
5.6 Future Directions of Optical Coherence Tomography in Glaucoma 119
References 120

6 Anterior Segment OCT Imaging 125
Alexandre Denoyer, Antoine Labbé, and Christophe Baudouin
6.1 Introduction 125
6.2 Principles of Anterior Segment OCT 126
6.3 Clinical Applications 127
6.3.1 Corneal Thickness Assessment 127
6.3.2 Corneal Grafts 127
6.3.3 Corneal Refractive Surgery 128
6.4 Phakic Intraocular Lens 131
6.5 Glaucoma 132
6.6 Conclusion 137
References 137

7 Optical Coherence Tomography: A Concept Review 139
Pedro Serranho, António Miguel Morgado and Rui Bernardes
7.1 Introduction 139
7.2 The Genesis of OCT 141
7.3 The Physical Principle 142
7.4 Low-Coherence Light Interferometer 144
7.5 Time-Domain Optical Coherence Tomography 146
7.6 Spectral-Domain Optical Coherence Tomography 147
7.7 Swept-Source Optical Coherence Tomography 148
7.8 Polarization-Sensitive Optical Coherence Tomography 149

7.9 Full-Field Optical Coherence Tomography 150
7.10 Quantum Optical Coherence Tomography 151
7.11 Latest Developments and Future Directions 152
References 153

8 Evaluation of the Blood–Retinal Barrier with Optical Coherence Tomography 157
Rui Bernardes and José Cunha-Vaz
8.1 Introduction 157
8.2 Blood–Retinal Barrier 158
8.3 Clinical Evaluation of the Blood–Retinal Barrier 159
8.4 Quantitative Evaluation of the Blood–Retinal Barrier 159
8.4.1 Confocal Scanning Laser Ophthalmoscopy 160
8.5 Retinal Leakage Analyzer 161
8.6 Noninvasive Evaluation of Blood–Retinal Barrier with OCT 163
8.6.1 Data of Healthy Volunteers vs. Patient Data 163
8.6.2 Differences Within the Same Eye and Scan 164
8.6.3 Automatic Classification 168
8.7 Conclusions 172
References 173

9 Polarization Sensitivity 175
U. Schmidt-Erfurth, F. Schlanitz, M. Bolz, C. Vass, J. Lammer, C. Schütze, M. Pircher, E. Götzinger, B. Baumann, and C.K. Hitzenberger
9.1 Polarization-Sensitive Optical Coherence Tomography 175
9.1.1 Principles of Polarization-Sensitive OCT 176
9.1.2 Polarization Properties of Ocular Structures 179
9.1.3 Segmentation Algorithms 182
9.2 Imaging of the Eye with Polarization-Sensitive OCT 186
9.2.1 Age-Related Macular Degeneration 187
9.2.2 Glaucoma 193
9.2.3 Diabetic Maculopathy 196
9.3 Conclusion and Future Directions 198
References 199

10 Adaptive Optics in Ocular Optical Coherence Tomography 209
Enrique Josua Fernández and Pablo Artal
10.1 Introduction 209
10.2 The Eye as an Optical System and Adaptive Optics 211
10.3 Transverse Resolution in OCT 212
10.4 Effect of Aberration Correction in Contrast and Resolution 216
10.5 Correction of Ocular Monochromatic Aberrations in OCT 219
10.5.1 Time Domain 219
10.5.2 Frequency Domain 221

10.6 Aberration Pancorrection in UHR OCT: Toward Isotropic and Subcellular Resolution ... 226
10.6.1 Pancorrection: Full Aberration Correction ... 226
10.6.2 UHR OCT and Pancorrection of Aberrations ... 229
10.7 Summary ... 231
References ... 232

11 The SL SCAN-1: Fourier Domain Optical Coherence Tomography Integrated into a Slit Lamp ... 237
F.D. Verbraak and M. Stehouwer
11.1 Introduction ... 237
11.2 The SL SCAN-1 Device ... 238
11.3 Scanning the Anterior and the Posterior Segment ... 239
11.4 Scanning the Periphery with the SL SCAN ... 239
11.5 Scanning Through a Three-Mirror Contact Lens with the SL SCAN ... 239
11.6 Scanning with the Fundus Viewer, an Add-On Lens ... 240
11.7 Examples of Subjects Studied ... 241
11.8 Results ... 241
11.8.1 Scans of the Posterior Pole Obtained Through a Handheld Lens (Posterior Mode) ... 242
11.8.2 Scans of the Peripheral Retina Obtained Through a Handheld Lens or Three-Mirror Lens (Posterior Mode) ... 242
11.8.3 Scans of the Anterior Segment Obtained Directly (Anterior Mode) and Through a Three-Mirror Lens (Posterior Mode) ... 243
11.8.4 Thickness Measurements of the Posterior Segment with the Fundus Viewer and FastMap (Posterior Mode) ... 244
11.9 Discussion ... 246
References ... 248

Index ... 251

Contributors

Pablo Artal Laboratorio de Óptica, Centro de Investigación en Óptica y Nanofísica (CiOyN), Universidad de Múrcia, Campus de Espinardo, E-30071 Murcia, Spain, pablo@um.es

Christophe Baudouin Department of Ophthalmology, Versailles-Saint-Quentin-en-Yvelines Medical School, Centre National d'Ophtalmologie des Quinze-Vingts, Paris, France, chrbaudouin@aol.com

UMRS 968, INSERM UPMC, Institut de la Vision; UMR 7210, CNRS, Paris, France

Department of Ophthalmology, Ambroise Paré Hospital, AP-HP, Boulogne and University of Versailles-Saint Quentin en Yvelines, France

B. Baumann Center for Medical Physics and Biomedical Engineering, Medical University Vienna, Währinger Gürtel 18-20, 1090 Vienna, Austria

Fatmire Berisha Universtity Medical Center Mainz, Johannes Gutenberg University, Lagenbeckstr. 1, 55101 Mainz, Germany, fatmire.berisha@gmx.de

Rui Bernardes IBILI-Institute for Biomedical Research in Light and Image, Faculty of Medicine, University of Coimbra, Azinhaga de Santa Comba, Celas, 3000-548 Coimbra, Portugal, rcb@aibili.pt

AIBILI-Association for Innovation and Biomedical Research on Light and Image, Azinhaga de Santa Comba, Celas, 3000-548 Coimbra, Portugal

M. Bolz Department of Ophthalmology and Optometry, Medical University Vienna, Währinger Gürtel 18-20, 1090 Vienna, Austria

Suk Ho Byeon Yonsei University Health System (YUHS), 50 Yonsei—Ro, Seodaemun-Gu, Seoul 120-752, Korea

José Cunha-Vaz AIBILI-Association for Innovation and Biomedical Research on Light and Image, Azinhaga Santa Comba, 3000-548 Coimbra, Portugal, cunhavaz@aibili.pt

Alexandre Denoyer Department of Ophthalmology III, Centre National d'Ophtalmologie des Quinze-Vingts, Paris, France

UMRS 968, INSERM UPMC, Institut de la Vision; UMR 7210, CNRS, Paris, France

Enrique Josua Fernández Laboratorio de Óptica, Centro de Investigación en Óptica y Nanofísica (CiOyN), Universidad de Múrcia, Campus de Espinardo, E-30071 Murcia, Spain, enriquej@um.es

Monika Fleckenstein Department of Ophthalmology, University of Bonn, Ernst Abbe Strasse 2, Bonn D53127, Germany, monika.fleckenstein@ukb.uni-bonn.de

E. Götzinger Center for Medical Physics and Biomedical Engineering, Medical University Vienna, Währinger Gürtel 18-20, 1090 Vienna, Austria

C.K. Hitzenberger Center for Medical Physics and Biomedical Engineering, Medical University Vienna, Währinger Gürtel 18-20, 1090 Vienna, Austria

Esther M. Hoffmann Universtity Medical Center Mainz, Johannes Gutenberg University, Lagenbeckstr. 1, 55101 Mainz, Germany

Frank G. Holz Department of Ophthalmology, University of Bonn, Ernst Abbe Strasse 2, Bonn D53127, Germany, frank.holz@ukb.uni-bonn.de

Min Kim Yonsei University Health System (YUHS), 50 Yonsei—Ro, Seodaemun-Gu, Seoul 120-752, Korea

Oh Woong Kwon Yonsei University Health System (YUHS), 50 Yonsei—Ro, Seodaemun-Gu, Seoul 120-752, Korea, owkwon0301@yuhs.ac

Antoine Labbé Department of Ophthalmology III, Centre National d'Ophtalmologie des Quinze-Vingts, Paris, France

UMRS 968, INSERM UPMC, Institut de la Vision; UMR 7210, CNRS, Paris, France

Department of Ophthalmology, Ambroise Paré Hospital, AP-HP, Boulogne and University of Versailles-Saint Quentin en Yvelines, France

J. Lammer Department of Ophthalmology and Optometry, Medical University Vienna, Währinger Gürtel 18-20, 1090 Vienna, Austria

Conceição Lobo AIBILI-Association for Innovation and Biomedical Research on Light and Image Azinhaga Santa Comba, 3000-548 Coimbra, Portugal, clobo@aibili.pt

António Miguel Morgado IBILI-Institute for Biomedical Research in Light and Image, Faculty of Medicine, University of Coimbra, Azinhaga de Santa Comba, Celas, 3000-548 Coimbra, Portugal, mmorgado@ibili.uc.pt

Department of Physics, University of Coimbra, Portugal

Norbert Pfeiffer Universtity Medical Center Mainz, Johannes Gutenberg University, Lagenbeckstr. 1, 55101 Mainz, Germany, norbert.pfeiffer@unimedizin-mainz.de

M. Pircher Center for Medical Physics and Biomedical Engineering, Medical University Vienna, Währinger Gürtel 18-20, 1090 Vienna, Austria

Isabel Pires AIBILI-Association for Innovation and Biomedical Research on Light and Image, Azinhaga Santa Comba, 3000-548 Coimbra, Portugal

Adzura Salam Department of Ophthalmology, Kulliyyah of Medicine, International Islamic University Malaysia, 25200 Bandar Indera Mahkota, Pahang, Malaysia

F. Schlanitz Department of Ophthalmology and Optometry, Medical University Vienna, Währinger Gürtel 18-20, 1090 Vienna, Austria

U. Schmidt-Erfurth Department of Ophthalmology and Optometry, Medical University Vienna, Währinger Gürtel 18-20, 1090 Vienna, Austria, ursula.schmidt-erfurth@meduniwien.ac.at

Steffen Schmitz-Valckenberg Department of Ophthalmology, University of Bonn, Ernst Abbe Strasse 2, Bonn D53127, Germany, steffan.schmitz-valckenberg@ukb.uni-bonn.de

C. Schütze Department of Ophthalmology and Optometry, Medical University Vienna, Währinger Gürtel 18-20, 1090 Vienna, Austria

Pedro Serranho Department of Science and Technology, Open University, Campus TagusPark, Av. Dr. Jacques Delors, 2740-122 Porto Salvo, Oeiras, Portugal

IBILI-Institute for Biomedical Research in Light and Image, Faculty of Medicine, University of Coimbra, Portugal, pserranho@uab.pt

M. Stehouwer Department of Ophthalmology, Academisch Medisch Centrum, Postbus 22660, 1100 DD Amsterdam

C. Vass Department of Ophthalmology and Optometry, Medical University Vienna, Währinger Gürtel 18-20, 1090 Vienna, Austria

F.D. Verbraak Department of Ophthalmology, Academisch Medisch Centrum, Postbus 22660, 1100 DD Amsterdam, f.d.verbraak@amc.nl

Department of Biomedical Engineering and Physics, Academisch Medisch Centrum, Postbus 22660, 1100 DD Amsterdam

Sebastian Wolf Universitätsklinik für Augenheilkunde, Inselspital, University Bern, 3010 Bern, Switzerland, sebastian.wolf@insel.ch

Ute Ellen Kathrin Wolf-Schnurrbusch Universitätsklinik für Augenheilkunde, Inselspital, University Bern, 3010 Bern, Switzerland

Chapter 1
Diabetic Macular Edema

Conceição Lobo, Isabel Pires, and José Cunha-Vaz

Abstract The optical coherence tomography (OCT), a noninvasive and noncontact diagnostic method, was introduced in 1995 for imaging macular diseases.

In diabetic macular edema (DME), OCT scans show hyporeflectivity, due to intraretinal and/or subretinal fluid accumulation, related to inner and/or outer blood–retinal barrier breakdown. OCT tomograms may also reveal the presence of hard exudates, as hyperreflective spots with a shadow, in the outer retinal layers, among others.

In conclusion, OCT is a particularly valuable diagnostic tool in DME, helpful both in the diagnosis and follow-up procedure.

1.1 Introduction

The World Health Organization (WHO) estimates that more than 180 million people worldwide have diabetes, and this number is expected to increase and to rise to epidemic proportions within the next 20 years [1]. Diabetic retinopathy, one of the most frequent complications of diabetes, remains a major public health problem with significant socioeconomic implications, affecting approximately 50% of diabetic subjects, and remains the leading cause of blindness in working-age populations of industrialized countries.

Diabetic macular edema (DME) is the largest cause of visual acuity loss in diabetes [2]. It affects central vision from the early stages of retinopathy, and it is the

C. Lobo (✉)
AIBILI-Association for Innovation and Biomedical Research on Light and Image,
Azinhaga Santa Comba, 3000-548 Coimbra, Portugal
e-mail: clobo@aibili.pt

R. Bernardes and J. Cunha-Vaz (eds.), *Optical Coherence Tomography*,
Biological and Medical Physics, Biomedical Engineering,
DOI 10.1007/978-3-642-27410-7_1,

most frequent sight-threatening complication of diabetic retinopathy, particularly in older type 2 diabetic patients. Its role in the process of vision loss in diabetic patients and its occurrence in the evolution of the retinal disease are being increasingly recognized.

DME leads to distortion of visual images and may cause a significant decrease in visual acuity even in the absence of severe retinopathy.

Although macular edema is a common and characteristic complication of diabetic retinopathy and shows apparent association with the systemic metabolic alterations of diabetes, it does not necessarily fit the regular course of diabetic retinopathy progression. It may occur at any stage of diabetic retinopathy, whether nonproliferative, moderate, or severe, or even at the more advanced stages of the retinopathy [3].

These facts are particularly important regarding the relevance of DME in the natural history of diabetic retinopathy. Diabetic retinopathy often progresses for many years without vision loss, making it sometimes challenging for the physician to counsel the patient for the need for treatment when progression occurs.

1.2 Pathophysiology of Retinal Edema

Retinal edema occurs when there is an increase of water in the retinal tissue, resulting in an increase in its thickness. This increase in water content of the retinal tissue may be initially intracellular (cytotoxic edema) or extracellular (vasogenic edema) [4]. In DME, extracellular edema resulting from breakdown of the blood–retinal barrier (BRB) is generally present.

In the retina, there is a specialized structure, the BRB, that regulates fluid movements into and out of the retinal tissue. If the BRB breaks down, as occurs in diabetes, it results in an "open BRB", which enables increased movements of fluids and molecules into the retina, with extracellular accumulation of fluid and deposition of macromolecules.

1.3 Extracellular Edema

Extracellular edema is directly associated with a situation of "open BRB" [4]. In this situation, the increase in tissue volume is due to an increase in the retinal extracellular space. Breakdown of the BRB is well identified by fluorescein leakage, which can be detected in a clinical environment by fluorescein angiography or vitreous fluorometry measurements. In this type of edema, Starling's law governing the movements of fluids applies [5].

1.4 Starling's Law

In extracellular edema, the "force" driving water across the capillary wall is the result of a hydrostatic pressure difference ΔP and an effective osmotic pressure difference $\Delta\pi$. The equation regulating fluid movements across the BRB is, therefore

$$\text{Driving\_force} = \text{Lp}\,[(\text{Pplasma} - \text{Ptissue}) - \sigma\,(\pi\text{plasma} - \pi\text{tissue})]$$

where Lp is the membrane permeability of the BRB, σ is an osmotic reflection coefficient, Pplasma is blood pressure, and Ptissue is the retinal tissue osmotic pressure.

An increase in ΔP, contributing to retinal edema, may be due to an increase in Pplasma and/or a decrease in Ptissue. An increase in Pplasma due to increased systemic blood pressure contributes to retinal edema formation only after loss of autoregulation of retinal blood flow and alteration of the structural characteristics of the BRB. A decrease in Ptissue is an important component that has not been given sufficient attention. Any loss in the cohesiveness of the retinal tissue due to pathologies such as cyst formation, vitreous traction, or pulling at the inner limiting membrane will lead to a decrease in Ptissue. A decrease in Ptissue, i.e., increased retinal tissue compliance, may lead to fluid accumulation, edema formation, and an increase in retinal thickness.

A decrease in $\Delta\pi$, contributing to retinal edema, may occur due to increased protein accumulation in the retina after breakdown of the BRB. Extravasation of proteins will draw more water into the retina. This is the main factor provoking a decrease in $\Delta\pi$, as a reduction in plasma osmolarity high enough to contribute to edema formation is an extremely rare event.

After a breakdown of the BRB, the progression of retinal edema depends directly on the ΔP and $\Delta\pi$ gradients. In these situations, tissue compliance becomes more important, influencing directly the rate of edema progression.

Thus, in the presence of retinal edema, it is essential to recognize the presence of an "open BRB" [6].

1.5 Incidence and Prevalence of DME

The incidence and prevalence of DME have been reported in different epidemiologic studies with significant variations, depending on the type (type I or II), treatment (insulin, oral hypoglycemic agents, or diet only), and the mean diabetes duration. Although DME can develop at any stage of DR, it is frequently related with increase in duration and severity of DR.

DME prevalence, indicated in the Wisconsin Epidemiologic Study in Diabetic Retinopathy (WESDR), is only about 3% in mild nonproliferative diabetic retinopathy (NPDR), but increases to 38% in moderate to severe NPDR and to 71% in eyes with proliferative diabetic retinopathy (PDR). In this study, the incidence of clinically significant DME was 4.3% in type I diabetic patients and 5.1% in type II with insulin and 1.3% in those without insulin. At 10 years, the rate of developing DME was 20.1% in patients with diabetes type I and 25.4% in type II diabetic patients needing insulin and 13.9% in those without insulin [7].

1.6 Clinical Evaluation of Macular Edema

Clinical evaluation of macular edema has been characterized by its difficulty and subjectivity. Direct and indirect ophthalmoscopy may only show an alteration of the foveal reflexes. Stereoscopic fundus photographs (SFP) and slit-lamp fundus stereo biomicroscopy have been the standard clinical methods to evaluate changes in retinal volume in the macular area, but they are dependent on the observer experience, and the results do not offer a reproducible measurement of the volume change [5]. Nevertheless, together they are useful to visualize signs correlated with retinal thickening, such as hard and soft exudates, hemorrhages, and microaneurysms.

The introduction of imaging methods, such as optical coherence tomography (OCT), made macular edema evaluation more precise and reliable.

1.7 Diagnosis and Classification

The Early Treatment Diabetic Retinopathy Study Group (ETDRS) defined DR severity stages [8] and DME [9] based on clinical grounds by SFP. DME is an increase in retinal thickness at or within 1 disk diameter of the foveal center, whether focal or diffuse, with or without hard exudates, sometimes associated with cysts. In this study, the term "clinically significant macular edema" (CSME) was introduced to characterize the severity of the disease and to provide a threshold level to apply laser photocoagulation. Three different CSME situations can occur:

1. Increase in retinal thickness $\leq 500\,\mu m$ of the center of the fovea
2. Hard exudates $\leq 500\,\mu m$ of the center of the fovea with increased retinal thickness
3. Increase in retinal thickness ≥ 1 disk diameter with at least one part within 1 disk diameter at the center of the fovea

Fluorescein angiography (FA) has been an important method to evaluate DME, and although not considered a screening exam, it provides important information about retinal perfusion, blood–retinal barrier integrity, and new vessel growth.

Angiographic classifications of DME have included noncystoid and cystoid macular edema (CME) [10] and focal or diffuse DME [11,12]. Focal macular edema is characterized by the presence of localized areas of retinal thickening associated with focal leakage of individual microaneurysms or clusters of microaneurysms or dilated capillaries. Diffuse macular edema is a more generalized and chronic form of edema, visualized as widespread macular leakage and pooling of dye in cystic spaces [13].

Ophthalmoscopy, SFP, and FA have been for many years the traditional methods used to evaluate diabetic retinopathy and DME, although they do not provide neither quantitative measurements of retinal thickness nor information about cross-sectional retinal morphology.

Recently, one methodology capable of measuring objective changes in retinal thickness and giving morphological and topographic surface images became available, the OCT, changing dramatically the landscape of DME diagnostic and follow-up.

OCT is a noninvasive and noncontact diagnostic method, well tolerated by patients, that provides important information about the retina. OCT imaging is analogous to B-scan ultrasound imaging, except that it uses infrared light reflections instead of ultrasound. It produces reliable, reproducible, and objective cross-sectional images of the retinal structures and the vitreoretinal interface and allows quantitative measurements of retinal thickness (RT). Since its commercial introduction in 1995, it enhanced the ability to diagnose and guide treatment decisions in retinal pathology, namely, macular holes, DME, epiretinal membranes, choroidal neovascularization, and vitreomacular traction, and thus became a powerful tool, widely used, for research and clinical evaluation of retinal disease.

OCT brought new insights about morphological changes of the retina in diabetic retinopathy and DME. It showed that macular edema may assume different morphologic patterns [14, 15]. In addition, a quantitative characterization of macular edema became feasible, as determined by measurements of retinal thickness and volume. OCT has been demonstrated to be more sensitive than slit-lamp biomicroscopy in detecting small changes in retinal thickness [16–19] and is clearly less subjective. In cases of DME, OCT scans may demonstrate diffuse thickening of the neurosensory retina and loss of the foveal depression; cystic retinal changes, which manifest as areas of low intraretinal reflectivity; and serous retinal detachment, alone or combined.

Over the last decade, the development of OCT instrumentation progressed rapidly. The first and second generations of commercial OCT instrument time domain (TD) (OCT1, OCT2) had an axial resolution of 10–15 μm. Third generation OCT (OCT3, Stratus; Carl Zeiss Meditec, Dublin, California, USA) provided an axial resolution of 8–10 μm. The recently available spectral-domain OCT (SD-OCT) has an axial resolution of 5–6 μm and has almost 100-fold improvement in acquisition speed over conventional time-domain OCT scanners since the moving reference arm is eliminated and all data points can be analyzed simultaneously. With increased imaging speed and greater signal-to-noise ratio, SD-OCT scanners produce more detailed and brighter images, with greater detail. Consequently, it is possible to decrease the motion artifacts and obtain a more precise registration from a larger area to be scanned. It is also possible to obtain an in vivo three-dimensional (3D) imaging that allows correlation between OCT images and clinical fundus features.

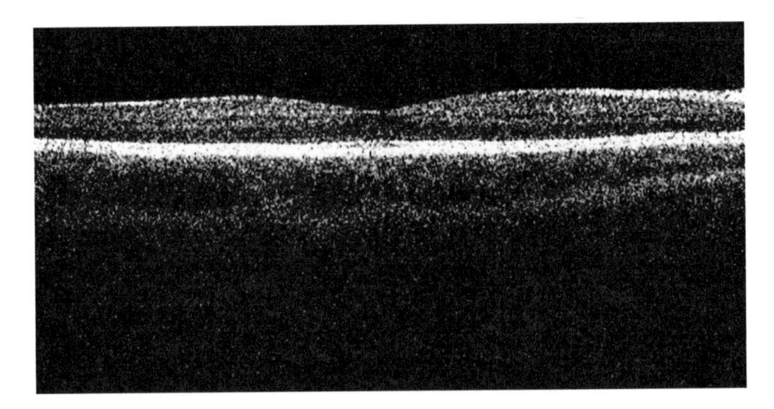

Fig. 1.1 TD-OCT, Stratus: normal cross-sectional macular image

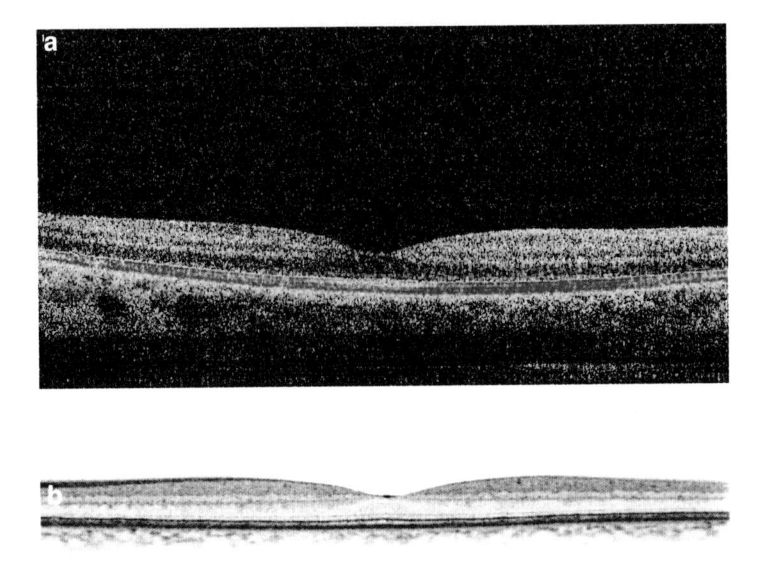

Fig. 1.2 SD-OCT, Cirrus: normal cross-sectional macular image; (**a**) false color; (**b**) gray scale

Cross-sectional images resemble closely the histological appearance of the retina [20] (Figs. 1.1–1.4). The top of the image corresponds to the vitreous cavity, which is optically silent, in a normal patient, or may show the posterior hyaloidal face, if there is a posterior vitreous detachment [21]. Central foveal depression is visible in normal eyes. The anterior boundary of the retina corresponds to the internal limiting membrane (ILM), at the vitreoretinal interface, hyperreflective and well defined, because of the contrast between the nonreflective vitreous and the backscattering of the retina.

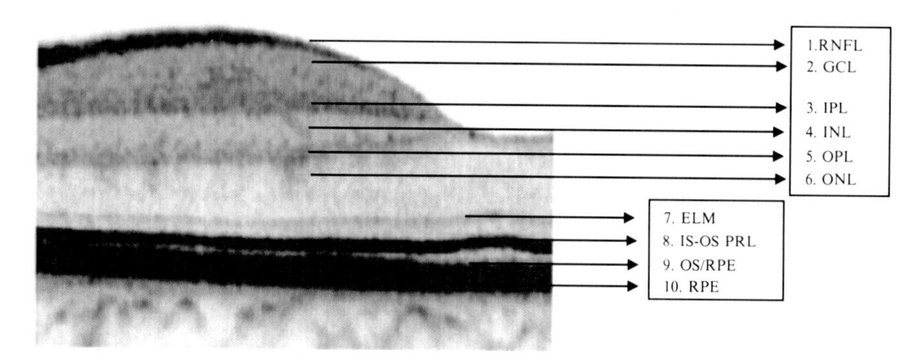

Fig. 1.3 SD-OCT, Spectralis: normal cross-sectional macular image (gray scale) and anatomic correlation. (1) RNFL, (2) GCL, (3) IPL, (4) INL, (5) OPL, (6) ONL, (7) ELM, (8) IS-OS PRL, (9) OS/RPE junction, and (10) RPE

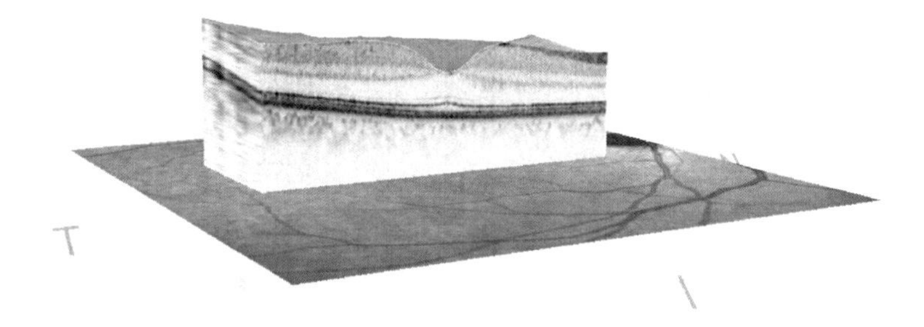

Fig. 1.4 SD-OCT, Spectralis: 3D image of RT from a healthy volunteer

The internal structure of the retina has heterogeneous reflections and distinct bands, and an anatomic correlation with the layers of the human retina has been proposed [22] (Fig. 1.3). Retinal nerve fiber layer (RNFL) is aligned horizontally, demonstrating higher tissue signal strength and appears thicker closer to the optic nerve, as expected. Axially aligned cellular layers—ganglion cell layer, inner nuclear layer, and outer nuclear layer (GCL, INL, and ONL, respectively)—have lower tissue signal compared with horizontally aligned layers, ILM, RNFL, and plexiform layers, which have higher tissue signal. Typically, nuclear layers appear hyporeflective, while plexiform layers (inner plexiform layer and outer plexiform layer—IPL and OPL, respectively) and axonal layers are relatively hyperreflective.

In the outer retina, different hyperreflective structures (bands) are visualized. TD Stratus OCT image shows the outer retinal layers as two hyperreflective bands, the photoreceptor's outer segments (inner) and the RPE/choriocapillaris complex (outer). On the other hand, SD-OCT scans of the outer retina allow the visualization of more bands than the TD-OCT. With this high resolution technology, 3 or 4 distinct strongly reflective bands are apparent, although their histological correlation remains a matter of discussion. According to Pircher et al. [23], the first (inner) band may correspond to the external limiting membrane (ELM), the second to the interface of the inner and outer segments (IS-OS) of the photoreceptor layer (IS-OS PRL), the third band may represent the outer segment-RPE junction (OS-RPE), and the fourth (outer) is assumed to represent the RPE (Fig. 1.3). The separation between the third and the fourth hyperreflective lines may not always be visible [23]. The analysis of structural changes in the outer retinal layers, particularly affecting photoreceptors and their interface, is now possible, using SD-OCT scanners [24]. In fact, the disruption of the photoreceptor IS/OS junction appears to be an important indicator of photoreceptor integrity or impairment and visual acuity outcome, as highlighted in several recent studies on distinct retinal diseases, such as branch retinal vein occlusion [25], central serous chorioretinopathy [26], retinitis pigmentosa [27], type 2 idiopathic macular telangiectasia (IMT) [28, 29], and DME [30, 31]. A recent report showed a strong correlation of photoreceptor's outer segments length, measured with Cirrus SD-OCT, and visual acuity, in DME [32].

Initial studies found a good correlation between macular thickening, assessed with OCT, and visual acuity [15, 33–35]. However, recent reports show only a moderate correlation between central retinal thickness and visual acuity, in patients with DME [36,37], implying that visual acuity may depend mainly on the disruption of the retinal architecture or direct photoreceptor damage.

The possibility to quantify retinal thickness by OCT is based in the distance between the anterior and posterior highly reflective boundaries of the retina, using appropriate algorithms [38]. All instruments identify the vitreoretinal interface as the inner retinal border; however, the segmentation of the outer retinal border differs widely among different OCT instruments. While the Stratus OCT system uses the photoreceptor's outer segments band for segmentation, spectral OCT devices use the second or the fourth hyperreflective lines as the outer border of the retina. As a consequence, Stratus OCT generates lower values of retinal thickness, while spectral-domain technology gives higher thickness values.

Since the commercialization of OCT systems, several types of software to quantify macular thickness became available. The first version (OCT 1, OCT 2—Carl Zeiss Meditec, Humphrey Division, Dublin, CA, USA) and, later, Stratus OCT (Carl Zeiss Meditec, Humphrey Division, Dublin, CA, USA) comprised automated macular thickness acquisition protocols consisting of six radial line scans, approximately 6 mm long each, obtained in a spoke-like pattern centered on the fovea (radially). Retinal thickness is measured along these six intersecting lines, with the advantage of concentrating measurements in the central fovea [38]. The high-density scanning protocol of the Stratus OCT (software version 4.0) uses six radial lines of 6 mm (B-scan), each containing 128 A-scans, with a total of 768 A-scans. Acquisition time for the scan is 1.5 s. Mean macular retinal thickness (RT) is displayed as a two-dimensional false color-coded map, where bright colors (e.g., red and white) represent thick areas and dark colors (e.g., blue and black) represent thin areas, and as a numerical map, for nine ETDRS-type areas (Fig. 1.5). Average RT is calculated automatically for each of the nine quadrants. Because data point density is greater centrally than peripherally, interpolated thickness measurements

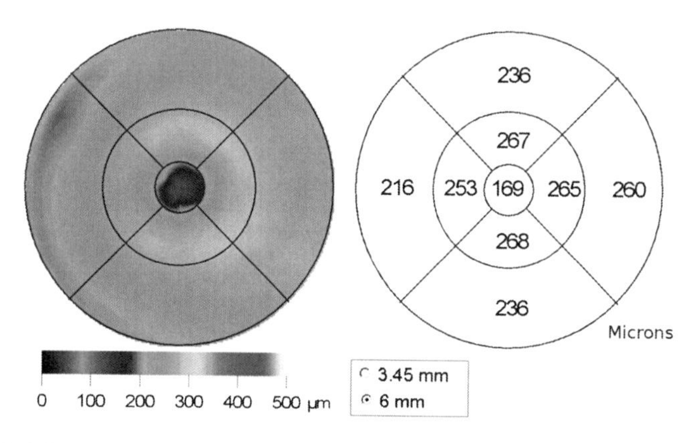

Fig. 1.5 TD-OCT, Stratus: two-dimensional color-coded RT map and its numerical representation from a healthy volunteer

of regions farther from the fovea are determined from fewer measurements and thus may be less accurate than those in central regions.

SD-OCT images are obtained by scanning multiple parallel lines in a rectangular scan pattern, providing a more homogeneous distribution of measured points within the macular area [39,40]. Nevertheless, the number of A-scans per line varies among different OCT scanners. It is now possible to obtain a three-dimensional image of the normal (Fig. 1.6) and diseased retina.

To evaluate and quantify macular thickening, patient's data should be compared with a normative database (defining threshold values), established in eyes with normal retinal status. However, retinal thickness measurements obtained by various OCT instruments differ, mainly due to different segmentation algorithms. Normal retinal thickness in the macular area, measured with time-domain devices, has been calculated to be 200–250 µm, and physiological foveal depression has a mean thickness of 170 µm [17, 41]. Based on these reference values, retinal thickness is considered normal, borderline, or thickened, in the setting of edema [41, 42] (Table 1.1).

Recently, Wolf-Schnurrbusch et al. [43] compared central retinal thickness (CRT) measurements (central 1,000-µm diameter area) in healthy eyes, obtained by six different commercially available OCT instruments, including time-domain and spectral-domain OCT. These authors also compared the intersession reproducibility of such measurements. Table 1.2 shows (different) CRT values of the right eye (µm ± SD).

OCT scans should be analyzed in two steps, regarding qualitative and quantitative information. Qualitative assessment relies on the characterization of the reflectivity profiles and morphological properties of the intraocular structures, whether normal or abnormal, visualized in the scans; quantitative evaluation refers to the possibility to measure these structures. After this analysis, the data should be integrated and correlated with clinical data, with previous exams and, if necessary, with

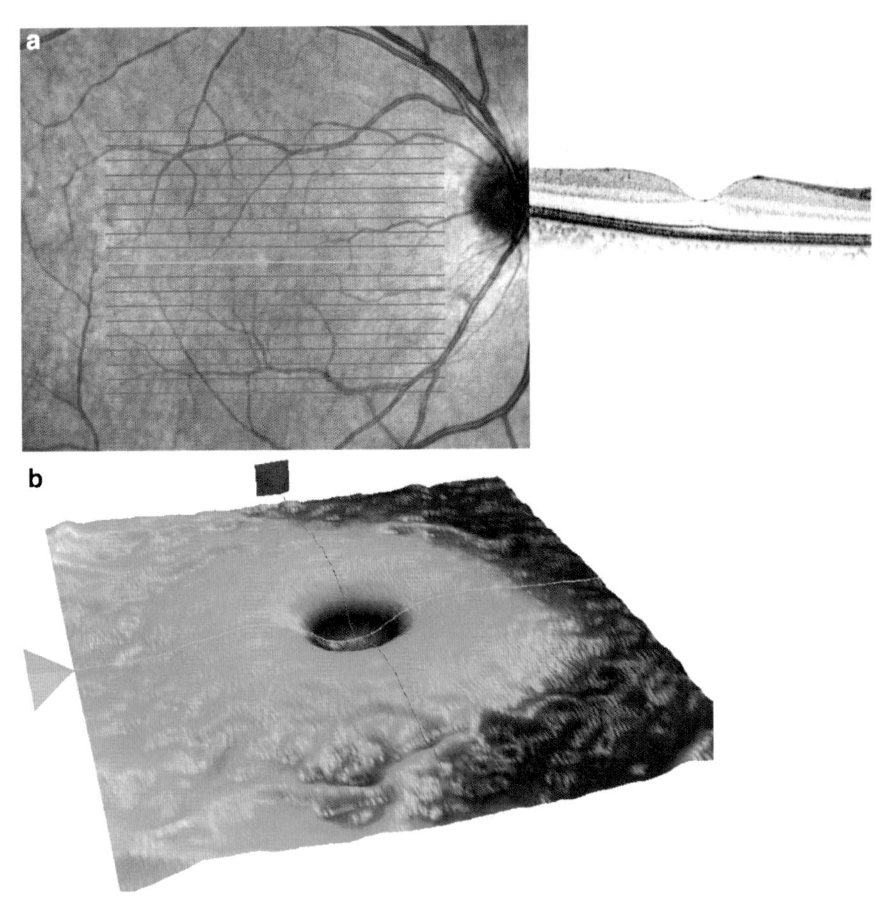

Fig. 1.6 (**a**) SD-OCT, Spectralis: acquisition protocol (parallel lines); (**b**) SD-OCT, Cirrus: 3D image of RT from a healthy volunteer

Table 1.1 Retinal thickness (RT) and standard deviation (SD) values, measured by OCT2 [41]

	RT [μm] (average ± SD)		
	Fovea	Central area (1.0 mm diameter)	Perifoveal and peripheral areas
Normal	150 ± 20	170 ± 20	230 ± 20
Borderline	170–210	190–230	250–290
Edema	≥210	≥230	≥290

information obtained by other diagnostic tools, namely, fluorescein angiography, aiming for a better correct diagnosis [17].

Reflections of the low coherent light from the ocular tissues should be differentiated between hyperreflectivity, hyporeflectivity, and shadowing effects (Table 1.3) [17].

Table 1.2 Mean CRT values (±standard deviation) obtained in central 1,000-μm diameter area [43]

Instrument	CRT [μm] (average ± SD)
Stratus OCT	212 ± 19
Spectralis HRA + OCT	289 ± 16
Cirrus HD-OCT	277 ± 19
Spectral OCT/SLO	243 ± 25
SOCT Copernicus	246 ± 23
RTVue—100	245 ± 28

Table 1.3 OCT qualitative interpretation [17]

Hyperreflectivity	Hyporeflectivity	Shadow effect
Hard exudates	Intraretinal edema	Hemorrhages
Cotton wool spots	Exudative retinal detachment	Exudates
	Cystoid macular edema	Retinal vessels

Cross-sectional images and retinal thickness measurements from various OCT devices show differences, due to different resolution, program segmentation, and/or alignment algorithms (Fig. 1.7).

Nowadays, OCT is increasingly used in the management of DME. Cross-sectional images of the retinal structures and thickness maps provide an objective and reproducible baseline characterization of the retinal disease. OCT imaging seems to be more sensitive than slit-lamp biomicroscopy to detect small changes in retinal thickness [16–18] and to visualize infraclinical foveolar detachments [19]. OCT scans also allow an accurate evaluation of disease progression, over time, and particularly after treatment.

Optical coherent tomography images of DME depict the presence of low intraretinal reflectivity, due to fluid accumulation in the extracellular space of the retina. The process begins as a diffuse retinal thickening with sponge-like appearance of the retinal layers, showing increase in the extracellular spaces advancing to the typical image of cystoid spaces [44, 45]. The hyporeflective cystoid-like cavities within the neurosensory retina are separated by highly reflective septa bridging retinal layers (Fig. 1.8a). They can progress to large and confluent hyporeflective (cystoid) spaces, involving the full thickness of the retina, with atrophy of the surrounding layers (Fig. 1.8b). Therefore, in newly developed CME, cystoid spaces are primarily located in the plexiform layer, while in well-established CME, cystoid spaces become confluent, and large cystoid cavities appear. Micropseudocysts, sometimes in the inner retina, may be identified with spectral-domain devices, even when retinal thickening is moderate, demonstrating the increase in extracellular space. They appear as small round hyporeflective lacunae with high signal elements bridging retinal layers (Fig. 1.8c) [44]. These small lesions are not visualized by time-domain OCT.

Increased thickening of the retina may also appear as an accumulation of serous fluid under the neurosensory retina, leading to a serous retinal detachment [46, 47]

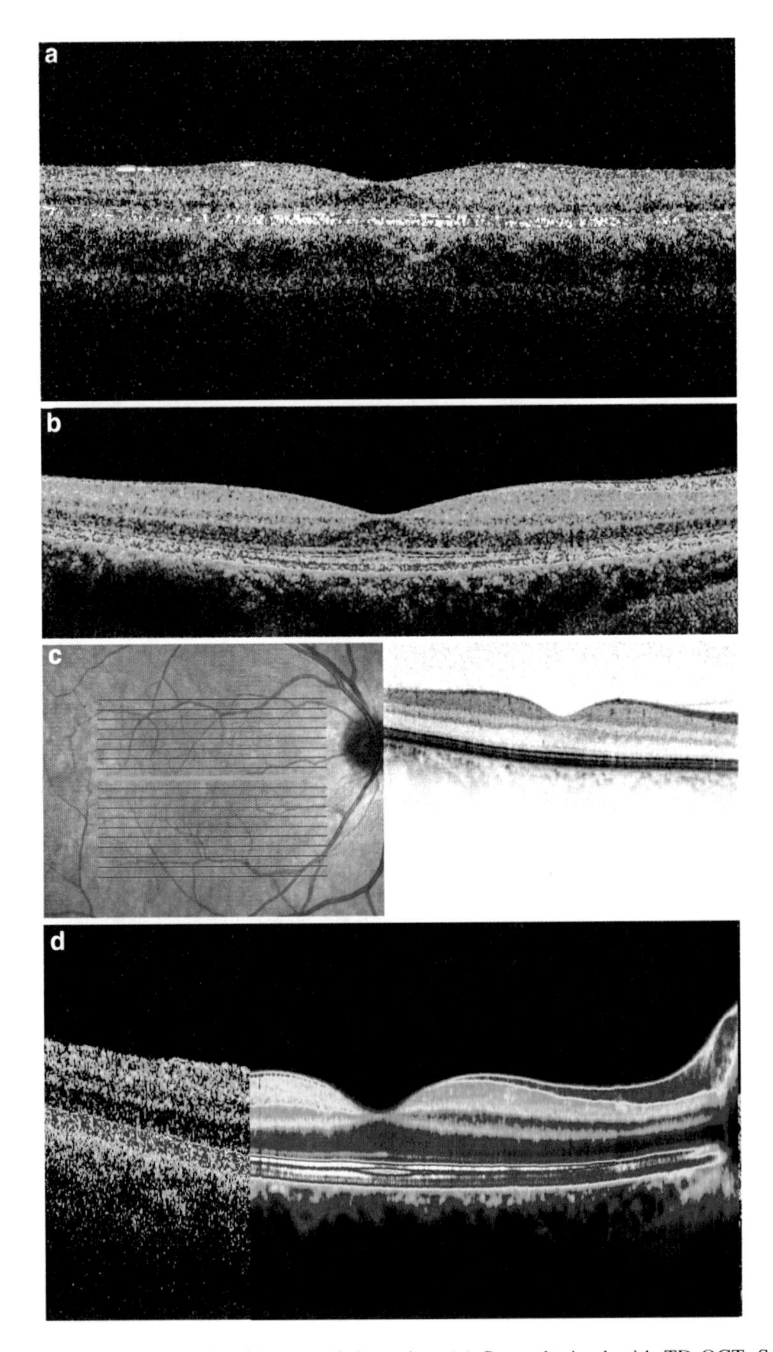

Fig. 1.7 Normal cross-sectional image of the retina. (**a**) Scan obtained with TD-OCT, Stratus; (**b**) scan obtained with SD-OCT, Cirrus; (**c**) scan obtained with SD-OCT, Spectralis (gray scale); and (**d**) comparison between scan obtained with TD-OCT (Stratus) and SD-OCT (Spectralis) (false color scale)

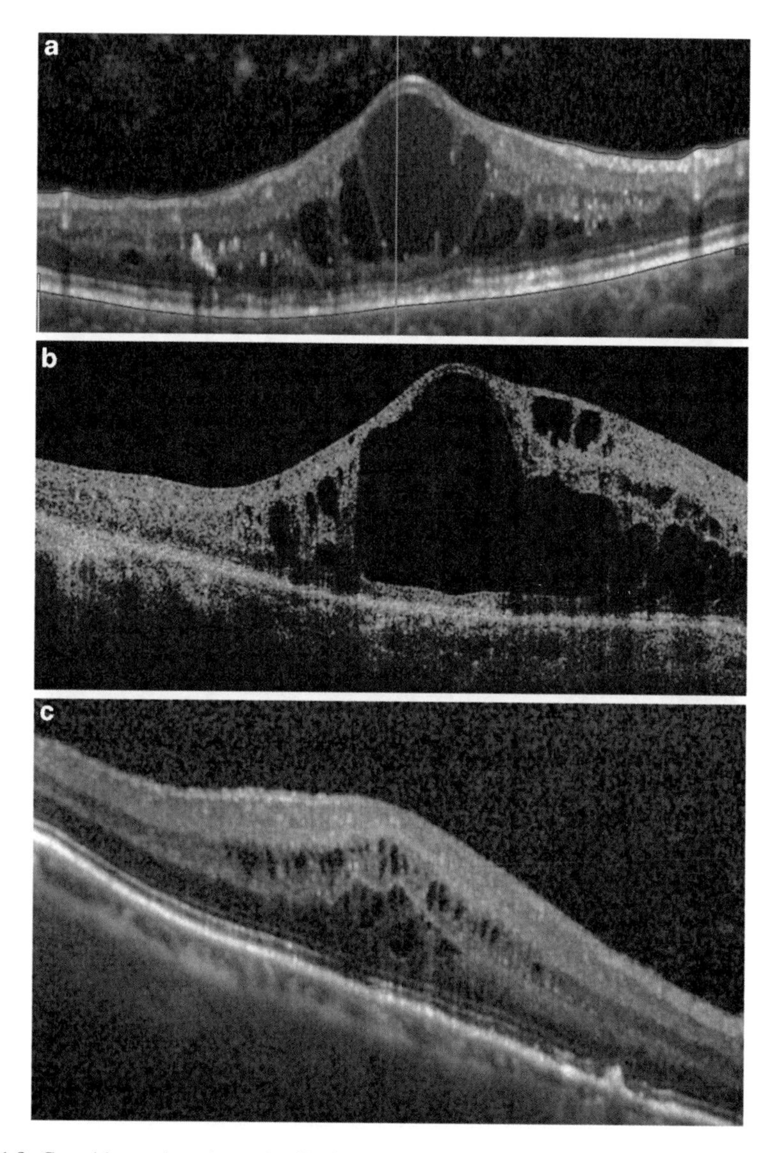

Fig. 1.8 Cystoid macular edema. (**a**) SD-OCT, Spectralis; (**b**) SD-OCT, Cirrus: confluent cystoid spaces (long-standing CME); and (**c**) SD-OCT, Spectralis: small cysts in the inner nuclear layer

(Fig. 1.9). This finding is not visible on biomicroscopy, but can be detected by OCT, as a hyporeflective area under the macula elevating the neurosensory retina above. It corresponds to an optically clear space between the retina and RPE, with a distinct outer border of the detached retina [48]. The distinction between subretinal and intraretinal fluid is possible by the identification of the highly reflective posterior

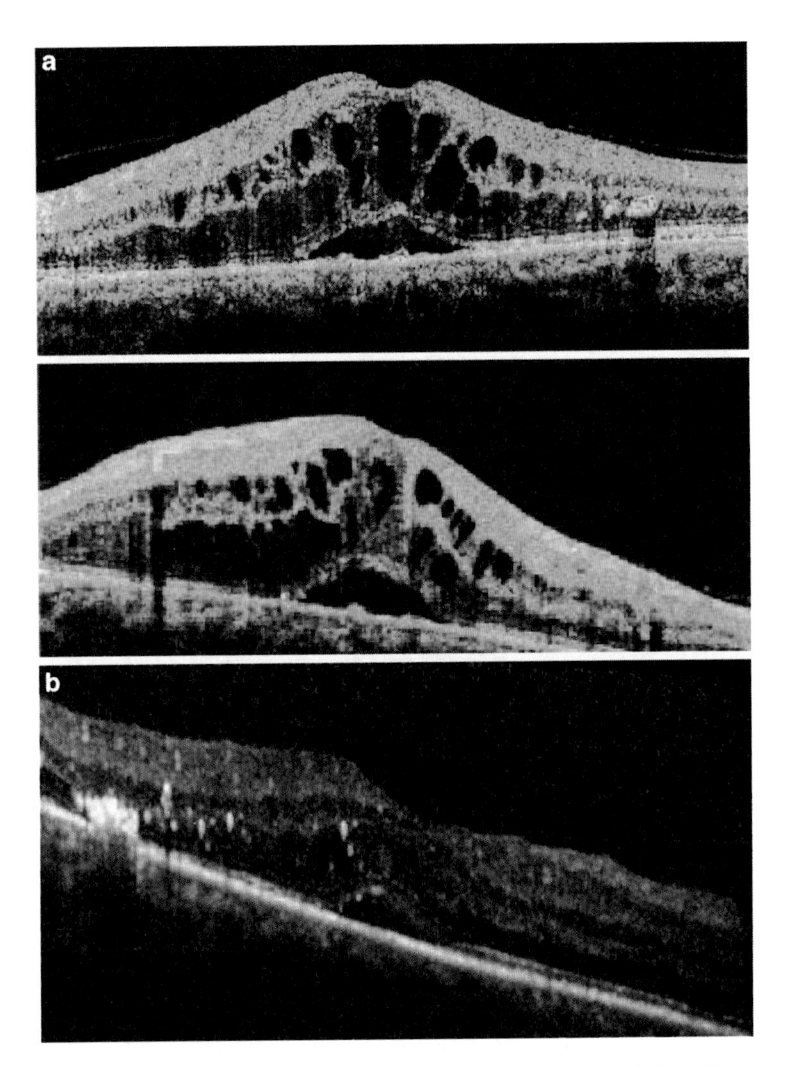

Fig. 1.9 DME with serous fluid detachment. (**a**) SD-OCT, Cirrus (*right* and *left* eye from the same patient); and (**b**) SD-OCT, Spectralis

border of the detached retina. Although there are different pathogenic mechanisms proposed for this type of edema [49–52], it may be associated with a dysfunction of the outer BRB [53, 54].

Another finding in DME is an outer retinal thickening, characterized by an ill-defined, widespread hyporeflective area, which can be distinguished from serous retinal detachment, by the absence of a highly reflective anterior boundary (Fig. 1.10).

Hard exudates are visualized as spots of high reflectivity with low reflective areas behind them (shadow effect), located in the outer retinal layers [55]

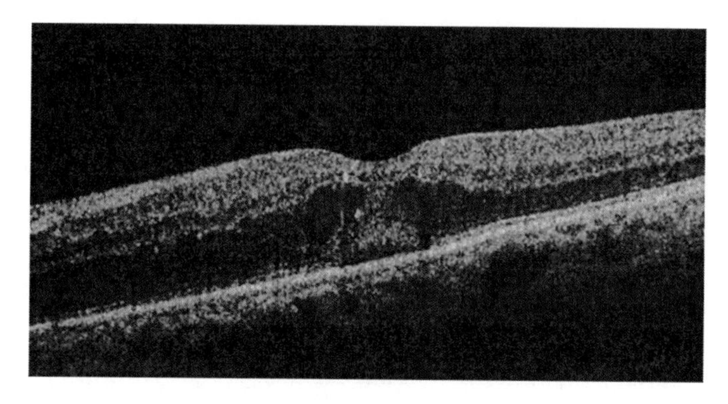

Fig. 1.10 SD-OCT, Cirrus. DME with outer retinal thickening

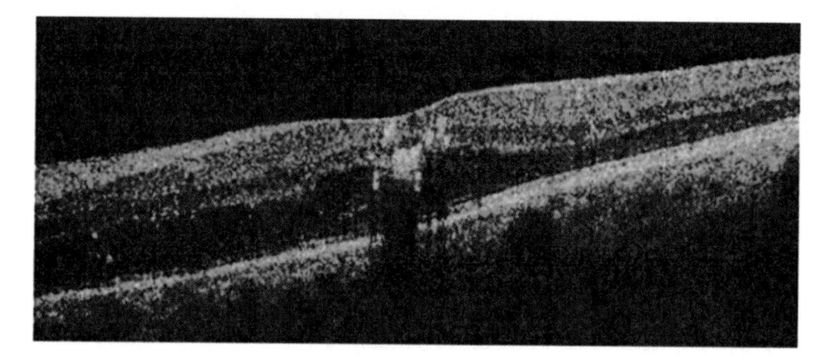

Fig. 1.11 SD-OCT, Cirrus. Hard exudates

(Figs. 1.9 and 1.11). They are due to protein or lipid deposition, secondary to the breakdown of the BRB. Hemorrhages also block the reflections from deeper retinal layers. Areas of previous focal laser treatment show an increased retinal reflectivity in the outer retinal layers (outer nuclear layer, photoreceptor layer, and retinal pigment epithelium). These morphologic changes are often associated with a decrease in central RT and thickening at the lesion site, especially in the photoreceptor layer [56].

Another advantage of the OCT is the possibility to analyze the vitreomacular interface. It is possible to determine the status of the posterior hyaloids when it is only slightly detached from the macular surface [57]. The concept of vitreoretinal traction is now considered of major relevance in the OCT classification of DME [51, 58] as it is considered to have an important role in the occurrence of ME. It appears in the OCT as a peak-shaped detachment of the retina with an area of low signal underlying the highly reflective border of the neurosensory retina, accompanied by posterior hyaloidal traction. The posterior hyaloidal traction is visible in the OCT as highly reflective signal arising from the inner retinal surface, partially detached from the posterior pole, toward to the optic nerve or peripherally, with adjacent vitreomacular traction (Figs. 1.12 and 1.13). The vitrectomy is considered beneficial in these cases [59–62].

DME can assume different morphologic patterns on OCT [15, 44, 45]. Kim et al. [15] proposed five morphologic patterns. A modification of their classification based on OCT interpretation and relationship with predominant inner or outer BRB breakdown is followed by our group:

Patterns of macular edema:

I. – Edema of the inner retinal layers

 – Breakdown of inner/outer BRB

II. – Cystoid spaces in the retina. Overall involvement

 – Breakdown of inner/outer BRB

III. –Subretinal fluid accumulation

 – Breakdown of outer BRB

IV. – Tractional retina edema

 – Breakdown of inner BRB

V. – Combination of patterns I, II, III, IV

In summary, OCT is today the only method that allows an objective follow-up of the major characteristics of DME. It allows a clear identification of the intraretinal fluid distribution and the presence or absence of vitreous traction. It is an excellent method to document these findings. Furthermore, OCT allows a quantitative diagnosis of ME, as it is used to obtain numerical representation of the retinal thickness.

CSME may be diagnosed using only biomicroscopy, but CSME with minimal increase in retinal thickness is difficult to recognize without OCT. Different studies demonstrated that OCT may identify DME in patients with normal biomicroscopy [17, 18, 63]. In diabetic patients with increased retinal thickness between 200 and 300 μm, considering abnormal values if they are above 200 μm, only 14% are detected by ophthalmoscopy. It corresponds to a subclinical form of macular edema [64].

Macular thickening is usually topographically correlated with leakage, in fluorescein angiography [39, 64, 65]; thus, it is considered an indicator of permeability of the BRB in the macular area. This correlation is better in the area within 1,500–3,000 μm around the fovea and less clear in the central 500–1,000 μm [66]. Recent work by our group has shown that it is possible to detect the alteration of the BRB, noninvasively with OCT measurements [67].

Clinical evaluation of macular edema should include the following parameters: extension of macular edema (i.e., thickened area); location of the edema in the macular area and, particularly, central foveal involvement (central area, 500 μm wide); presence or absence of vitreous traction; and chronicity of the edema (i.e., time elapsed since initial diagnosis and response to therapy). If considered

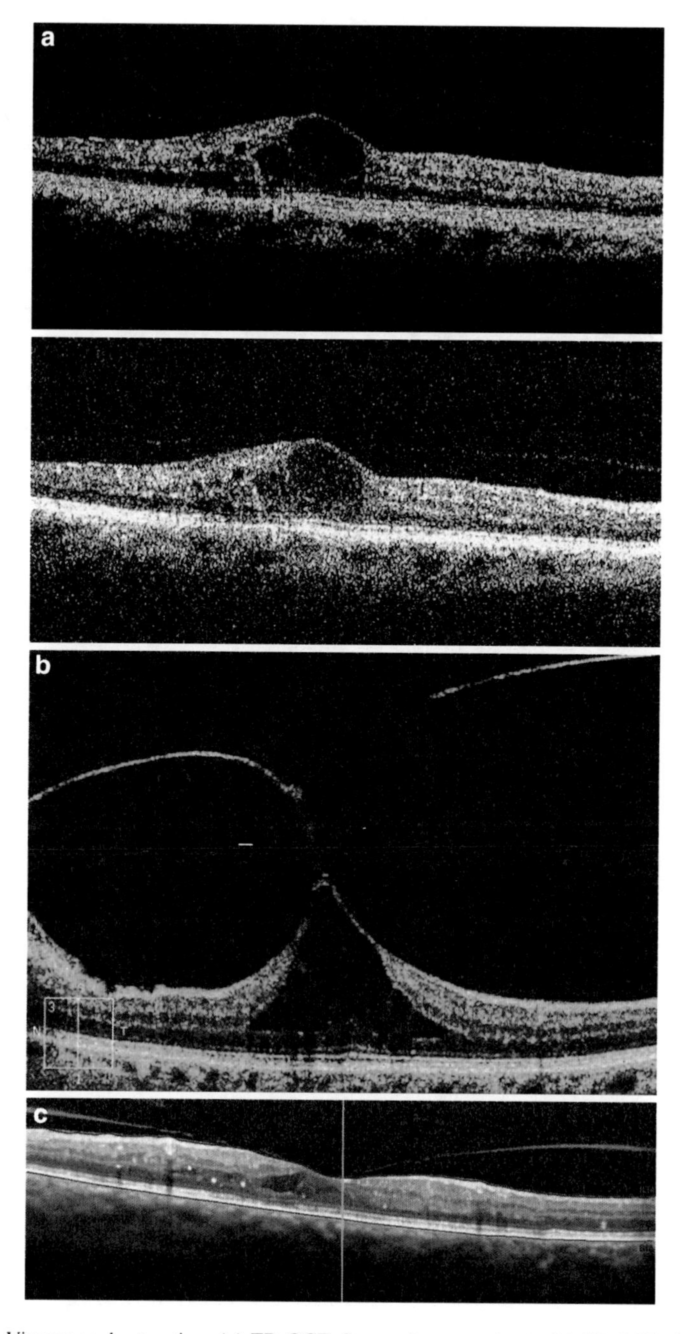

Fig. 1.12 Vitreomacular traction. (**a**) TD-OCT, Stratus (same patient); (**b**) SD-OCT, Cirrus; and (**c**) SD-OCT, Spectralis

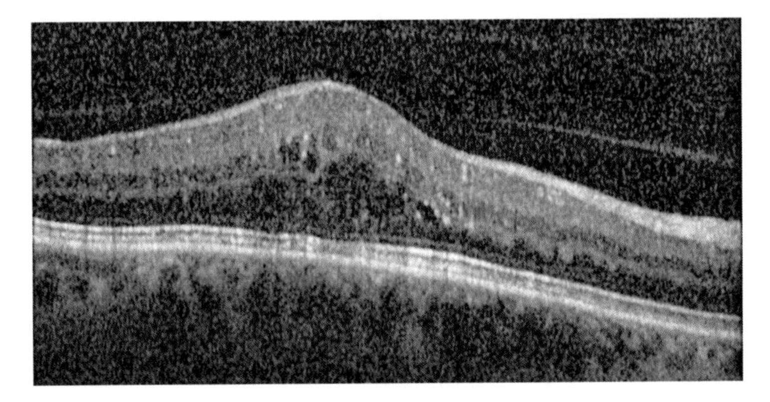

Fig. 1.13 Detachment of posterior hyaloids with vitreomacular traction. SD-OCT, Spectralis

necessary to complement the information obtained with OCT, information about the presence of ischemia may be obtained by performing fluorescein angiography.

In conclusion, OCT is a particularly valuable diagnostic tool in DME, helpful both in the diagnosis and follow-up procedure. DME classification systems should be based on OCT evaluation and measurements.

References

1. H. King, R.E. Aubert, W.H. Herman, Global burden of diabetes, 1995–2025: prevalence, numerical estimates, and projections. Diabetes Care **21**(9), 1414–1431 (1998)
2. L.P. Aiello, T.W. Gardner, G.L. King, G. Blankenship, J.D. Cavallerano, F.L.I.I.I. Ferris, R. Klein, Diabetic retinopathy. Technical review. Diabetes Care **21**, 143–156 (1998)
3. B.E. Klein, R. Klein, K.E. Lee, Components of the metabolic syndrome and risk of cardiovascular disease and diabetes in beaver dam. Diabetes Care **25**(10), 1790–1794 (2002)
4. J. Cunha-Vaz, Diabetic macular edema. Eur. J. Ophthalmol. **8**(3), 127–130 (1998)
5. J. Cunha-Vaz, A. Travassos, Breakdown of the blood-retinal barriers and cystoid macular edema. Surv. Ophthalmol. **28**(Suppl), 485–492 (1984)
6. C. Lobo, R. Bernardes, J.R. Faria de Abreu, J. Cunha-Vaz, Novel imaging techniques for diabetic macular edema. Doc. Ophthalmol. **97**, 341–347 (1999)
7. R. Klein, B. Klein, S. Moss, K. Cruickshanks, The Wisconsin Epidemiologic study of Diabetic Retinopathy XV. Ten year incidence and progression of diabetic retinopathy. Arch. Ophthalmol. **112**, 1217–1288 (1994)
8. Early Treatment Diabetic Retinopathy Study Research Group, Grading diabetic retinopathy from stereoscopic color fundus photographs—an extension of the modified Airlie House classification. ETDRS report number 10. Ophthalmology **98**(5 Suppl), 786–806 (1991)
9. Early Treatment Diabetic Retinopathy Study Group, Photocoagulation for diabetic macular edema: early treatment diabetic retinopathy study report no 1. Arch. Ophthalmol. **103**(12), 1796–1806 (1985)
10. G. Richard, G. Soubrane, L.A. Yannuzzi, *Fluorescein and ICG Angiography*,, 2nd edn. (Thieme, Stuttgart, 1998), chap. 2, pp. 15–16
11. A. Girach, H. Lund-Andersen, Diabetic macular edema: a clinical review. J. Clin. Pract. **61**, 88–97 (2007)

12. R. Klein, M.D. Knudtson, K.E. Lee, R. Gangnon, B.E. Klein, The Wisconsin Epidemiologic study of Diabetic Retinopathy XXIII: the twenty-five-year incidence of macular edema in persons with type I diabetes. Ophthalmology **116**, 497–503 (2009)
13. N. Bhagat, R.A. Grigorian, A. Tutela, M.A. Zarbin, Diabetic macular edema: pathogenesis and treatment. Surv. Ophthalmol. **54**(1), 1–32 (2009)
14. S. Yamamoto, T. Yamamoto, M. Hayashi, S. Takeuchi, Morphological and functional analyses of diabetic macular edema by optical coherence tomography and multifocal electroretinograms. Graefes Arch. Clin. Exp. Ophthalmol. **239**(2), 96–101 (2001)
15. B.Y. Kim, S.D. Smith, P.K. Kaiser, Optical coherence tomographic patterns of diabetic macular edema. Am. J. Ophthalmol. **142**(3), 405–412 (2006)
16. M.R. Hee, C.A. Puliafito, C. Wong, J.S. Duker, E. Reichel, B. Rutledge, J.S. Schuman, E.A. Swanson, J.G. Fujimoto, Quantitative assessment of macular edema with optical coherence tomography. Arch. Ophthalmol. **113**, 1019–1029 (1995)
17. G.E. Lang, Optical coherence tomography findings in diabetic retinopathy, in *Diabetic Retinopathy*, vol. 39, ed. by G.E. Lang. *Dev Ophthalmol* (Karger, Basel, 2007), pp. 31–47
18. C.S. Yang, C.Y. Cheng, F.L. Lee, W.M. Hsu, J.H. Liu, Quantitative assessment of retinal thickness in diabetic patients with and without clinically significant macular edema using optical coherence tomography. Acta Ophthalmol. Scand. **79**(3), 266–270 (2001)
19. P. Massin, A. Girach, A. Erginay, A. Gaudric, Optical coherence tomography: a key to the future management of patients with diabetic macular oedema. Acta Ophthalmol. Scand. **84**(4), 466–474 (2006)
20. R. Margolis, K. Kaiser, Diagnostic modalities in diabetic retinopathy, in *Diabetic Retinopathy—Contemporary Diabetes*, vol. 1, ed. by E.J. Duh (Humana Press, Totowa, NJ, 2008), pp. 109–133
21. J. Cunha-Vaz, G. Coscas, Diagnosis in macular edema. In: Steroids and management of macular edema. Ophthalmologica **224**(Suppl 1), 2–7 (2010)
22. W. Drexler, Cellular and functional optical coherence tomography of the human retina: the Cogan lecture 1. Invest Ophthalmol Vis Sci **48**, 5339–5351 (2007)
23. M. Pircher, E. Götzinger, O. Findl, S. Michels, W. Geitzenauer, C. Leydolt, U. Schmidt-Erfurth, C.K. Hitzenberger, Human macula investigated in vivo with polarization-sensitive optical coherence tomography. Invest. Ophthalmol. Vis. Sci. **47**, 5487–5494 (2006)
24. G. Coscas, *Optical Coherence Tomography in Age Related Macular Degeneration (OCT in AMD)*, (Springer; Edit., Heidelberg, 2009; 1389-Lamy, Publ., Marseille, 2009). ISBN 978-3-642-01468-0 (Print) 978-3-642-01467-3 (Online)
25. T. Murakami, A. Tsujikawa, M. Ohta, K. Miyamoto, M. Kita, D. Watanabe, H. Takagi, N. Yoshimura, Photoreceptor status after resolved macular edema in branch retinal vein occlusion treated with tissue plasminogen activator. Am. J. Ophthalmol. **143**(1), 171–173 (2007)
26. C.M. Eandi, J.E. Chung, F. Cardillo-Piccolino, R.F. Spaide, Optical coherence tomography in unilateral resolved central serous chorioretinopathy. Retina **25**(4), 417–421 (2005)
27. M.A. Sandberg, R.J. Brockhurst, A.R. Gaudio, E.L. Berson, The association between visual acuity and central retinal thickness in retinitis pigmentosa. Invest. Ophthalmol. Vis. Sci. **46**(9), 3349–3354 (2005)
28. L.A. Paunescu, T.H. Ko, J.S. Duker, A. Chan, W. Drexler, J.S. Schuman, J.G. Fujimoto, Idiopathic juxtafoveal retinal telangiectasis: new findings by ultrahigh-resolution optical coherence tomography. Ophthalmology **113**(1), 48–57 (2006)
29. A. Gaudric, G. Ducos de Lahitte, S.Y. Cohen, P. Massin, B. Haouchine, Optical coherence tomography in group 2A idiopathic juxtafoveal retinal telangiectasis. Arch. Ophthalmol. **124**(10), 1410–1419 (2006)
30. A.S. Maheshwary, S.F. Oster, R.M. Yuson, L. Cheng, F. Mojana, W.R. Freeman, The association between percent disruption of the photoreceptor inner segment-outer segment junction and visual acuity in diabetic macular edema. Am. J. Ophthalmol. **150**(1), 63–67 (2010)
31. H.J. Shin, S.H. Lee, H. Chung, H.C. Kim, Association between photoreceptor integrity and visual outcome in diabetic macular edema. Graefes Arch. Clin. Exp. Ophthalmol. **250**(1), 61–70 (2012)

32. F. Forooghian, P.F. Stetson, S.A. Meyer, E.Y. Chew, W.T. Wong, C. Cukras, C.B. Meyerle, F.L. Ferris 3rd, Relationship between photoreceptor outer segment length and visual acuity in diabetic macular edema. Retina **30**(1), 63–70 (2010)
33. A. Martidis, J.S. Duker, P.B. Greenberg, A.H. Rogers, C.A. Puliafito, E. Reichel, C. Baumal, Intravitreal triamcinolone for refractory diabetic macular edema. Ophthalmology **109**(5), 920–927 (2002)
34. S. Yamamoto, T. Yamamoto, M. Hayashi, S. Takeuchi, Morphological and functional analyses of diabetic macular edema by optical coherence tomography and multifocal electroretinograms. Graefes Arch. Clin. Exp. Ophthalmol. **239**(2), 96–101 (2001)
35. C. Strøm, B. Sander, N. Larsen, M. Larsen, H. Lund-Andersen, Diabetic macular edema assessed with optical coherence tomography and stereo fundus photography. Invest. Ophthalmol. Vis. Sci. **43**(1), 241–245 (2002)
36. S. Nunes, I. Pereira, A. Santos, R. Bernardes, J. Cunha-Vaz, Central retinal thickness measured with HD-OCT shows a weak correlation with visual acuity in eyes with CSME. Br. J. Ophthalmol. **94**(9), 1201–1204 (2010)
37. Diabetic Retinopathy Clinical Research Network, D.J. Browning, A.R. Glassman, L.P. Aiello, R.W. Beck, D.M. Brown, D.S. Fong, N.M. Bressler, R.P. Danis, J.L. Kinyoun, Q.D. Nguyen, A.R. Bhavsar, J. Gottlieb, D.J. Pieramici, M.E. Rauser, R.S. Apte, J.I. Lim, P.H. Miskala, Relationship between optical coherence tomography-measured central retinal thickness and visual acuity in diabetic macular edema. Ophthalmology **114**(3), 525–536 (2007)
38. M.R. Hee, C.A. Puliafito, C. Wong, J.S. Duker, E. Reichel, B. Rutledge, J.G. Coker, J.R. Wilkins, J.S. Schuman, E.A. Swanson, J.G. Fujimoto, Topography of diabetic macular edema with optical coherence tomography. Ophthalmology **105**(2), 360–370 (1998)
39. P. Massin, E. Vicaut, B. Haouchine, A. Erginay, M. Paques, A. Gaudric, Reproducibility of retinal mapping using optical coherence tomography. Arch.Ophthalmol. **119**, 1135–1142 (2001)
40. A. Polito, M. Del Borrello, M. Isola, N. Zemella, F. Bandello, Repeatability and reproducibility of fast macular thickness mapping using stratus optical coherence tomography. Arch. Ophthalmol. **123**(10), 1330–1337 (2005)
41. G. Panozzo, B. Parolini, E. Gusson, A. Mercanti, S. Pinackatt, G. Bertolo, S. Pignatto, Diabetic macular edema: an OCT-based classification. Semin. Ophthalmol. **19**(1–2), 13–20 (2004)
42. R. Brancato, B. Lumbroso, *Guide to Optical Tomography Interpretation* (Innovation-News Communication, Rome, 2004)
43. U. Wolf-Schnurrbusch, L. Ceklic, C.K. Brinkmann, M. Iliev, M. Frey, S. Rothenbuehler, V. Enzmann, S. Wolf, Macular Thickness measurements in healthy eyes using six different optical coherence tomography instruments. Invest. Ophthalmol. Vis. Sci. **50**(7), 3432–3437 (2009)
44. T. Otani, S. Kishi, Y. Mauyama, Patterns of diabetic macular edema with optical coherence tomography. Am. J. Ophthalmol. **127**(6), 688–693 (1999)
45. H. Alkuraya, D. Kangave, A.M. Abu El-Asrar, The correlation between coherence tomography features and severity of retinopathy, macular thickness and visual acuity in diabetic macular edema. Int. Ophthalmol. **26**(3), 93–99 (2005)
46. H. Ozdemir, M. Karacorlu, S. Karacorlu, Serous macular detachment in diabetic cystoid macular oedema. Acta Ophthalmol. Scand. **83**, 63–66 (2005)
47. A. Catier, R. Tadayoni, M. Paques, A. Erginay, B. Haouchine, A. Gaudric, P. Massin, Optical coherence tomography characterization of macular edema according to various etiology. Am. J. Ophthalmol. **140**(2), 200–206 (2005)
48. G.H. Bresnick, Diabetic macular edema: a review. Ophthalmology **93**(7), 989–997 (1986)
49. T. Otani, S. Kishi, Topographic assessment of vitreous surgery for diabetic macular edema. Am. J. Ophthalmol. **129**, 487–494 (2000)
50. T. Nagaoka, N. Kitaya, R. Sugawara, Alteration of choroidal circulation in the foveal region in patients with type II diabetes. Br. J. Ophthalmol. **88**(8), 1060–1063 (2004)
51. S. Kang, C. Yon Park, D. Ham, The correlation between fluorescein angiography and optical coherence tomography features in clinically significant macular edema. Am. J. Ophthalmol. **137**(2), 313–322 (2004)

52. W. Soliman, B. Sancher, T. Martini, Enhanced optical coherence patterns of diabetic macula edema and the correlation with the pathophysiology. Acta Ophthalmol. Scand. **85**(6), 613–617 (2007)
53. D. Gaucher, C. Sebah, A. Erginay, B. Haoucine, R. Tadayoni, A. Gaudric, P. Massin, Optical coherence tomography features during the evolution of serous retinal detachment in patients with macular edema. Am. J. Ophthalmol. **145**(2), 289–296 (2008)
54. F. Bandello, M. Battaglia Parodi, P. Lanzetta, A. Loewenstein, P. Massin, F. Menchini, D. Veritti, Diabetic macular edema, in *Macular Edema*, vol. 47, ed. by G. Coscas. *Dev Ophthalmol* (Karger, Basel, 2010), pp. 73–110
55. M. Bolz, U. Schmidt-Erfurth, G. Deak, G. Mylonas, K. Kriechbaum, C. Scholda, Optical coherence tomographic hyperreflective foci: a morphologic sign of lipid extravasation in diabetic macular edema. Ophthalmology **116**(5), 914–920 (2009)
56. M. Bolz, K. Kriechbaum, C. Simader, G. Deak, J. Lammer, C. Treu, C. Scholda, C. Prünte, U. Schmidt-Erfurth, Diabetic Retinopathy Research Group Vienna. In vivo retinal morphology after grid laser treatment in diabetic macular edema. Ophthalmology **117**(3), 538–554 (2010)
57. D. Gaucher, R. Tadayoni, A. Erginay, B. Haouchine, A. Gaudric, P. Massin, Optical coherence tomography assessment of the vitreoretinal relationship in diabetic macular edema. Am. J. Ophthalmol. **139**(5), 807–813 (2005)
58. G. Panozzo, E. Gusson, B. Parolini, A. Mercanti, Role of OCT in diagnosis and follow up of diabetic macular edema. Semin. Ophthalmol. **18**(2), 74–81 (2003)
59. H. Lewis, G.W. Abrams, M.S. Blumen Krans, R.V. Campo, Vitrectomy for diabetic macular traction and edema associated with posterior hyaloid traction. Ophthalmology **99**, 753–759 (1992)
60. P. Massin, G. Duguid, A. Erginay, B. Haouchine, A. Gaudric, Optical coherence tomography for evaluating diabetic macular edema before and after vitrectomy. Am. J. Ophthalmol. **135**(2), 169–177 (2003)
61. D. Thomas, C. Bunce, C. Moorman, A.H. Laidlaw, Frequency and associations of a taut thickened posterior hyaloid, partial vitreomacular separation, and subretinal fluid in patients with diabetic macular edema. Retina **25**(7), 883–888 (2005)
62. S.D. Pendergast, T.S. Hassan, G.A. Williams, M.S. Cox, R.R. Margherio, P.J. Ferrone, B.R. Garretson, M.T. Trese, Vitrectomy for diffuse diabetic macular edema associated with taut premacular posterior hyaloid. Am. J. Ophthalmol. **130**(2), 178–186 (2000)
63. U.H. Schaudig, C. Glaefke, F. Scholz, G. Richard, Optical coherence tomography for retinal thickness measurement in diabetic patient without clinical significant macular edema. Ophthalmic Surg. Lasers. **31**(3), 182–186 (2000)
64. J.C. Brown, S.D. Solomon, S.B. Bressler, A.P. Schachat, C. DiBernardo, N. Bressler, Detection of diabetic foveal edema, contact lens biomicroscopy compared with optical coherence tomography. Arch Ophthalmol. **122**(3), 330–335 (2004)
65. W. Goebel, T. Kretzchmar-Gross, Retinal thickness in diabetic retinopathy. A study using optical coherence tomography (OCT). Retina **22**(6), 759–767 (2002)
66. A. Neubauer, C. Chryssafis, S. Priglinger, C. Haritoglou, M. Tiel, V. Welge-Luben, A. Kampik, Topography of diabetic macular edema compared with fluorescein angiography. Acta Ophthalmol. Scand. **85**(1), 32–39 (2007)
67. R. Bernardes, T. Santos, P. Serranho, C. Lobo, J. Cunha-Vaz, Noninvasive evaluation of retinal leakage using optical coherence tomography. Ophthalmologica **226**(2), 29–36 (2011)

Chapter 2
Ischemia

Suk Ho Byeon, Min Kim, and Oh Woong Kwon

Abstract "Ischemia" implies a tissue damage derived from perfusion insufficiency, not just an inadequate blood supply. Mild thickening and increased reflectivity of inner retina and prominent inner part of synaptic portion of outer plexiform layer are "acute retinal ischemic changes" visible on OCT. Over time, retina becomes thinner, especially in the inner portion.

Choroidal perfusion supplies the outer portion of retina; thus, choroidal ischemia causes predominant change in the corresponding tissue.

2.1 Retinal Ischemia

Ischemia refers to a pathological situation derived from an inadequate blood flow to a tissue, resulting in a failure to meet cellular energy demand. It does not necessarily mean a complete loss of perfusion to a tissue.

Ischemia can be divided into two types according to the rate of progression [1]. Acute ischemia develops from occlusion of large retinal vessels, such as emboli in the arterioles or thrombosis in the venules. Causes and mechanisms of chronic ischemia are, however, usually obscure.

Funduscopic examination shows the typical findings of damaged area by acute retinal ischemia as pale inner retina or cotton wool spots. In central retinal artery occlusion (CRAO), the macula shows diffusely edematous area with a "cherry-red spot". Fundus appearance in the affected area depends on whether the occlusion is total or partial, acute or chronic, and whether it involves arteriole or venule [1].

O.W. Kwon (✉)
Yonsei University Health System (YUHS), 50 Yonsei—Ro, Seodaemun-Gu,
Seoul 120–752, Korea
e-mail: owkwon0301@yuhs.ac

R. Bernardes and J. Cunha-Vaz (eds.), *Optical Coherence Tomography*,
Biological and Medical Physics, Biomedical Engineering,
DOI 10.1007/978-3-642-27410-7_2,

With time, acute retinal ischemic changes disappear, and occluded artery is sometimes recanalized [2, 3]. The predominant late changes after ischemic damage are atrophic retinal thinning.

2.2 Brief Anatomy of Retinal Circulation

The retina receives its blood supply from two blood flow systems [4, 5].

The inner portion of the retina nerve fiver layer (NFL), ganglion cell layer (GCL), inner plexiform layer (IPL), and inner portion of inner nuclear layer (INL) is supplied by the central retinal blood vessels [6]. The outer portion, which includes the retinal layers from outer portion of INL to retinal pigment epithelial cell (RPE), is nourished by the choriocapillaris of choroid.

Central retinal artery (CRA) is a direct branch of the ophthalmic artery and constitutes an intraretinal end artery microvascular system. Posterior ciliary arteries, which are also branches of the ophthalmic artery, supply choroidal circulation. Cilioretinal arteries (present in about one-third of the population) contribute to the retinal circulation and are originated from posterior ciliary vessels [7]. Central retinal vein is drained into the ophthalmic vein and cavernous sinus.

Main retinal arteries are end arteries without interanastomosis. However, retinal perfusion system has some collateral circulations including cilioretinal capillary anastomosis, pial, and intraneural anastomosis of CRA.

Histologically, main retinal vessels lie superficially in the nerve fiber layer. The vein usually lies deeper than the artery [8]. Macular vasculature consists of two or three capillary plexuses. Superficial capillary plexus lies in NFL or GCL, and deep capillary plexus is located within INL [9] (Fig. 2.1). These two capillary plexuses have anastomosis [9]. In adult primate fovea, there is a central avascular area known as foveal avascular zone (FAZ). Capillary plexus is situated on the slope of fovea, not in the foveola [10] (Fig. 2.2).

The macula has three layers of capillaries, the peripheral retina has one to two, and the peripapillary retina has four.

2.3 Fluorescein Angiography (FA) and Optical Coherence Tomography (OCT): Anatomic Correlations

In fluorescein angiography (FA), superficial capillary plexus is mainly visible as a network, but deeper capillary plexus is seen as blurred outlined structure or only background fluorescence [11] (Fig. 2.1).

Fovea is a highly specialized area for high-resolution vision. In the foveal center, both GCL and INL are absent from the foveola [9]. Inner retinal neuron and process show lateral displacement from the foveal center [6]. In early phase of high-resolution FA, we can find capillary plexus defining FAZ. In histologic

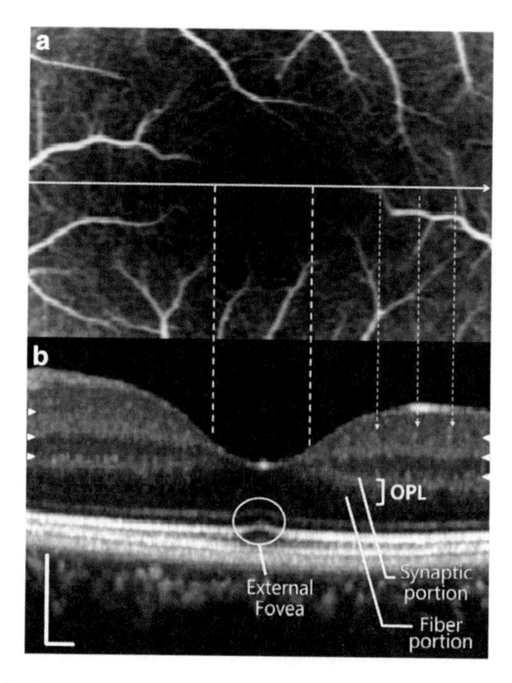

Fig. 2.1 Normal retinal vasculature visible on fluorescein angiography (FA) and detailed anatomy on optical coherence tomography (OCT); 39-year-old male with −5 diopter myopia. (**a**) In the early-phase FA, fine arterioles and venules in the macula are visible. Macular capillary networks (mainly superficial capillary plexus) are discriminable as fine mesh against background macular hypofluorescence [11, 16]. Foveal avascular zone (FAZ) is defined by foveal capillary plexus. (**b**) Corresponding horizontal OCT scan shows each retinal layers and multiple hyperreflective dots. The capillary plexuses of the human macula consist of superficial and deep capillary plexus. They are visible as multiple rows of hyperreflective dots on OCT (*white triangles*). Arterioles and venules are also visible as larger hyperreflective dots. These are well matched to the area where these vessels are visible on OCT (*broken arrows*). The capillary plexuses are visible on the upper foveal slope, and the size of foveal floor visible on OCT is well matched to that of FAZ on FA (*broken lines*). Histologically, the outer plexiform layer (OPL) can be divided into three zones, but these zones cannot be distinguished from each other on the current OCT. Thus, we propose more simplified terms, which simply divide the OPL into two zones, synaptic portion (synapses between bipolar cells and photoreceptor cells) and fiber portion (axons of photoreceptor) [17]. This synaptic portion of OPL is thought to be the outermost territory that normal retinal circulation can supply. In foveolar center, both ELM and IS/OS lines are displaced vertically due to the special morphology of foveal cones. On high-resolution OCT, this "external fovea" is an important landmark for finding true anatomic center of fovea (*Circle*)

exam, the blood vessels are visible on the upper foveal slope [10]. This particular anatomical structure of fovea is thought to be a result of a morphological adaptation to the available blood flow [12]. Thus, oxygen supply and consumption of macula are "on a knife-edge." Any small damage in capillary plexus may result in the damage of supplied corresponding tissue [13, 14].

Size of normal FAZ is quite variable in each individual from 0 to 1,200 μm [15]. The size of foveal floor visible on OCT is well matched to that of FAZ on FA [14] (Fig. 2.1). Size and shape of foveal slope and floor are also variable accordingly.

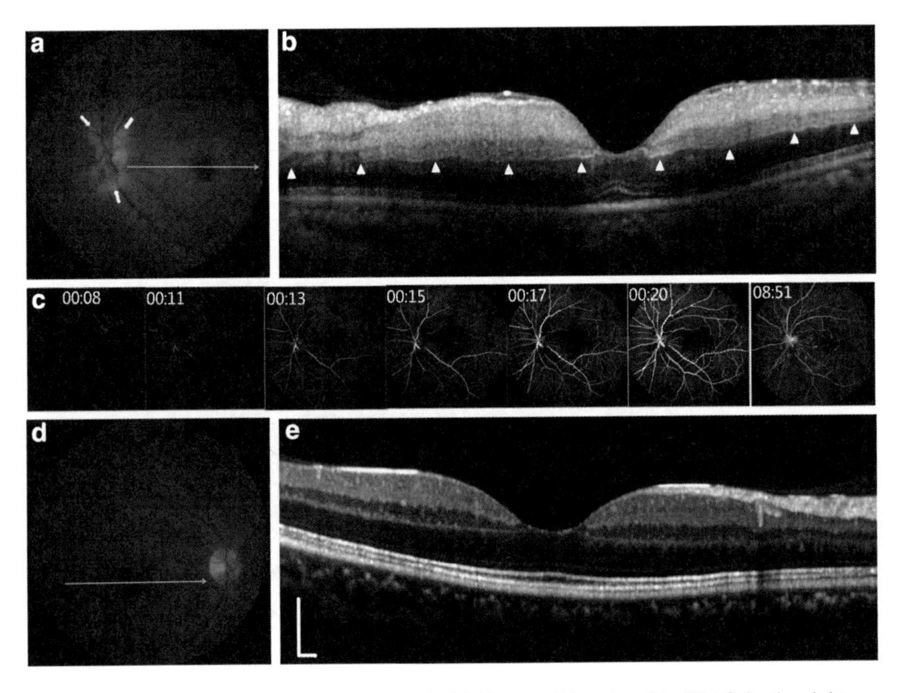

Fig. 2.2 Acute retinal ischemic changes on OCT. 48-year-old male with CRAO in the right eye (1 week after the onset of disease). (**a**) Fundus examination reveals an extensively pale retina with *cherry-red* spot in fovea in comparison with the fellow eye (D). Retinal opacification is more prominent in the posterior pole, becoming less apparent in the periphery. *Yellowish white* peripapillary area which represents axoplasmic material accumulation can be noticed (*white arrows*). (**b**) OCT shows increased reflectivity and diffuse mild thickening (edema) in the inner retinal area. "Prominent middle limiting membrane" is a hyperreflective line visible near the synaptic portion of outer plexiform layer (*white triangles*) in acute ischemic inner retinal area. The outer retina shows hyporeflective change, which is caused by shadowing from the abnormally hyperreflective inner retina. In the transparent foveolar area, because of the relative absence of inner retinal tissue, light attenuation resulting from inner retinal opacification is less severe. (**c**) FA shows delayed AV transit time, especially at the superior temporal arcade. Capillary plexus in the macular area is severely damaged, and the late-phase FA shows some dye leakage from the damaged capillary plexus. (**d**, **e**) Fundus photo and OCT of normal fellow eye

This implies that the extent of centrifugal displacement of GCL and NFL is variable from subjects to subjects; thus, the topographical variability of retinal circuitry exists in each individual [14].

2.4 "External Fovea" in OCT

Foveal cone has special morphology of long, slender cell body and longer outer segment, thus displaying special feature on OCT. In foveolar center, ELM is displaced vertically. In histological terminology, this central hump had been named as "external fovea" [6]. On high-resolution OCT, this "external fovea" is an important landmark for determining the true anatomic center of fovea (Fig. 2.1).

2.5 Detection of Ischemic Retinal Area: Comparisons Between Fluorescein Angiography and OCT

FA is still the best method to evaluate the retinal vasculature and circulation status [18, 19]. However, FA itself is an invasive technique, and it cannot provide ultrastructural anatomic details [20]. Noninvasive retinal imaging modalities such as OCT or autofluorescence play an increasingly important role in current practice [11, 21].

A lack of filling of the capillary (capillary dropout) or delayed arterial filling or an alteration of blood-retinal barrier (BRB) (dye leakage) is a diagnostic clue to detect ischemic areas in FA. Arterial narrowing, widening, or staining of retinal vessels can also be noticed [19].

Ischemia does not always imply a complete loss of perfusion to a tissue [4]. Also, delayed arterial filling is a specific feature of perfusion insufficiency, but not a characteristic sensitive enough to diagnose ischemia. Retinal circulation on FA can revert to normal even after acute retinal arterial occlusion [2, 19, 22]. Even though residual circulation is seen on FA, ischemic damage can persist due to low oxygen tension in the blood [2, 11]. Damage in BRB can be restored after acute phase of ischemia. In many retinal vascular obstruction diseases, on several days or weeks after onset, the retinal circulation is restored to a variable extent via various collateral anastomosis [2].

Current OCT can provide us with detailed ultrastructural anatomy of retina. Direct visualization of retinal vessel and capillary plexus is possible on the current commercially available OCT; however, determination of damage or loss of vascular structure on the current OCT is not possible [23] (Fig. 2.1). Consequently, we cannot consistently determine the ischemic damage or loss of vessel on OCT.

When we evaluate the area of ischemic damage, we should take the following points into consideration. The extent of tissue damage is determined not only by perfusion abnormalities in vasculature but also by the anatomy of blood supply (collateral) [2]. Thus, the anatomic complexity of retinal blood supply influences the outcome of ischemic retinal injury. Retinal tissue consists of various cell types of neurons or glial cells. Each cell type or structure has variable oxygen consumption rate; thus, perfusion abnormality itself cannot directly reflect the extent of tissue damage [19]. Detection of functional or ultrastructural changes in ischemia-damaged tissues or cells would provide the most definite information [24, 25].

2.6 Pathologic Changes and Relevant OCT Findings in Ischemic Tissues

Acute ischemic damage in retina appears as infarction, which implies that all elements of retinal tissue are lost without gliosis [26, 27].

In CRAO, a typical acute inner retinal ischemic disease, coagulative necrosis of inner layer occurs. Neurons become rapidly edematous within the first few hours [28, 29]. Clinically, we can observe grayish retinal opacity (marked edema of inner half of the retina is noted several hours after arterial obstruction and becomes maximal within 24 h) [26, 28, 30].

With time, homogenous, diffuse, and acellular zone begins to replace ischemic tissue, and inner half of the retina becomes homogenized. This neuronal cell loss is an ongoing process that may last up to 3 months after ischemic insult [31]. Gliosis usually does not occur. The boundaries among the different retinal layers in the inner half of the retina become obliterated. The inner retinal layer turns into an indistinguishably homogenized zone.

Similar histological changes can be visualized on OCT. In acute retinal ischemia, increased reflectivity and mild edematous change in the inner retina, including GCL, IPN, and INL, can be found. The outer retina shows hyporeflective change, which is caused by shadowing from the abnormally hyperreflective inner retina [23, 25, 27, 32–36]. The structures of photoreceptor layer are preserved; thus, IS/OS junction and outer nuclear layer (ONL) thickness are maintained (Fig. 2.2). Some authors reported a highly reflective deposit between the outer segments and the RPE in ischemic retinal area [37, 38]. However, these findings should be differentiated from secondary changes by concomitant choroidal insufficiency or by exudation from severe inner BRB damage.

2.7 Prominent Middle Limiting Membrane

The plexiform layer is a dominant oxygen consumption region among several retinal layers; thus, they are thought to be the most vulnerable to hypoxic insult [13]. In acute phase of inner retinal ischemia, a prominent line near the synaptic portion of OPL in speckle-noise-reduced OCT can be noticed (Fig. 2.2). We would like to newly reintroduce the concept of "middle limiting membrane" (MLM) to the description of OCT finding, which is an old terminology and no longer in use. Middle limiting membrane is not a true membrane and is located in the synaptic portion of OPL, which is regarded as the outermost extent of normal retinal vascular supply [39].

Increased inner retinal reflectivity with mild thickening of inner retina is an early sign of inner retinal ischemia. However, "prominent MLM" is quite a specific finding to indicate acute retinal ischemic damage on OCT (Byeon SH, unpublished data).

These OCT findings of acute retinal ischemia confirm the diagnosis of RAO in some cases of unexplained loss of visual function simulating an optic neuropathy [40].

After a long time of ischemic damage, retinal thinning and loss of foveal depression from retinal atrophic change can be found [32,33,41,42]. However, gross retinal thinning itself is a nonspecific finding and less informative [42]. Layer-by-layer analysis of retina shows inner retinal thinning in retinal vascular insufficiency and outer retinal thinning in choroidal vascular insufficiency [14].

2.8 Retinal Artery Occlusion

Retinal artery obstructive disease can be classified into CRAO, branched retinal artery occlusion (BRAO), ophthalmic artery occlusion, combined central vein occlusion (CRVO) with CRAO, and cilioretinal artery occlusion according to the level of occlusion [43]. In a broader sense, cotton wool spots may be included in this category.

2.9 Central Retinal Artery Occlusion

The earliest characteristic fundus change in CRAO is the retinal opacity, especially in the posterior pole. Cotton wool spot, optic disc edema, attenuated or box-carrying arteries or veins, sheathed arteries, and emboli can also be visualized in fundus examination [2, 18, 30]. The retinal opacity is essentially due to opacification of the retinal ganglion cells caused by acute ischemia [2,30]. Consequently, nonedematous transparent foveolar retina is discriminated in the center of opaque retinal area because there are no ganglion cells in the foveola. "Cherry-red spot" is often considered to be present due to the underlying normally perfused choroid and visible RPE [44]. Thus, some authors insisted that "cherry-red spot" is not visible in cases of ophthalmic artery occlusion [45]. However, fundus examination of recently died patients shows similar findings despite the absence of both choroidal and retinal blood supply [17]. We would rather presume that the visibility of "cherry-red spot" may be more related to individual variability of the size of FAZ [14]. The size of transparent foveola, which is closely related with the size of FAZ, is variable in each individual (Fig. 2.3).

In FA, arterial filling delay is a highly specific finding in CRAO including marked delay in the transit time from the arm to the retina [2, 18, 19]. A delay in retinal AV transit time, however, is the most common finding [19]. Choroid fills normally in CRAO on the contrary to ophthalmic artery occlusion [46]. After several days or weeks, the retinal circulation is restored to a variable extent; thus, narrowing and irregularity in the arteries are the only FA findings in the late phase of CRAO in some cases [2, 18].

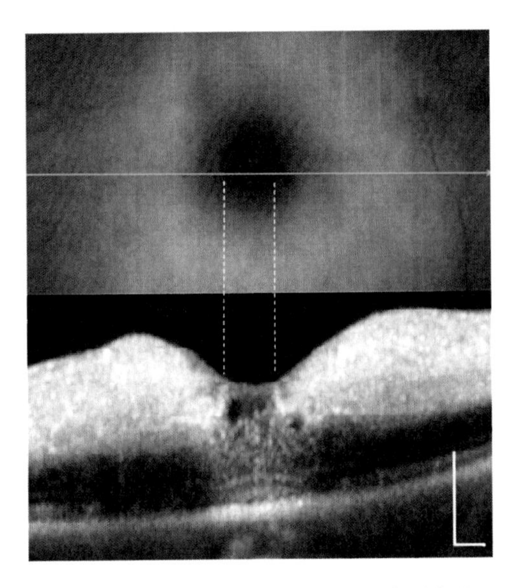

Fig. 2.3 *Cherry-red* spot in CRAO. 59-year-old female with CRAO (1 week after the onset of disease). (**a**) *Whitish yellow* change (pallor or swelling) is visible due to ischemic swelling and opacification of the inner retinal layers. (**b**) The *whitish yellow* change is most prominent in the outer margin of fovea, where ganglion cell layer (GCL) comprises the thickest portion of retinal layers. The *whitish* opacification is predominantly confined to the area where ganglion cells predominantly exist. The transparent fovea is visible as a "*red* spot" due to the absence of opaque inner retina and visibility of RPE or choroid (*cherry-red* spot). Definite *red* spot in fundus is well matched to the transparent foveolar area, where GCL and inner nuclear layer are absent on OCT (*broken lines*). In previous report, we found that the size of transparent foveola is related with the size of foveal avascular zone; thus, the size of "*red* spot" in *cherry-red* spot lesion may be related with the size of foveola in each individual

With time, arterial narrowing with or without venous narrowing may persist, but retinal circulation in FA can revert to normal [30]. Retinal thinning, especially in the inner retina at posterior pole, develops; thus, foveal annular light reflex disappears in fundus examination [47]. Optic disc also becomes pale [48].

Acute ischemic inner retinal change including increased reflectivity with mild edema of inner retina and "prominent MLM" can be found on OCT [23, 32] (Fig. 2.2). A "prominent MLM" is noticed in the early phase (usually within 1 month) of inner retinal ischemia, but sometimes the inner aspect of this prominent line is fused with dense inner retinal hyperreflectivity (Fig. 2.4). The outer retina shows hyporeflective change, which is caused by shadowing from the inner retinal change. The outer retina seems to expand; however, that is not a true expansion in most cases. Inner boundary of outer retina (synaptic portion of OPL) seems to be displaced inward because "prominent MLM" is located at rather inner aspect of synaptic portion of OPL (Fig. 2.2). Fluid accumulation in the outer retina is rarely noticed.

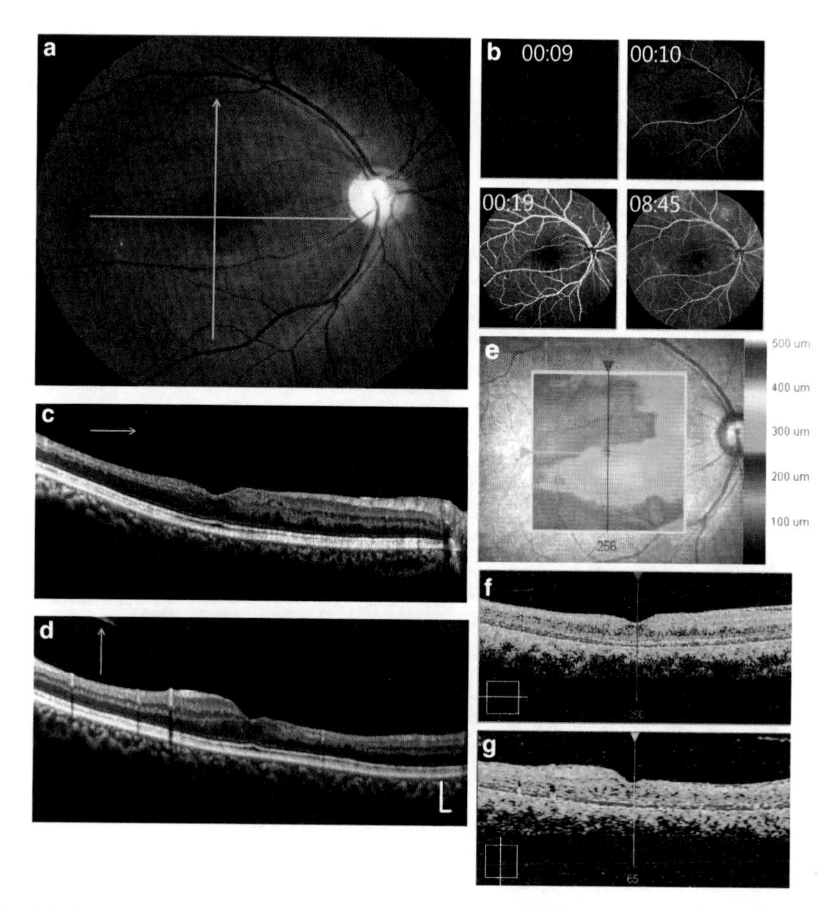

Fig. 2.4 Late change of ischemic damage. 38-year-old hypertensive and diabetic patients with BRAO in the *right* eye (8 months after the onset of disease). (**a**) Fundus examination shows irregular shaped foveal annular reflex due to atrophic change of the superior macula. (**b**) In fluorescein angiography, delayed dye filling in the superior temporal retinal artery and a delay in dye filling of the retinal vein in the corresponding area are compared with inferior temporal vascular arcade. However, AV transit time is within normal limits even in the affected vascular arcade. In many cases of retinal vascular obstruction, the retinal circulation is restored to a variable extent with time via various anastomosis. (**c**) On the horizontal speckle-noise-reduced OCT scan, ischemic damaged area at temporal macula shows retinal thinning with indistinguishably homogenized change in the inner layers. However, the outer retinal layer remains relatively intact. On this section of OCT, irregular inner retinal thinning is also visible in the nasal fovea. (**d**) On the vertical OCT scan, the superior retina, especially in the macular area, shows indistinguishably homogenized change in the inner layer. The inferior retina on this section shows incidental thinning in both the ganglion cell layer and the nerve fiber layer. (**e**) Retinal thickness map through cube scan shows superior macular thinning from arterial occlusion and retinal thinning at the inferior arcade from nerve fiber layer/ganglion cell layer thinning. (**c, d, f, g**) Speckle-noise-reduced OCT can provide more detailed image for layer-by-layer analysis compared to cube scan OCT image

With time, homogenous, diffuse, and acellular zone replaces ischemic tissue; thus, inner half of the retina becomes homogenized on OCT. The boundaries among the different retinal layers in the inner half of the retina become obliterated (Fig. 2.4).

2.10 CRAO with Cilioretinal Artery Sparing

Cilioretinal artery supplies the macular region in about 32% of eyes [7]. In case of CRAO, the presence of cilioretinal blood supply to the macula can help preserve central vision [7, 49] (Fig. 2.5).

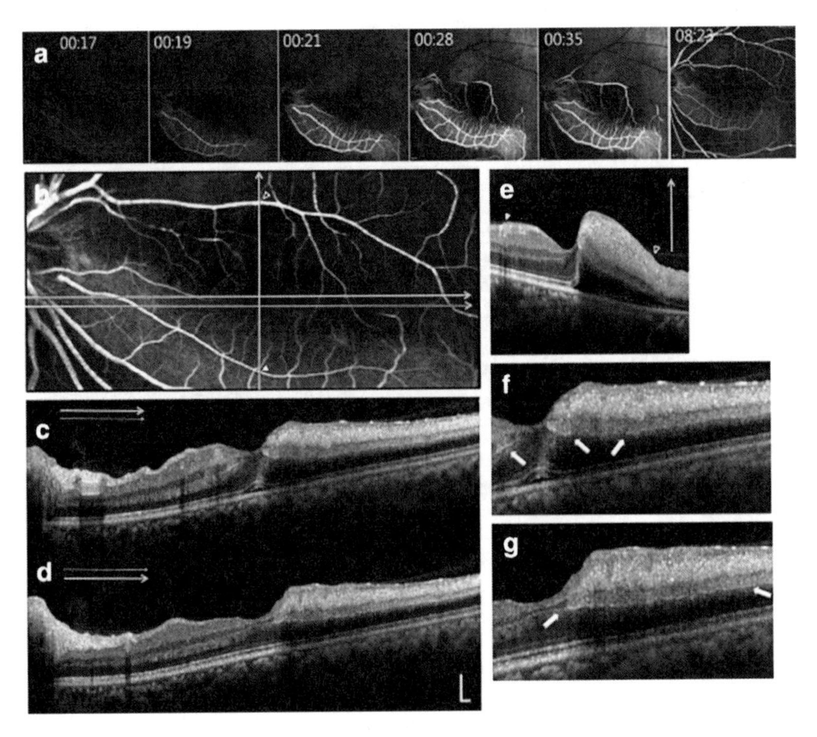

Fig. 2.5 Acute CRAO with cilioretinal artery sparing. 52-year-old male with CRAO with acute cilioretinal artery sparing. (**a**) In the early-phase fluorescein angiography, a large inferotemporal cilioretinal artery shows early-phase dye filling and subsequent dye filling in adjacent vein. Retrograde filling of an arteriole at the superior macula is noticed. (**c–g**) On OCT, increased reflectivity and mild thickening of the retina are noticed at nonperfused area. There is sharp demarcation between perfused area and nonperfused ischemic area. However, "prominent MLM," another important sign of inner retinal ischemia, is sometimes helpful in determining ischemic damage in ambiguous area (**f, g**, *thick arrows*). Large retinal vessels are visible on OCT, and perfused cilioretinal artery is visible as vacant round structure on the vertical OCT section (**b, e**, *solid triangle*). Collapse of the downstream circulation and increased reflectivity of adjacent tissue sometimes obscure the contour of vessels on OCT (**b, e**, *vacant triangle*). However, evaluation of perfusion status by direct analysis of the vessel is not possible in the current commercially available OCT

2.11 Cilioretinal Artery Occlusion

Cilioretinal artery occlusion, which accounts for about 5% of retinal arterial occlusion diseases, involves sudden visual loss [43].

Clinically, cilioretinal artery occlusion can be divided into three distinctive patterns, including isolated occlusion, central retinal vein occlusion (CRVO) with cilioretinal occlusion, and anterior ischemic optic neuropathy with cilioretinal occlusion [49–51].

Inner retinal whitening corresponding to the distribution of cilioretinal artery can be noticed in fundus examination. In FA, cilioretinal artery, which normally fills just before the CRA, shows delayed perfusion or lack thereof [7].

Signs of acute inner retinal ischemia can also be noticed on OCT at the corresponding ischemic area. In particular, acute ischemic retinal sign of "prominent MLM" provides an additional diagnostic clue, especially in cases showing subtle fundus changes.

2.12 Branched Retinal Artery Occlusion

Fundus examination shows an area of retinal whitening along the distribution of the obstructed arterial branch. Right eye is more commonly affected in BRAO because cardiac emboli travel more readily through right carotid artery [52]. Emboli are visible in fundus exam up to 68% [53].

In FA, the characteristics of BRAO are similar to those of CRAO. Arterial filling delay is visible in the affected artery. In some cases, retrograde filling of the occluded vessel from adjacent vessel can be seen [2, 54].

OCT shows sharp distinction between acute ischemic area and normally perfused area [33, 34]. Acute ischemic signs (increased reflectivity, mild thickening, and "prominent MLM") can be noticed in ischemic area [33, 34] (Fig. 2.6).

Over time, sectoral thinning with homogenized changes of inner half of the retina takes place, which can be observed on OCT [23, 41] (Fig. 2.3).

2.13 Ophthalmic Artery Occlusion

Ophthalmic artery splits into the central retinal artery and the ciliary artery. The ciliary artery supplies uveal perfusion both posteriorly (posterior ciliary artery) and anteriorly (anterior ciliary artery). Ophthalmic artery occlusion, which is proximal to the branching of the central retinal artery and the ciliary artery, causes a global ischemia of the entire globe.

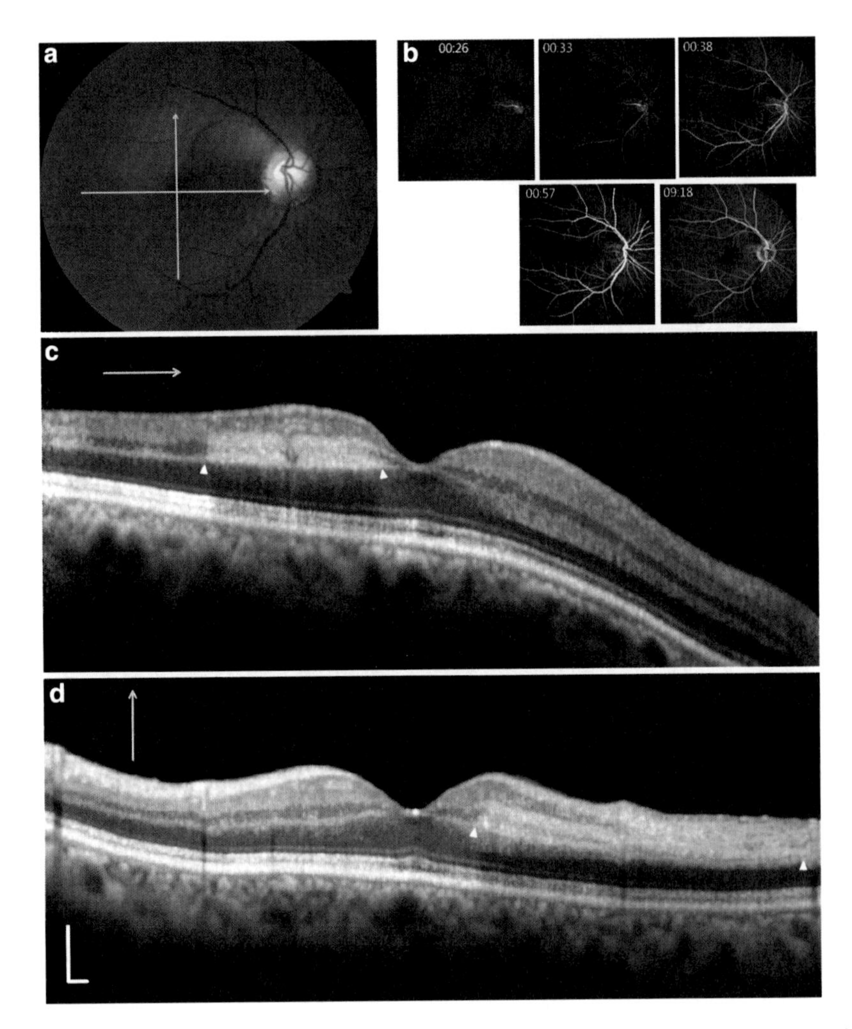

Fig. 2.6 Acute BRAO. 64-year-old female with sudden onset of visual disturbance 2 weeks ago. (**a**) In fundus examination, geographic pale change of the retina is noticed at the superior macular area. (**b**) In the early-phase fluorescein angiography, dye filling of cilioretinal artery at the peripapillary is noticed. Dye filling is delayed at the superior temporal artery compared to that at the inferior temporal artery. Delayed AV transit time at the superior temporal arcade is noticed. (**c, d**) On the horizontal and vertical OCT, increased reflectivity and mild thickening in the inner retina are noticed at the corresponding area of pale retina in fundus. There is sharp demarcation between perfused area and ischemia area. Attenuation of the outer retinal reflectivity may be caused by shadowing from increased inner retinal reflectivity (*white triangles*). In this case, the inner aspect of "prominent middle limiting membrane" is fused with highly increased reflectivity in the inner nuclear layer (*white triangles*)

Ophthalmic artery occlusion results in complete retinal ischemia and infarction. The most common cause is giant cell arteritis [46].

Initial retinal opacity of the entire retina can be seen in a similar manner to that of CRAO. "Cherry-red spot" is visible in only about a half of cases, which

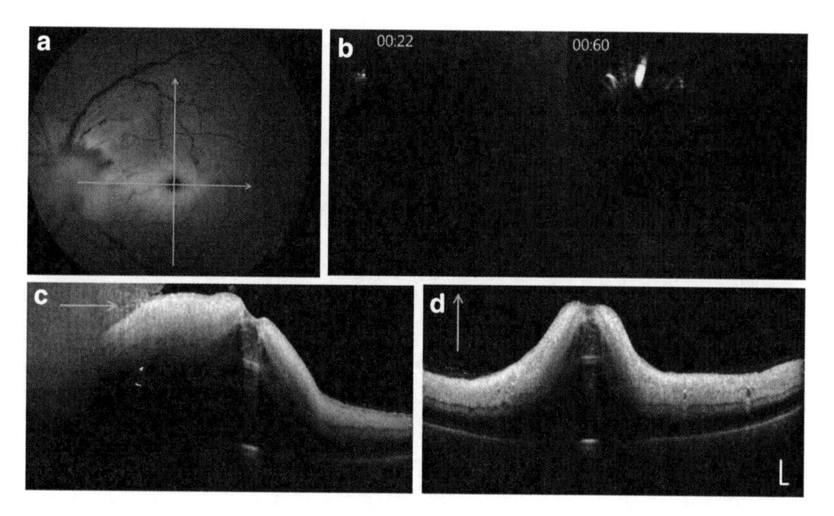

Fig. 2.7 Ophthalmic artery occlusion. 57-year-old female with sudden visual loss of no light perception in the eye. (**a**) Fundus examination shows extensive *whitish yellow* change of the retina and severely attenuated arteries. Severe optic disc swelling and retinal hemorrhage, especially in the peripapillary area, are noticed. Exudative materials are seen in front of the optic disc. (**b**) In fluorescein angiography, marked dye filling abnormalities in both the choroid and the retinal vessels are noticed. (**c, d**) Severe thickening with increased reflectivity was noted at the inner retinal area. Swelling of the outer retina, especially in the foveal area, is also noticed. At the foveal center, little light attenuation by relatively transparent foveola is responsible for the "*cherry-red* spot" appearance in fundus exam as in cases with CRAO. In cases with choroidal circulation insufficiency, the outer retinal changes are expected. However, due to severe light attenuation by increased reflectivity in the inner retina, the outer retinal changes cannot be evaluated in this case. Combined severe outer retinal thickening might be related with choroidal perfusion insufficiency

is believed to be due to the lack of choroidal perfusion [45] (Fig. 2.7). However, there is controversy over the explanation for the visibility of central red spot in the ophthalmic artery occlusion case (Fig. 2.4).

The differences from CRAO are that optic atrophy can develop with time. Also, varying amounts of pigmentation can develop due to the lack of perfusion of the RPE [48] (Fig. 2.8). This pigment is generally diffusely scattered throughout the posterior pole, but it may also be seen in greater amount in the periphery [46].

2.14 Cotton Wool Spots

Cotton wool spots represent microinfarction of nerve fiber layer resulting from focal arteriole occlusion or capillary obstruction. Focal hypoxia, which obstructs axoplasmic flow, results in accumulation of axoplasmic debris in the nerve fiber layers of the retina [55].

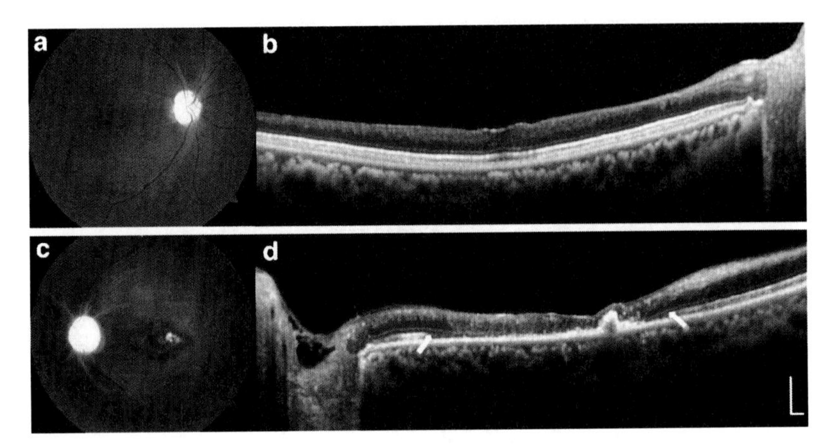

Fig. 2.8 Late OCT changes in central retinal artery occlusion (CRAO) vs. ophthalmic artery occlusion. 60-year-old female with 12-month history of vision loss from a central retinal artery occlusion (**a, b**). (**a**) In CRAO, fundus examination shows a loss of foveal annular light reflex at posterior pole and pale optic disc. (**b**) On OCT, severe retinal thinning, especially in the inner retina, is noticed. Inner retinal layers show indistinguishably homogenized change. The boundaries between the different retinal layers in the inner half of the retina become obliterated. The outer retina remains intact. 57-year-old female in 12 months after ophthalmic artery occlusion (Fig. 2.7) (**c, d**). (**c**) In ophthalmic artery occlusion, loss of foveal annular reflex and pigmentary atrophic change due to choroidal perfusion insufficiency in the posterior pole are visible in fundus examination. Compared to CRAO, the optic disc shows more severe atrophic change. (**d**) On OCT, atrophic changes in both the inner and outer retina are noticed; especially severe atrophic change is more prominent in the posterior pole. The inner retina shows indistinguishably homogenized change, and the outer retina shows thinning and loss of IS/OS line (*arrow*) with irregular contour of RPE line due to pigmentary proliferation

Cotton wool spots clinically disappear within 4–12 weeks; however, they can still be detected as hyperreflective nodules on OCT [56]. After a long time, thinning of the inner retinal layer occurs, but the overall thickness of the retina remains relatively intact with dragging of the outer retina at the corresponding area of cotton wool spots [55].

2.15 Combined CRAO and CRVO

In fundus examination, characteristics of both diseases including whitening of the retina associated with cherry-red spots, optic disc edema, retinal hemorrhages, and venous engorgement can be observed [36, 57]. Later, optic atrophy and extreme narrowing of the retinal vessels can develop [58].

On acute-phase OCT, the features of acute retinal ischemia such as mild macular thickening can be noticed. Late OCT findings include inner retinal ischemic atrophy identical to that seen in long-standing CRAO [57].

2.16 Ocular Ischemic Syndrome

Ocular ischemic syndrome (OIS) is a hypoperfusion retinopathy related with carotid artery insufficiency [59, 60].

Dot and blot hemorrhages develop predominantly in the midperiphery. Microaneurysms are frequently visible in midperipheral area sparing macula. Retinal veins are dilated, but they show minimal tortuosity [59]. "Cherry-red spot" occurs in 12% of eyes [61]. Optic disc usually shows no severe swelling in OIS, in contrast to CRVO.

In FA, slow arm to retina or delayed AV transit time indicates ophthalmic perfusion insufficiency [59]. Microaneurysm or capillary nonperfusion area can be noticed in the midperiphery; however, they usually do not involve the posterior pole.

Retinal thinning on OCT, especially the thinning of inner retina, might result from chronic perfusion abnormality, but this is not a specific finding of OIS. Distinctive features usually develop in midperipheral area; thus, OCT alone is not suitable for diagnosis of OIS. Occasionally, such features can be found in posterior pole, and cystoid macular edema can sometimes be accompanied [62]. Capillary nonperfusion area, visible as severe atrophy of inner retina on OCT, can be seen in some cases [63] (Fig. 2.9).

2.17 Evaluation of Retinal Perfusion State on OCT

The term "ischemia" implies a hypoperfusion state, in which perfusion is down to the level of causing a tissue damage, and not just a hypoperfusion [4]. In evaluating retinal perfusion in FA, a "definite" delay in dye filling is specific for ischemic damage; however, determining the critical point of hypoperfusion state which causes tissue damage is not possible. Also, retinal circulation can revert to normal even after causing permanent ischemic damage. On the other hand, identifying actual changes or loss of retinal tissue on OCT seems to be objective and more reliable indicator for confirming ischemic damage (Fig. 2.10).

2.18 Central retinal vein occlusion (CRVO)

Depending on the site of the occlusion, one can differentiate central retinal vein occlusion (CRVO) and branch retinal vein occlusion (BRVO). Retinal venous occlusion shows variable clinical features according to extent or degree of obstruction and existence of collateral circulation [64].

In many acute CRVO cases, retinal hemorrhages and macular edema hinder the early evaluation of perfusion status in FA or ischemic change on OCT [64]. After

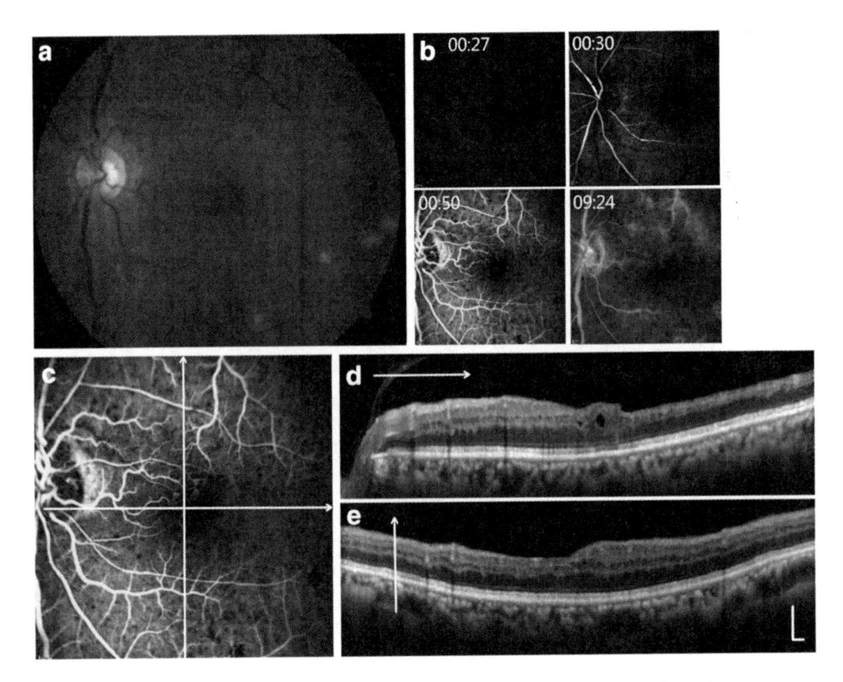

Fig. 2.9 Ocular ischemic syndrome with capillary dropout. 80-year-old male with severe stenosis of left carotid artery. (**a**) Fundus shows multiple small retinal hemorrhages and attenuated arterioles. Retinal vein shows dilation but minimal tortuosity. (**b**) In fluorescein angiography (FA), prolonged and delayed patch dye filling in choroid with marked delay of retinal arterial filling is noticed. AV transit time is also much delayed. Capillary nonperfusion area involving fovea is located at temporal macular area. The late-phase FA shows staining of retinal vessels in both arterioles and venules. Optic disc shows hyperfluorescence with staining. (**c**) The early-phase FA shows capillary dropout area at the temporal macula. (**d**) The horizontal OCT scan shows severe inner retinal thinning with obliterated boundaries between the different retinal layers at the corresponding capillary dropout area. But the overall retinal thickness, especially at the inner retina, seems to be thinner. Incidental vitreous traction at the fovea is also noted. (**e**) The vertical OCT scan shows retinal thinning at the inferior foveal area where capillary damage is visible near the foveal avascular zone area

retinal hemorrhage is absorbed, images of adequate quality can be obtained and nonperfusion area can be evaluated by FA [64]. Acute ischemic sign or retinal thickness changes on OCT can be detected [65].

Late sequela of Central retinal vein occlusion (CRVO) is ischemic atrophy of inner retina. However, outer retinal atrophy is not uncommonly observed, especially in macular area. This might be caused by accompanied choroidal perfusion insufficiency, but chronic macular edema seems to be more responsible for the late atrophic change of outer retina [66]. We cannot clearly differentiate whether the retinal damage results from chronic edema or ischemia by OCT.

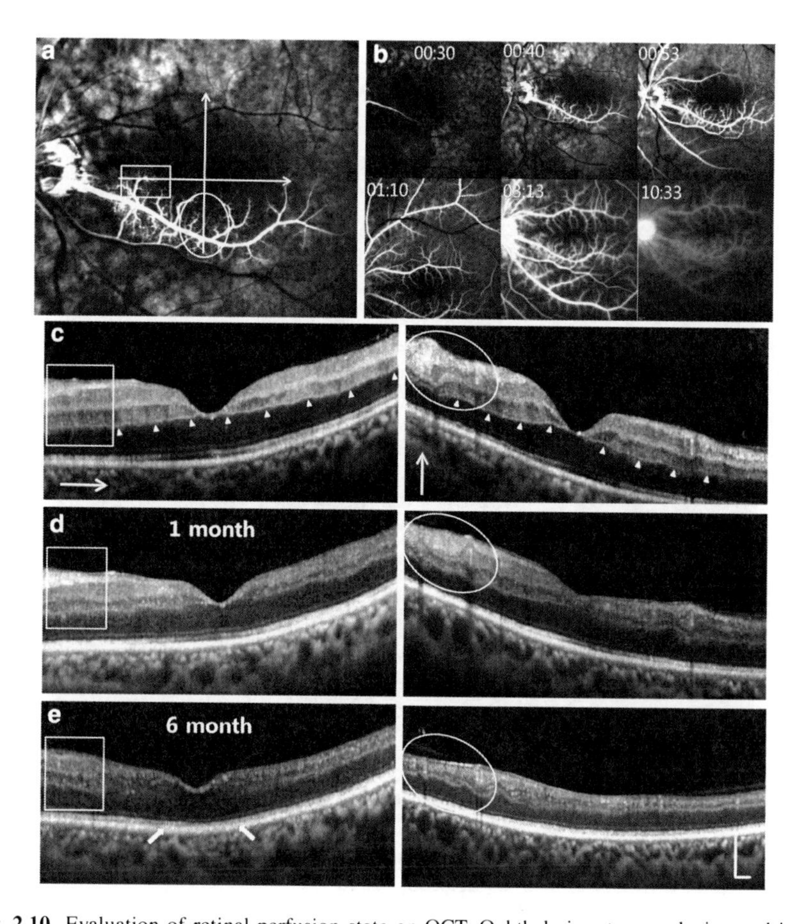

Fig. 2.10 Evaluation of retinal perfusion state on OCT. Ophthalmic artery occlusion and insufficiency of remained cilioretinal artery. (**a**) The early-phase FA shows delayed choroidal dye filling of patch pattern, and the cilioretinal artery is visible at the inferior macula area. (**b**) In fluorescein angiography (FA), a delay in choroidal dye filling of abnormal pattern and much delayed arterial filling are noted. Before the dye filling of retinal arteries, cilioretinal vessel at the inferior macular area shows dye filling. Cilioretinal artery seems to fill completely; however, filling time is much delayed compared to normal cilioretinal circulation. At 30 s after injection, the cilioretinal vessel starts to fill, implying insufficient perfusion state. (**c**) On both horizontal and vertical OCT, increased reflectivity with mild inner retinal thickening is noticed. "Prominent middle limiting membrane" (MLM) is also well defined at the ischemic area on both the horizontal and vertical OCT (*white triangles*). The presumed perfused areas from the cilioretinal artery on FA are marked as *white square* or *ellipse* at the corresponding areas on both FA and OCT images. At initial visit, the area perfused by cilioretinal artery shows irregular but mild increased reflectivity and vague MLM compared to nonperfused area (**c**, *white square*). On the vertical OCT section, the retinal area perfused by cilioretinal artery shows increased reflectivity with prominent MLM (**c**, *white ellipse*). (**d, e**) During the follow-up period, progressive inner retinal thinning occurs, and boundaries between the different retinal layers in the inner retina become obliterated with time. However, the area perfused by cilioretinal artery shows relatively less thinning and less morphologic distortion compared to the adjacent retina (**d**, **e**, *white ellipse*). Acute ischemic sign including "prominent MLM" reflects the extent of the ischemic damage in the early phase (**c**). The outer retina shows progressive thinning and loss of IS/OS junction, especially at the foveal area, which may result from the accompanying choroidal insufficiency (**e**, *white arrow*)

2.19 Central retinal vein occlusion

CRVO is usually classified into two subtypes, ischemic or nonischemic, based on the clinical and angiographic evidence of the degree of retinal ischemia [2, 64].

Changes caused by retinal infarction as in acute arterial obstruction show relatively consistent histological changes; however, hypoxic retinal changes in ischemic CRVO are quite variable in each case [64, 66] (Fig. 2.11).

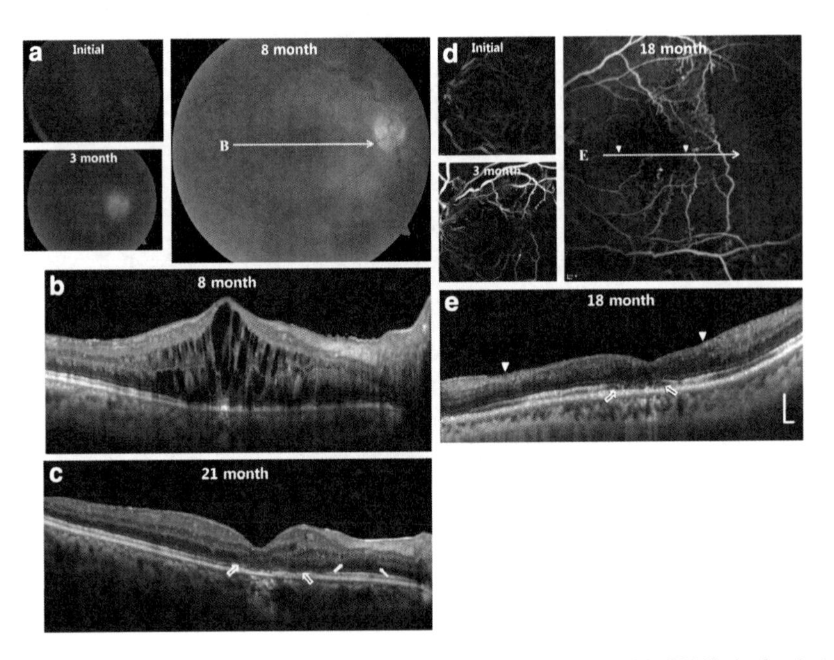

Fig. 2.11 Central retinal vein occlusion (CRVO). 59-year-old female with CRVO (**a, b, c**). (**a**) At initial visit, severe splint retinal hemorrhage with disc swelling was noticed. Three months later, retinal hemorrhage was still quite severe, and mild vitreous opacity was developed. Eight months after initial visit, media opacity is cleared. Retinal veins are dilated and became tortuous. Optociliary shunt vessels are visible at the optic disc. (**b**) OCT shows that severe cystoid changes are mainly located in the inner nuclear layer, the outer plexiform layer, and the outer nuclear layer. (**c**) After 21 months, macular edema was resolved. Atrophic changes of the retina are prominent in the outer nuclear and the outer plexiform layers (*vacant arrows*). This outer retinal changes may be associated with the damage from chronic macular edema. At the nasal macular area, retinal thinning including INL and GCL is noticed. This inner retinal thinning is thought to be related with ischemic damage. 52-year-old female with CRVO (**d, e**). (**d**) Early-phase fluorescein angiography (FA) shows severe tortuosity of retinal vessels, but evaluation of capillaries is not possible due to hemorrhage and retinal opacity. Three months later, capillary dropout area can be detected after absorption of hemorrhage [76]. Extensive capillary dropout area and severely damaged foveal avascular zone are shown in FA after 18 months. (**e**) Horizontal OCT scan dissecting the foveal center shows severe inner retinal atrophy with homogenous obliteration of different inner retinal layers in the corresponding capillary nonperfusion area (**d, e**, *white arrow*). Atrophic retinal changes of outer plexiform layer and outer nuclear layer in the foveal area may be more related with the damage from chronic macular edema (*vacant arrows*)

Current OCT is not suitable for evaluation of peripheral nonperfusion area. However, macular perfusion status is one of the important prognostic factors for both predicting visual outcome and differentiating ischemic from nonischemic type [66, 67]. In recent onset CRVO, cystoid macular edema or retinal hemorrhages are frequently accompanied, thereby prohibiting a detailed examination of retinal layers on OCT [68] (Fig. 2.11). However, detection of acute ischemic changes of the macula is possible on OCT, which is helpful in determining the extent of ischemic retinal damage. With time, atrophic change of the inner retina is noticed on OCT [65, 69, 70]. But differentiating the damage by concomitant macular edema from true ischemic damage on OCT is a challenge [69–75] (Fig. 2.11).

2.20 Diabetic Retinopathy

Diabetic retinopathy is still one of the most common causes of blindness, and it is clinically manifested by microangiopathy of retinal circulation. Retinal vascular changes start with telangiectatic or microaneurysmal changes of capillaries, and capillary closures develop afterward. Vascular leakage and resultant macular edema, ischemic change, and secondary vasoproliferation are main pathologies responsible for the permanent visual loss.

Capillary closure or ischemia causes new vascular proliferation. Secondary hemorrhage or fibrosis leads to severe visual loss. Capillary closure tends to occur on the arterial side of the microaneurysm. Initially, this vascular closure starts in isolated capillaries, eventually extending to arterioles [77]. Capillary dropout usually starts in the midperipheral retina, although the central retina is also often involved [78]. Since current OCT is not useful for analyzing peripheral lesion, FA becomes valuable for the detection of ischemic damage in the periphery.

2.21 Diabetic Ischemic Maculopathy

Diabetic macular ischemia usually starts with the enlargement of FAZ [77, 79]. This condition is not quite uncommon, but it draws little attention to it. Even with some extent of FAZ damage, visual acuity is still maintained normal until severe damage occurs. Also, this condition is difficult to detect without special diagnostic imaging tool (FA). Most importantly, there is a lack of treatment options until now.

However, diabetic macular ischemia is one of the important causes of visual loss and poor response to treatment of diabetic macular edema [38, 80]. Even with normal visual acuity, visual function is deteriorated in the early phase of FAZ enlargement [38, 81–84].

In funduscopic examination, macular ischemia shows "featureless" areas of ischemic retina. Dilated capillaries or microaneurysms are sometimes visible at the

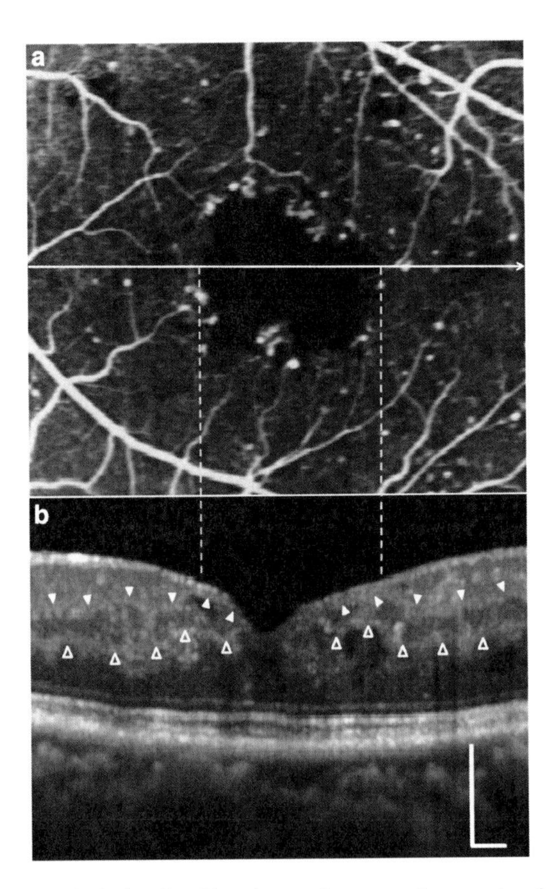

Fig. 2.12 Diabetic macular ischemia—Foveal avascular zone enlargement and ganglion cell layer loss. 53-year-old female with diabetic macular ischemia. (**a**) In fluorescein angiography, irregular foveal avascular zone (FAZ) damage with capillary dilation and microaneurysms at the margin is visible. (**b**) OCT shows thinning of ganglion cell layer (GCL) (*solid triangle*; inner plexiform layer). Thickness of inner nuclear layer (INL) is relatively maintained compared to that of GCL (*vacant triangle*; synaptic portion of outer plexiform layer). Superficial capillary plexus, which supplies GCL and nerve fiber layer, is mainly visible in fluorescein angiography; thus, "enlargement of FAZ" is primarily related with loss of foveal GCL. However, when arterioles are involved, both GCL and INL become thin

boundary surrounding the ischemic area [85]. Focal depressions or loss of annular light reflex in the macula are detectable due to atrophy from ischemic infarcts [47].

In FA, retinal ischemia area shows relative hypoflourescence with nonfilling of macular capillaries. Enlargement and irregularity of the FAZ, and intervening capillary loss (increased perifoveal intercapillary area; widened, nonuniform spaces between macular capillaries) are the hallmarks of macular ischemia [77, 79, 86] (Fig. 2.12). In advanced cases, arterioles are also damaged, thereby showing focal narrowing, prunting, staining, or blurred contour in FA [86, 87].

Neurons in the GCL and INL are totally dependent on oxygen derived from retinal circulation. Any small capillary loss leads to neural tissue loss in the corresponding area [14]. GCL, which is nourished by superficial capillary plexus, is damaged first in diabetic macular ischemia. Enlargement of FAZ can be visualized as thinning or loss of foveal GCL [14] (Fig. 2.12). In the area where arterioles are damaged, both GCL and INL thinning can be visualized [14, 88].

Grading or evaluation of FAZ enlargement with OCT can provide additional information as compared with FA alone [14, 89]. Considering variability of the size of FAZ in each individual, evaluation of the size of predamaged FAZ is important for grading of FAZ damage [90, 91] (Fig. 2.13). In FA, the center of FAZ is usually considered as the anatomic center of fovea [86, 91–93]. Using OCT, we can accurately determine the true anatomic center of fovea in damaged FAZ; thus, we can more exactly estimate the size of predamaged FAZ [14] (Fig. 2.13).

As the size of FAZ is variable in each individual, the extent of centrifugal displacement of GCL and NFL is also variable from subjects to subjects (Fig. 2.1). This implies that there is topographical variability of foveal retinal circuitry in each individual [14]. Thus, this variability of retinal circuitry and lower reliability of current grading system for ischemic damage seem to be responsible for unpredictability of visual function in diabetic ischemic maculopathy [14, 77]. Observation of actual tissue loss on OCT would be more promising in grading of macular ischemia.

OCT provides us with more objective information about actual tissue damage, even when FA cannot provide enough information due to unusual damage pattern or image quality problem [14]. Also, diabetic macular ischemia usually progresses incidentally. In FA, both loss of capillaries and nonperfusion area are visible as hypofluorescence. OCT is helpful in differentiating the acute ischemic damage from old lesion using acute ischemic signs (increased inner retinal reflectivity or prominent MLM and loss of tissue (retinal layer) on OCT).

2.22 Choroidal Ischemia

Appearance of the ischemic choroidal lesion varies, depending on the size of the vessel occluded and the severity of ischemia.

The occlusion of larger choroidal artery can be detected as a wedge-shaped whitish retina. Subsequently, pigmentation may develop in the corresponding area of lesion. The occlusion of small terminal choroidal arteriole results in small patch area of whitening, with subsequent overlying pigmentation (Elschnig's spots) [94, 95].

Severe ischemia results in the infarction of choroid, RPE, and outer retina. Mild ischemia produces ischemic dysfunction of the RPE and blood–retinal barrier, resulting in serous retinal detachment. Various systemic diseases, such as malignant hypertension and preeclampsia, are related with development of serous retinal detachment [95, 96].

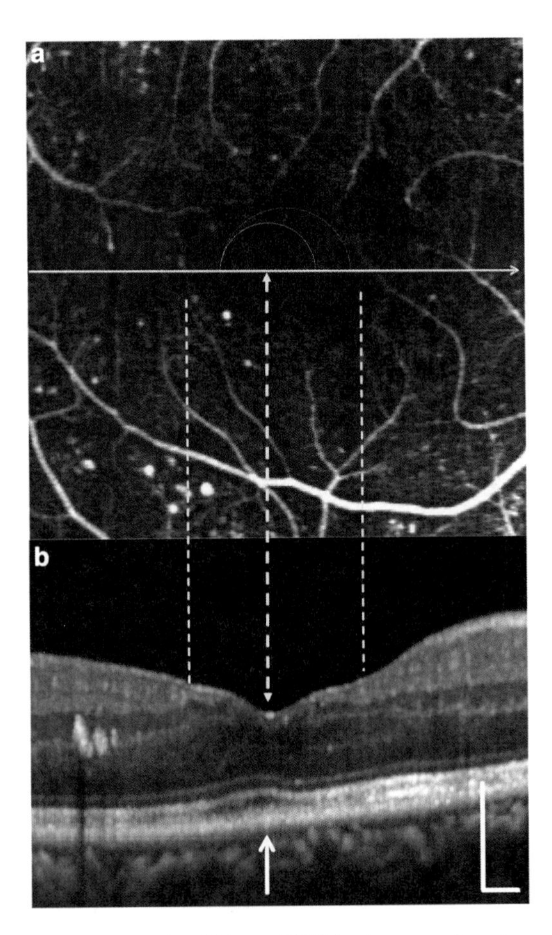

Fig. 2.13 Diabetic macular ischemia—Value of OCT in the evaluation of macular ischemia. 61-year-old male with diabetic macular ischemia. (**a**) In fluorescein angiography (FA), irregular enlargement of foveal avascular zone (FAZ) with suspicious arteriolar damage near FAZ is noticed. (**b**) FAZ visible in FA is well matched to the ganglion cell layer (GCL)-damaged area on OCT. On OCT, thinning of both GCL and inner nuclear layer (INL) at fovea, especially on the nasal side, is noticed. INL thinning implies there might be accompanying arteriolar damage. FAZ damage does not always progress symmetrically; thus, the estimation of the size of predamaged FAZ with only FA using remaining visible capillaries is not accurate (*large semicircle*). Determination of true anatomic center of fovea is very important for exact estimation of the size of the predamaged FAZ. We can precisely determine the true center of fovea using "external fovea" on OCT (*arrow*). We can more precisely assume the predamaged FAZ size (*small semicircle*). However, topographical comparison with the remained INL on OCT reveals that predamaged FAZ size might be much smaller than that estimated by FA in this case

On OCT, disrupted choriocapillaries, atrophic change of RPE, outer retinal pigment creeping, or outer retinal thinning can be found in the ischemic choroidal area [94]. Choroidal thickness had been measured using enhanced depth imaging (EDI) SD-OCT and was found to be variable topographically within the posterior pole. The thickness of the choroid showed a negative correlation with age [76]. However, choroidal thickness is quite variable in each individual, and choroidal thickness itself cannot reflect the extent of choroidal ischemia. Current EDI SD-OCT image is not so detailed enough to identify the ischemic choroidal changes. With further development of SD-OCT using longer wavelength, we could explore the ischemic and other pathologic changes of choroid in vivo in the near future.

References

1. T. Bek, Inner retinal ischaemia: current understanding and needs for further investigations. Acta Ophthalmol. **87**(4):362–367 (2009)
2. S. Hayreh, Prevalent misconceptions about acute retinal vascular occlusive disorders. Prog. Retin. Eye Res. **24**(4), 493–519 (2005)
3. M. Yanoff, B.S. Fine, *Ocular Pathology: A Text and Atlas*, 3rd ed. (Lippincott, Philadelphia, 1988), pp. 383–394
4. N.N. Osborne, R.J. Casson, J.P.M. Wood, G. Childlow, M. Graham, J. Melena, Retinal ischemia: mechanisms of damage and potential therapeutic strategies. Prog. Retin. Eye Res. **23**(1), 91–147 (2004)
5. R.W. Green, Retinal ischemia; vascular and circulatory conditions and diseases, in *Ophthalmic Pathology: An Atlas and Textbook*, ed by W.H. Spencer (Saunders, Philadelphia, 1985), p. 655
6. M.J. Hogan, J.A. Alvarado, J.E. Weddell, *Histology of the Human Eye; An Atlas and Textbook* (Saunders, Philadelphia, 1971)
7. J. Justice Jr., R.P. Lehmann, Cilioretinal arteries. A study based on review of stereo fundus photographs and fluorescein angiographic findings. Arch. Ophthalmol. **94**(8), 1355–1358 (1976)
8. M.J. Hogan, J.A. Alvarado, J.E. Weddell, *Histology of the Human Eye*. 2nd ed. (Saunders, Philadelphia, 1971), pp. 448–471
9. J.M. Provis, Development of the primate retinal vasculature. Prog. Retin. Eye Res. **20**(6), 799–821 (2001)
10. J.M. Provis, A.E. Hendrickson, The foveal avascular region of developing human retina. Arch. Ophthalmol. **126**(4), 507–511 (2008)
11. R.S. Weinhaus, J.M. Burke, F.C. Delori, D.M. Snodderly, Comparison of fluorescein angiography with microvascular anatomy of macaque retinas. Exp. Eye Res. **61**(1), 1–16 (1995)
12. P.L. Penfold, M.C. Madigan, J.M. Provis, Antibodies to human leucocyte antigens indicate subpopulations of microglia in human retina. Vis. Neurosci. **7**(4), 383–388 (1991)
13. D.Y. Yu, S.J. Cringle, Oxygen distribution and consumption within the retina in vascularised and avascular retinas and in animal models of retinal disease. Prog. Retin. Eye Res. **20**(2), 175–208 (2001)
14. S.H. Byeon, Y.K. Chu, H. Lee, S.Y. Lee, O.W. Kwon, Foveal ganglion cell layer damage in ischemic diabetic maculopathy: correlation of optical coherence tomographic and anatomic changes. Ophthalmology **116**(10), 1949–1959.e8 (2009)
15. B. Sander, M. Larsen, C. Engler, H. Lund-Andersen, Absence of foveal avascular zone demonstrated by laser scanning fluorescein angiography. Acta Ophthalmol. **72**(5), 550–552 (1994)
16. D.M. Foreman, S. Bagley, J. Moore, G.W. Ireland, D. McLeod, M.E. Boulton, Three dimensional analysis of the retinal vasculature using immunofluorescent staining and confocal laser scanning microscopy. Br. J. Ophthalmol. **80**(3), 246–251 (1996)
17. S. Duke-Elder, *System of Ophthalmology*, vol. 2, The Anatomy of the Visual System (Kimpton, London, 1961)
18. S.S. Hayreh, T.A. Weingeist, Experimental occlusion of the central artery of the retina. I. Ophthalmoscopic and fluorescein fundus angiographic studies. Br. J. Ophthalmol. **64**(12), 896–912 (1980)

19. N.J. David, E.W.D. Noron, J.D. Gass, J. Beauchamp, Fluorescein angiography in central retinal artery occlusion. Arch. Ophthalmol. **77**(5), 619–629 (1967)
20. K.A. Kwiterovich, M.G. Maquire, R.P. Murphy, A.P. Schachat, N.M. Bressler, S.L. Fine, Frequency of adverse systemic reactions after fluorescein angiography. Results of a prospective study. Ophthalmology **98**(7), 1139–1142 (1991)
21. F.G. Holz, R.F. Spaide, *Medical Retina: Focus on Retinal Imaging* (Springer, Heidelberg, 2010)
22. D. Gold, Retinal arterial occlusion. Trans. Sect. Ophthalmol. Am. Acad. Ophthalmol. Otolaryngol. **83**(3 Pt 1), OP392–OP408 (1977)
23. R.K. Murthy, S. Grover, K.V. Chalam, Sequential spectral domain OCT documentation of retinal changes after branch retinal artery occlusion. Clin. Ophthalmol. **26**(4), 327–329 (2010)
24. K. Shinoda, K. Yamada, C.S. Matsumoto, K. Kimoto, K. Nakatsuka, Changes in retinal thickness are correlated with alterations of electroretinogram in eyes with central retinal artery occlusion. Graefes Arch. Clin. Exp. Ophthalmol. **246**(7), 949–954 (2008)
25. D. Schmidt, T. Kube, N. Feltgen, Central retinal artery occlusion: findings in optical coherence tomography and functional correlations. Eur. J. Med. Res. **11**(6), 250–252 (2006)
26. M.C. Kincaid, Uveal melanoma. Cancer Control **5**(4), 299–309 (1998)
27. A.J. Kroll, Experimental central retinal artery occlusion. Arch. Ophthalmol. **79**(4), 453–469 (1968)
28. R.L. Radius, D.R. Anderson, Morphology of axonal transport abnormalities in primate eyes. Br. J. Ophthalmol. **65**(11), 767–777 (1981)
29. B.E.I.I. Dahrling, The histopathology of early central retinal artery occlusion. Arch. Ophthalmol. **73**(4), 506–510 (1965)
30. S.S. Hayreh, M.B. Zimmerman, Fundus changes in central retinal artery occlusion. Retina **27**(3), 276–289 (2007)
31. M.P. Lafuente, M.P. Villegas-Pérez, I. Sellés-Navarro, S. Mayor-Torroglosa, J. Miralles de Imperial, M. Vidal-Sanz, Retinal ganglion cell death after acute retinal ischemia is an ongoing process whose severity and duration depends on the duration of the insult. Neuroscience **109**(1), 157–168 (2002)
32. S.M. Falkenberry, M.S. Ip, B.A. Blodi, J.B. Gunther, Optical coherence tomography findings in central retinal artery occlusion. Ophthalmic Surg. Lasers Imaging **37**(6), 502–505 (2006)
33. M. Karacorlu, H. Ozdemir, S. Arf Karacorlu, Optical coherence tomography findings in branch retinal artery occlusion. Eur. J. Ophthalmol. **16**(2), 352–353 (2006)
34. W. Cella, M. Avila, Optical coherence tomography as a means of evaluating acute ischaemic retinopathy in branch retinal artery occlusion. Acta Ophthalmol. Scand. **85**(7), 799–801 (2007)
35. K. Arashvand, Images in clinical medicine. Central retinal artery occlusion. N Engl J Med. **356**(8), 841 (2007)
36. S.G. Schwartz, M. Hickey, C.A. Puliafito, Bilateral CRAO and CRVO from thrombotic thrombocytopenic purpura: OCT findings and treatment with triamcinolone acetonide and bevacizumab. Ophthalmic Surg. Lasers Imaging **37**(5), 420–422 (2006)
37. T. Bek, T. Ledet, Vascular occlusion in diabetic retinopathy. A qualitative and quantitative histopathological study. Acta Ophthalmol. Scand. **74**(1), 36–40 (1996)
38. N. Unoki, K. Nishijima, A. Sakamoto, M. Kita, D. Watanabe, M. Hangai, T. Kimura, N. Kawagoe, M. Ohta, N. Yoshimura, Retinal sensitivity loss and structural disturbance in areas of capillary nonperfusion of eyes with diabetic retinopathy. Am. J. Ophthalmol. **144**(5), 755–760 (2007)
39. D.M. Kozart, Anatomic correlates of the retina, in *Duane's Clinical Ophthalmolgy*, ed. by W. Tasman, E.A. Jaeger (Lippincott, Philadelphia, 1991), pp. 1–18
40. N.G. Ghazi, E.P. Tilton, B. Patel, R.M. Knape, S.A. Newman, Comparison of macular optical coherence tomography findings between postacute retinal artery occlusion and nonacute optic neuropathy. Retina **30**(4), 578–585 (2010)
41. H. Takahashi, H. Iijima, Sectoral thinning of the retina after branch retinal artery occlusion. Jpn. J. Ophthalmol. **53**(5), 494–500 (2009)
42. C.K. Leung, C.C. Tham, S. Mohammed, E.Y. Li, K.S. Leung, W.M. Chan, D.S. Lam, In vivo measurements of macular and nerve fibre layer thickness in retinal arterial occlusion. Eye (Lond.) **21**(12), 1464–1468 (2007)

43. G.C. Brown, J.A. Shields, Cilioretinal arteries and retinal arterial occlusion. Arch. Ophthalmol. **97**(1), 84–92 (1979)
44. R.M. Burde, M.E. Smith, J.T. Black, Retinal artery occlusion in the absence of a cherry red spot. Surv. Ophthalmol. **27**(3), 181–186 (1982)
45. L.A. Yannuzzi, *The Retinal Atlas* (Elsevier, London, 2010) (Published: JUN-2010 ISBN 10: 0-7020-3320-0)
46. G.C. Brown, L.E. Magargal, R. Sergott, Acute obstruction of the retinal and choroidal circulations. Ophthalmology **93**(11), 1373–1382 (1986)
47. S.S. Hayreh, J.B. Jonas, Ophthalmoscopic detectability of the parafoveal annular reflex in the evaluation of the optic nerve: an experimental study in rhesus monkeys. Ophthalmology **107**(5), 1009–1014 (2000)
48. S.S. Hayreh, J.B. Jonas, Optic disk and retinal nerve fiber layer damage after transient central retinal artery occlusion: an experimental study in rhesus monkeys. Am. J. Ophthalmol. **129**(6), 786–795 (2000)
49. G.C. Brown, J.A. Shields, Cilioretinal arteries and retinal arterial occlusion. Arch. Ophthalmol. **97**(1), 84–92 (1979)
50. S.S. Hayreh, L. Fraterrigo, J. Jonas, Central retinal vein occlusion associated with cilioretinal artery occlusion. Retina **28**(4), 581–594 (2008)
51. S.S. Hayreh, Ischemic optic neuropathy. Prog. Retin. Eye. Res. **28**(1), 34–62 (2009)
52. M.A. Ros, L.E. Magargal, M. Uram, Branch retinal-artery obstruction: a review of 201 eyes. Ann. Ophthalmol. **21**(3), 103–107 (1989)
53. L.A. Wilson, C.P. Warlow, R.W. Russell, Cardiovascular disease in patients with retinal arterial occlusion. The Lancet. **313**(8111), 292–294 (1979)
54. D. Schmidt, A fluorescein angiographic study of branch retinal artery occlusion (BRAO)—the retrograde filling of occluded vessels. Eur. J. Med. Res. **4**(12), 491–506 (1999)
55. M.L. Gomez, F. Mojana, D.U. Bartsch, W.R. Freeman, Imaging of long-term retinal damage after resolved cotton wool spots. Ophthalmology **116**(12), 2407–2414 (2009)
56. I. Kozak, D.U. Bartsch, L. Cheng, W.R. Freeman, Sign in resolved cotton wool spots using high-resolution optical coherence tomography and optical coherence tomography ophthalmoscopy. Ophthalmology **114**(3), 537–543 (2007)
57. G.C. Brown, J.S. Duker, R. Lehman, R.C. Eagle Jr., Combined central retinal artery-central vein obstruction. Int. Ophthalmol. **17**(1), 9–17 (1993)
58. T.D. Duane, *Clinical Ophthalmology* (Harper & Row, Philadelphia, 2002)
59. G.C. Brown, L.E. Magargal, The ocular ischemic syndrome. Clinical, fluorescein angiographic and carotid angiographic features. Int. Ophthalmol. **11**(4), 239–251)(1988)
60. J.B. Mizener, P. Podhajsky, S.S. Hayreh, Ocular ischemic syndrome. Ophthalmology **104**(5), 859–864 (1997)
61. J.A. Pournaras, L. Konstantinidis, T.J. Wolfensberger, Sequential central retinal artery occlusion and retinal vein stasis as a result of ocular ischemic syndrome. Klin. Monbl. Augenheilkd. **227**(4), 338–339 (2010)
62. G.C. Brown, Macular edema in association with severe carotid artery obstruction. Am. J. Ophthalmol. **102**(4), 442–448 (1986)
63. G.D. Sturrock, H.R. Mueller, Chronic ocular ischaemia. Br. J. Ophthalmol. **68**(10), 716–723 (1984)
64. Baseline and early natural history report. The Central Vein Occlusion Study. Arch. Ophthalmol. **111**(8), 1087–1095 (1993)
65. D. Shroff, D.K. Mehta, R. Arora, R. Narula, D. Chauhan, Natural history of macular status in recent-onset branch retinal vein occlusion: an optical coherence tomography study. Int. Ophthalmol. **28**(4), 261–268 (2008)
66. The Central Vein Occlusion Study Group, Natural history and clinical management of central retinal vein occlusion. Arch. Ophthalmol. **115**(4), 486–491 (1997)
67. D.J. Browning, Patchy ischemic retinal whitening in acute central retinal vein occlusion. Ophthalmology **109**(11), 2154–2159 (2002)

68. N. Yamaike, A. Tsujikawa, M. Ota, A. Sakamoto, Y. Kotera, M. Kita, K. Miyamoto, N. Yoshimura, M. Hangai, Three-dimensional imaging of cystoid macular edema in retinal vein occlusion. Ophthalmology **115**(2), 355.e2–362.e2 (2008)
69. M. Karacorlu, H. Ozdemir, S.A. Karacorlu, Resolution of serous macular detachment after intravitreal triamcinolone acetonide treatment of patients with branch retinal vein occlusion. Retina **25**(7), 856–860 (2005)
70. R.F. Spaide, J.K. Lee, J.K. Klancnik Jr., N.E. Gross, Optical coherence tomography of branch retinal vein occlusion. Retina **23**(3), 343–347 (2003)
71. T.H. Williamson, A. O'Donnell, Intravitreal triamcinolone acetonide for cystoid macular edema in nonischemic central retinal vein occlusion. Am. J. Ophthalmol. **139**(5), 860–866 (2005)
72. R.A. Costa, R. Jorge, D. Calucci, L.A. Melo Jr., J.A. Cardillo, I.U. Scott, Intravitreal bevacizumab (Avastin) for central and hemicentral retinal vein occlusions: IBeVO study. Retina **27**(2), 141–149 (2007)
73. J. Hsu, R.S. Kaiser, A. Sivalingam, P. Abraham, M.S. Fineman, M.A. Samuel, J.F. Vander, C.D. Regillo, A.C. Ho, Intravitreal bevacizumab (Avastin) in central retinal vein occlusion. Retina **27**(8), 1013–1019 (2007)
74. T. Murakami, A. Tsujikawa, M. Ohta, K. Miyamoto, M. Kita, D. Watanabe, H. Takagi, N. Yoshimura, Photoreceptor status after resolved macular edema in branch retinal vein occlusion treated with tissue plasminogen activator. Am. J. Ophthalmol. **143**(1), 171–173 (2007)
75. D.J. Pieramici, M. Rabena, A.A. Castellarin, M. Nasir, R. See, T. Norton, A. Sanchez, S. Risard, R.L. Avery, Ranibizumab for the treatment of macular edema associated with perfused central retinal vein occlusions. Ophthalmology **115**(10), e47–e54 (2008)
76. R. Margolis, R.F. Spaide, A pilot study of enhanced depth imaging optical coherence tomography of the choroid in normal eyes. Am. J. Ophthalmol. **147**(5), 811–815 (2009)
77. G.H. Bresnick, R. Condit, S. Syrjala, M. Palta, A. Groo, K. Korth, Abnormalities of the foveal avascular zone in diabetic retinopathy. Arch. Ophthalmol. **102**(9), 1286–1293 (1984)
78. Early Treatment Diabetic Retinopathy Study Research Group, Fluorescein angiographic risk factors for progression of diabetic retinopathy. ETDRS report number 13. Ophthalmology **98**(5 Suppl), 834–840 (1991)
79. A.M. Mansour, A. Schachat, G. Bodiford, R. Haymond, Foveal avascular zone in diabetes mellitus. Retina **13**(2), 125–128 (1993)
80. E.J. Chung, M.I. Roh, O.W. Kwon, H.J. Koh, Effects of macular ischemia on the outcome of intravitreal bevacizumab therapy for diabetic macular edema. Retina **28**(7), 957–963 (2008)
81. O. Arend, S. Wolf, A. Harris, M. Reim, The relationship of macular microcirculation to visual acuity in diabetic patients. Arch. Ophthalmol. **113**(5), 610–614 (1995)
82. O. Arend, S. Wolf, F. Jung, B. Bertram, H. Pöstgens, H. Toonen, M. Reim, Retinal microcirculation in patients with diabetes mellitus: dynamic and morphological analysis of perifoveal capillary network. Br. J. Ophthalmol. **75**(9), 514–518 (1991)
83. D. Talwar, N. Sharma, A. Pai, R.V. Azad, A. Kohli, P.S. Virdi, Contrast sensitivity following focal laser photocoagulation in clinically significant macular oedema due to diabetic retinopathy. Clin. Exp. Ophthalmol. **29**(1), 17–21 (2001)
84. O. Arend, A. Remky, D. Evans, R. Stüber, A. Harris, Contrast sensitivity loss is coupled with capillary dropout in patients with diabetes. Invest. Ophthalmol. Vis. Sci. **38**(9), 1819–1824 (1997)
85. D. McLeod, A chronic grey matter penumbra, lateral microvascular intussusception and venous peduncular avulsion underlie diabetic vitreous haemorrhage. Br. J. Ophthalmol. **91**(5), 677–689 (2007)
86. Early Treatment Diabetic Retinopathy Study Research Group, Classification of diabetic retinopathy from fluorescein angiograms. ETDRS report number 11. Ophthalmology **98**(5 Suppl), 807–822 (1991)
87. N. Ashton, Arteriolar involvement in aiabetic retinopathy. Br. J. Ophthalmol. **37**(5), 282–292 (1953)

88. L.M. Jampol, Arteriolar occlusive diseases of the macula. Ophthalmology **90**(5), 534–539 (1983)
89. L. Yeung, V.C. Lima, P. Garcia, G. Landa, R.B. Rosen, Correlation between spectral domain optical coherence tomography findings and fluorescein angiography patterns in diabetic macular edema. Ophthalmology **116**(6), 1158–1167 (2009)
90. N. Drasdo, C.L. Millican, C.R. Katholi, C.A. Curcio, The length of Henle fibers in the human retina and a model of ganglion receptive field density in the visual field. Vision Res. **47**(22), 2901–2911 (2007)
91. J. Conrath, R. Giorgi, D. Raccah, B. Ridings, Foveal avascular zone in diabetic retinopathy: quantitative vs qualitative assessment. Eye (Lond.) **19**(3), 322–326 (2005)
92. J. Conrath, O. Valat, R. Giorgi, M. Adel, D. Raccah, F. Meyer, B. Ridings, Semi-automated detection of the foveal avascular zone in fluorescein angiograms in diabetes mellitus. Clin. Exp. Ophthalmol. **34**(2), 119–123 (2006)
93. Y. Zheng, J.S. Gandhi, A.N. Stangos, C. Campa, D.M. Broadbent, S.P. Harding, Automated segmentation of foveal avascular zone in fundus fluorescein angiography. Invest. Ophthalmol. Vis. Sci. **51**(7):3653–3659 (2010)
94. S.S. Hayreh, J.A. Baines, Occlusion of the posterior ciliary artery. II. Chorio-retinal lesions. Br. J. Ophthalmol. **56**(10):736–753 (1972)
95. S.S. Hayreh, Posterior ciliary artery circulation in health and disease. The Weisenfeld Lecture. Invest. Ophthalmol. Vis. Sci. **45**(3), 749–757 (2004)
96. P.G. Theodossiadis, A.K. Kollia, P. Gogas, D. Panagiotidis, M. Moschos, G.P. Theodossiadis, Retinal disorders in preeclampsia studied with optical coherence tomography. Am. J. Ophthalmol. **133**(5), 707–709 (2002)

Chapter 3
Optical Coherence Tomography and Visual Acuity: Photoreceptor Loss

Adzura Salam, Ute Ellen Kathrin Wolf-Schnurrbusch, and Sebastian Wolf

Abstract SD-OCT has improved the visualization of intraretinal morphologic features quantitatively and qualitatively. SD-OCT allows the retinal physician to evaluate the integrity of each retinal layer such as the external limiting membrane (ELM) and the junction between the inner and outer segments (IS/OS junction) of the photoreceptors

The presence and integrity of the external limiting membrane (ELM), the photoreceptor inner segment (IS), the outer segment (OS), and the retinal pigment epithelium (RPE) appears to be a good prognostic feature for visual improvement after treatment for various macular diseases.

Core Message

- Optical Coherence Tomography (OCT) plays a very important role in the development of the ophthalmic imaging field.
- OCT enables the Retina Physician to perform "optical biopsy," without the need to excise the retina.
- SD-OCT made both high resolution and fast scanning speed possible, thus improving the quality of images.
- SD-OCT has improved the visualization of intraretinal morphologic features, allowing to evaluate the integrity of each retinal layer such as the external limiting membrane (ELM) and the junction between the inner and outer segments (IS/OS junction) of the photoreceptors.

S. Wolf (✉)
Universitätsklinik für Augenheilkunde, Inselspital, University Bern, 3010 Bern, Switzerland
e-mail: sebastian.wolf@insel.ch

R. Bernardes and J. Cunha-Vaz (eds.), *Optical Coherence Tomography*, Biological and Medical Physics, Biomedical Engineering, DOI 10.1007/978-3-642-27410-7_3,

- The presence and integrity of the external limiting membrane (ELM), the photoreceptor inner segment (IS), the outer segment (OS), and the retinal pigment epithelium (RPE) appears to be a good prognostic feature for visual improvement after treatment for various macular diseases.

3.1 Introduction

The introduction of optical coherence tomography (OCT) in 1991 by Huang and associates [1] has become a tremendous turning point in the development of the ophthalmic imaging field. The clinical application of OCT in ophthalmology has undergo rapid evolution over the past decade especially with the introduction of spectral or Fourier-domain OCT.

In brief, OCT uses low-coherence interferometry to perform high-resolution, cross-sectional tomographic imaging of the retinal microstructure by measuring backscattered and back-reflected light [2]. The image resolution of 5–15 μm can be achieved. High-speed, real-time retinal imaging is made possible, and imaging can be performed in situ and nondestructively. OCT enables the retina physician to perform "optical biopsy," the imaging of retinal structure or pathology on resolution scale approaching that of histopathology without the need to excise the retina and process them as in standard excision biopsy and histopathology [3]. With this concept of nonexcisional "optical biopsy," more researches are being established in order to understand the pathophysiology of the retinal diseases and improved the ability of clinicians to individualize the approach to treatment.

3.2 Principle of SD-OCT

Spectral or Fourier-domain OCT refers to Fourier transformation of the optical spectrum of the low-coherence interferometer. SD-OCT made both high resolution and fast scanning speed possible, thus improving the quality of images. The most important advantage of the spectral domain technique over the conventional time domain OCT (TD-OCT) technique is the increase in scan speed [4]. SD-OCT is more than 100 times faster than TD-OCT technique.

Currently, various SD-OCT instruments are commercially available: Cirrus™ HD-OCT, Carl Zeiss Meditec, Inc.; RTVue-100 Fourier-Domain OCT, Optovue Corporation; Copernicus OCT, Reichert/Optopol Technology, Inc.; Spectral OCT/SLO, Opko/OTI, Inc.; Spectralis™ HRA + OCT, Heidelberg Engineering, Inc.; Topcon 3D OCT-1000 (Color + OCT), Topcon; and RS-3000 Retiscan, Nidek, and Tomey SS-1000, Tomey GmbH. All systems provide high-quality OCT line scans as well as special scan patterns for imaging the optic nerve fiber layer around the optic disk and for producing three-dimensional OCT images [5].

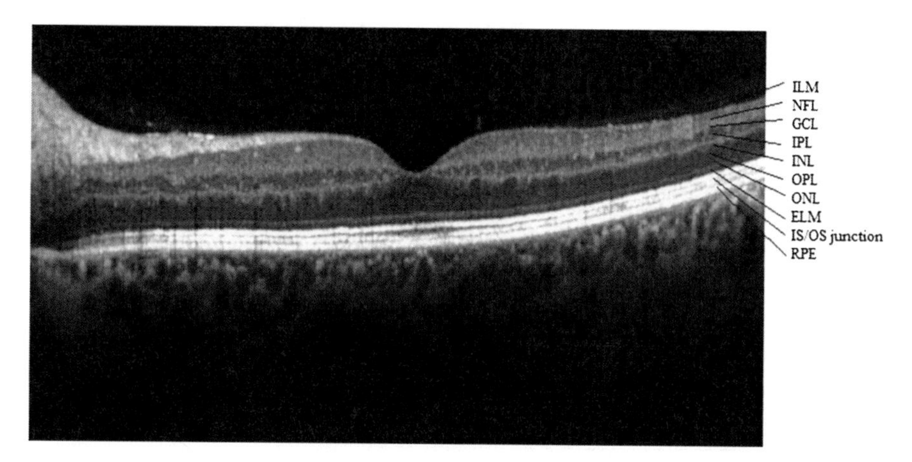

Fig. 3.1 The Spectralis™ HRA + OCT image shows normal retina layers from outermost to innermost: retinal pigment epithelium (RPE), junction between inner segment and outer (IS/OS junction), external limiting membrane (ELM), outer nuclear layer (ONL), outer plexiform layer (OPL), inner nuclear layer (INL), inner plexiform layer (IPL), ganglion cell layer (GCL), nerve fiber layer (NFL), and internal limiting membrane (ILM)

3.3 SD-OCT in Normal Eyes

SD-OCT imaging allows to distinguish the ganglion cell layer, the inner plexiform layer, the inner nuclear layer, the outer plexiform layer, the outer nuclear layer, the external limiting membrane (ELM), the photoreceptor inner segments (IS), the outer segments (OS), the retinal pigment epithelium (RPE), and, in pathologic cases, Bruch's membrane [6].

Most of the new SD-OCT systems image the outer retinal layers as three hyperreflective bands. The innermost of these hyperreflective bands has the lowest reflectivity. The bands may correspond to the external limiting membrane, the junction of the photoreceptor OS and IS and the RPE (Fig. 3.1). Some SD-OCT systems use the second inner hyperreflective band as outer border of the retina; others identify the most outer reflective band as the outer border of the retina. Wolf-Schnurrbusch et al. observed the discrepancies in macular thickness measurement in healthy eyes using six different OCT instruments [5]. These differences lead to differences in retinal thickness measurements of up to 70–80 μm. This implicates that the different OCT systems cannot be used interchangeably for the measurement of retinal thickness [5].

3.4 SD-OCT in Retinal Diseases

A very important feature of the SD-OCT system is it provides information on the retinal structures. SD-OCT has improved the visualization of intraretinal morphologic features, allowing to evaluate the integrity of each retinal layer.

Various macular diseases have been studied, and SD-OCT can aid in identifying, monitoring, and quantitatively assessing various posterior segment conditions including diabetic macular edema [7, 8], age-related macular degeneration [9, 10], macular edema due to retinal vein occlusion [11, 12] disease of the vitreomacular interface such as epiretinal membranes [13], full-thickness macular holes [14], pseudoholes, schisis from myopia or optic pits [15], central serous chorioretinopathy [16], and retinal detachment [17]. At the same time, it may be possible to explain why some patients respond to treatment while others do not. SD-OCT may become a valuable tool in determining the minimum maintenance dose of a certain drug in the treatment of diabetic macular edema (DME), macula edema (ME) due to retinal vein occlusion (CRVO), and wet age-related macular degeneration (wet AMD). It may demonstrate retinal changes that explain the recovery in some patients or nonimprovement in other patients without performing any invasive procedure like retina angiography.

3.5 Correlation Between Visual Acuity and Integrity of Photoreceptor in Retinal Diseases

3.5.1 Photoreceptor

Retina is a highly specialized light-sensitive tissue located in the innermost layer of the eye wall (Fig. 3.2). It consists of the photoreceptor cells, neurons, ganglions, and pigment cells. Human photoreceptors are a complex, layered structure with several layers of neurons interconnected by synapses (Fig. 3.3). The only neurons that are directly sensitive to light are the photoreceptor cells. These are mainly of two types: the rods and cones (Fig. 3.4). Rods function mainly in dim light and provide black-and-white vision, while cones support daytime vision and the perception of color. Important functional differences between rods and cones are summarized in Table 3.1.

SD-OCT has improved the visualization of intraretinal morphologic features, allowing to evaluate the integrity of each retinal layer [6].

3.5.2 Diabetic Macular Edema

Diabetic macular edema (DME), as defined by the Early Treatment of Diabetic Retinopathy Study (ETDRS) protocol, is clinically significant retinal thickening (CSME) or hard exudates (EH) associated with retinal thickening observed within 500 μm of the center of the macula, or if a zone of retinal thickening one-disk area in size is present within one-disk diameter of the center of the macula. DME is a

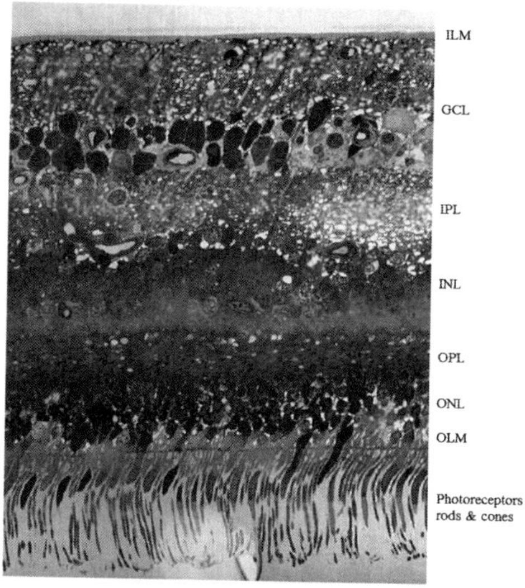

Fig. 3.2 Image of human retina by using electron microscopy

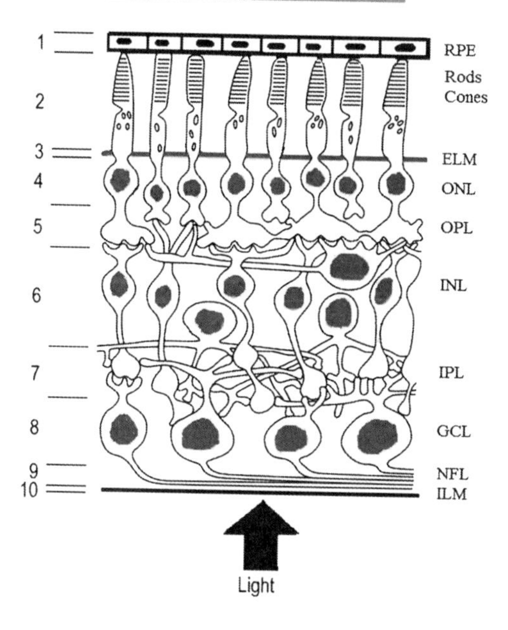

Fig. 3.3 Schematic representation of the human retinal layers: (1) retinal pigment epithelium, (2) layer of rods and cones, (3) external limiting membrane, (4) outer nuclear layer, (5) outer plexiform layer, (6) inner nuclear layer, (7) inner plexiform layer, (8) ganglion cell layer, (9) nerve fiber layer, and (10) internal limiting membrane

major cause of visual deterioration in diabetic retinopathy [19]. Historically, DME is diagnosed by fundus biomicroscopy, fundus stereophotography, or both [20]. Fluorescein angiography is also used to evaluate patients with DME; however, the procedure is invasive and usually performed with strong indication (Fig. 3.5). Various therapeutic modalities has been used to treat the DME including grid laser photocoagulation [21], triamcinolone acetonide [22, 23], vitrectomy [24], and intravitreal injection of ranibizumab [25].

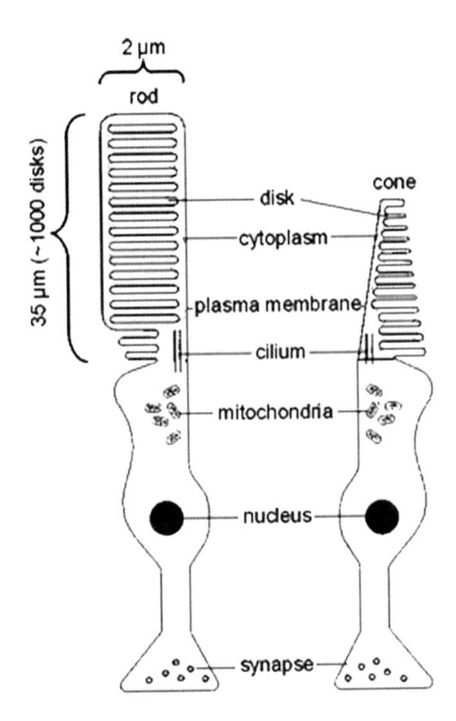

Fig. 3.4 Diagram shows human rod and cone photoreceptors

Table 3.1 The differences between rods and cones

Rods	Cones
Used for scotopic vision	Used for photopic vision
Very light sensitive; sensitive to scattered light	Not very light sensitive; sensitive to only direct light
Loss causes night blindness	Loss causes legal blindness
Low visual acuity	High visual acuity; better spatial resolution
Not present in fovea	Concentrated in fovea
Slow response to light, stimuli added over time	Fast response to light, can perceive more rapid changes in stimuli
Have more pigment than cones, so can detect lower light levels	Have less pigment than rods, require more light to detect images
Stacks of membrane-enclosed disks are unattached to cell membrane directly	Disks are attached to outer membrane
20 times more rods than cones in the retina	
One type of photosensitive pigment	Three types of photosensitive pigment in humans
Confer achromatic vision	Confer color vision

Source: From Eric Kandel et al. in *Principles of Neural Science* [19]

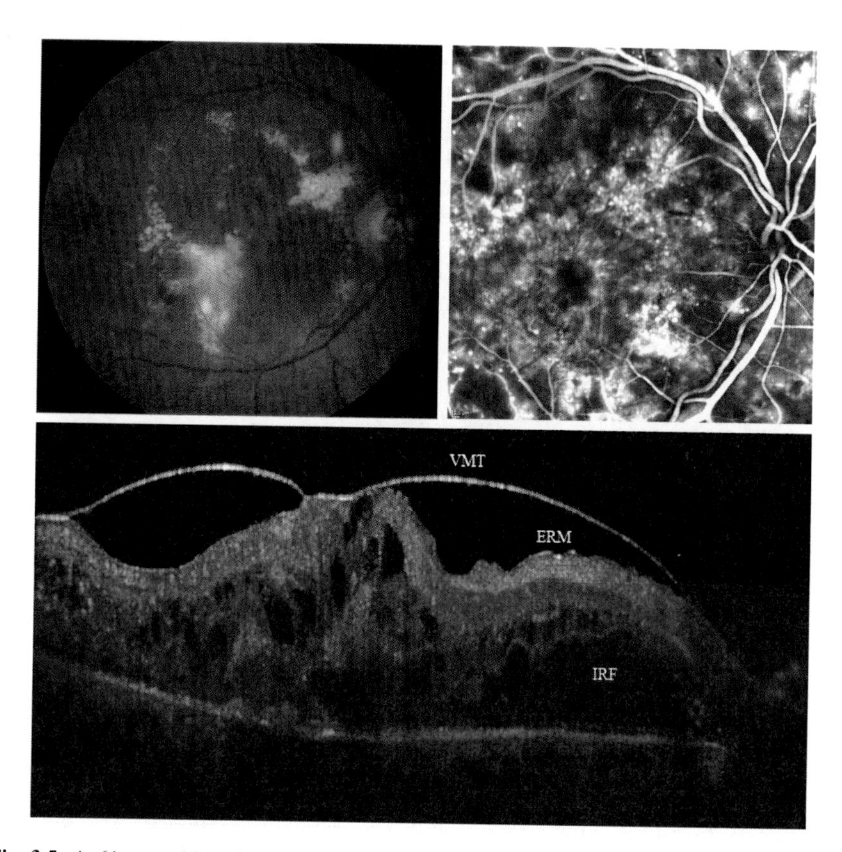

Fig. 3.5 A 61-year-old male presented with diabetic maculopathy in the right eye. The best corrected visual acuity was 20/100. Color picture of the right eye shows diffuse macula edema and presence of hard exudates in "succinate ring" shape. Fluorescein angiography (FA) demonstrates "petalloid pattern" that corresponded to cystoid macular edema (CME). The Spectralis™ HRA + OCT shows diffused CME, intraretinal fluid (IRF), disturbed IS/OS junction, and ELM. Note the presence of vitreomacular traction (VMT) and epiretinal membrane (ERM) which was not detected by FA

Advances in OCT have enabled us to identify the external limiting membrane (ELM) and the junction between the inner and outer segments (IS/OS junction) of the photoreceptors [26]. We conducted a cross section retrospective study to see the correlation between foveal photoreceptor changes and visual acuity (VA) in DME using SD-OCT [27]. We noted that patients with intact ELM and IS/OS junction has significantly better VA as compared to disturbed ELM and IS/OS junction. There was a moderate good correlation between VA, ELM, and IS/OS junction distance and weak correlation to CRT and CRV. The same study documented that presence of subretinal fluid or detachment, cystoid macular edema, and epiretinal membrane also resulted in reduced VA [27]. This finding is similar to earlier study by Otani et al. [28] who demonstrated that integrity of the ELM and IS/OS affects best corrected visual acuity (BCVA) more strongly than other OCT findings, particularly cystic changes and retinal thickness in DME. In another study, Maheshwary et al. [29] reported a statistically significant correlation between percentage disruption of the IS/OS junction and visual acuity in DME patients.

The causes of photoreceptor damage in DME was well described by Mervin et al. [30]. In experimental retinal detachment, hypoxia caused by the separation of the outer retina from its normal source of nutrients is a factor in inducing the photoreceptor death. In DME, subretinal detachment (SRD) is seen in 15–31% of eyes by OCT [7, 31]. Fluid leakage from abnormal vessels accumulates in the subretinal space. Although some oxygen and glucose is preserved in the subretinal fluid of DME, long-term SRD may cause the destruction of photoreceptors. In summary, macular edema, especially long-standing cases, and the association with ischemia may be prone to severe photoreceptor dysfunction. Atrophy causes deterioration of vision, irrespective of macular thickness, and exclusion of eyes with marked macular edema.

The standard treatment protocol for DME according to ETDRS is macular laser photocoagulation (MPL) [32]. However, 15% of patients in the ETDRS study experienced more than two lines of visual loss after 3 years [32]. Intravitreal triamcinolone acetonide (IVTA) and pars plana vitrectomy (PPV) are other treatment modalities investigated for DME; however, both IVTA and PPV have limited efficacy and/or significant side effects [33, 34]. Some eyes with DME have poor visual outcomes despite successful treatment and complete resolution of edema by PPV. Framme et al. [35] has documented postoperative RPE proliferation observed in both conventional retinal laser (CRL) and selective retina treatment (SRT) with a Nd:YLF. RPE atrophy appeared subsequently only in CRL lesions, whereas the neurosensory retina appeared unaffected following SRT. This could explain the reasons why some patients still have poor vision after MPL (Fig. 3.6).

Vascular endothelial growth factor (VEGF) has been shown to be extensively involved in the development and progression of DME [36]. VEGF promotes neovascularization and microvascular leakage [37]. Thus, drugs which inhibit VEGF may provide an alternative therapeutic approach in DME. Results from the RESOLVE study indicate that DME responds well to treatment with intravitreal ranibizumab over 1 year [25]. The same study shows improvements in BCVA and CRT over the 12-month study period combined with a good safety profile.

The advent of SD-OCT in DME has expanded the capabilities of OCT technology beyond diagnosis and disease-monitoring to prediction of treatment outcome. Einbock et al. were among the first to observe that patients, where the SD-OCT images demonstrated an intact outer limiting membrane, show better VA outcome after intravitreal anti-VEGF therapy as compared to patients with severely compromised outer retinal structures [38] (Fig. 3.7). Interestingly, Sakamoto et al. conducted a study looking for association between foveal photoreceptor status and visual acuity after resolution of DME by PPV [39]. The result shows significantly higher VA in intact IS/OS junction as compared with disturbed IO/OS junction after resolution of DME by PPV. In conclusion, the integrity of the outer retinal layers appears to be a predictive factor for the visual prognosis after DME treatment (Fig. 3.8).

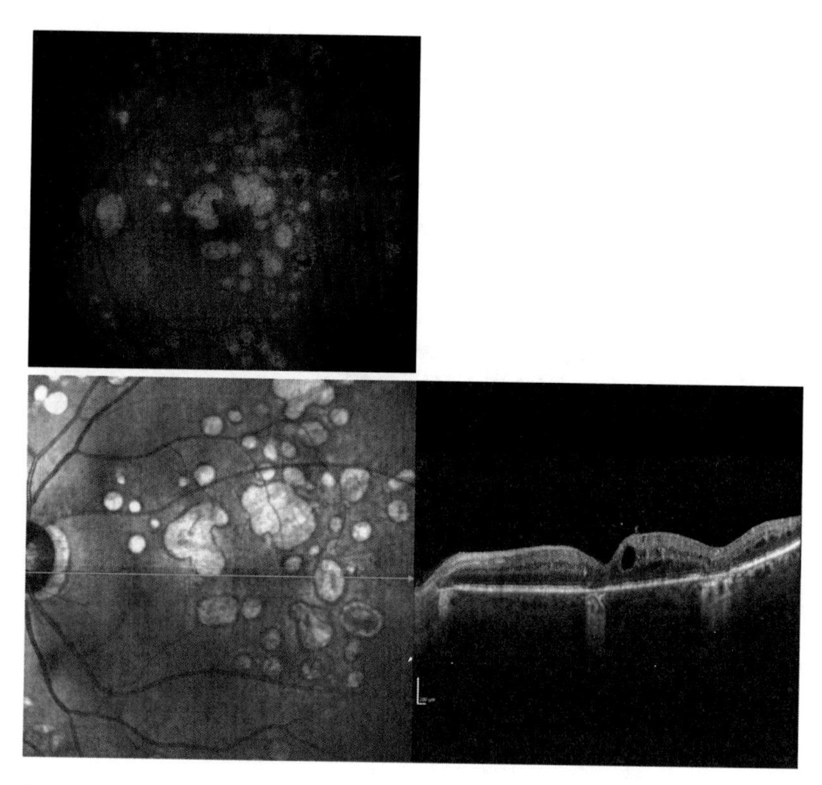

Fig. 3.6 A color fundus photograph of the left eye shows old laser scars in a 56-year-old female who presented with history of PDR and CSMO. Initially there was slight improvement of visual acuity 3 months after the treatment, then her vision was gradually worsened within 2 years after the lasers. The Spectralis^{TM} HRA + OCT shows loss of photoreceptor and ELM layers over the laser scars

3.5.3 Retinal Vein Occlusion

Retinal vein occlusion is a frequently seen retinal vascular disorder which often associated with severe decrease in VA. Macular edema (ME) is the most common vision-threatening complication associated with branch retinal vein occlusion (BRVO) and central retinal vein occlusion (CRVO) [40]. Various options for treatment of ME that is secondary to BRVO and CRVO include laser photocoagulation [40, 41], radial optic neurotomy [42, 43], and intravitreal injection of triamcinolone acetonide [44]. Intravitreal dexamethasone implant has recently been shown to be remarkably effective for the reduction of ME [45]. Indeed, the resolution of ME often leads to improvement in visual acuity. Even with such successful treatment, some patients still have a poor visual outcome despite complete resolution of the ME.

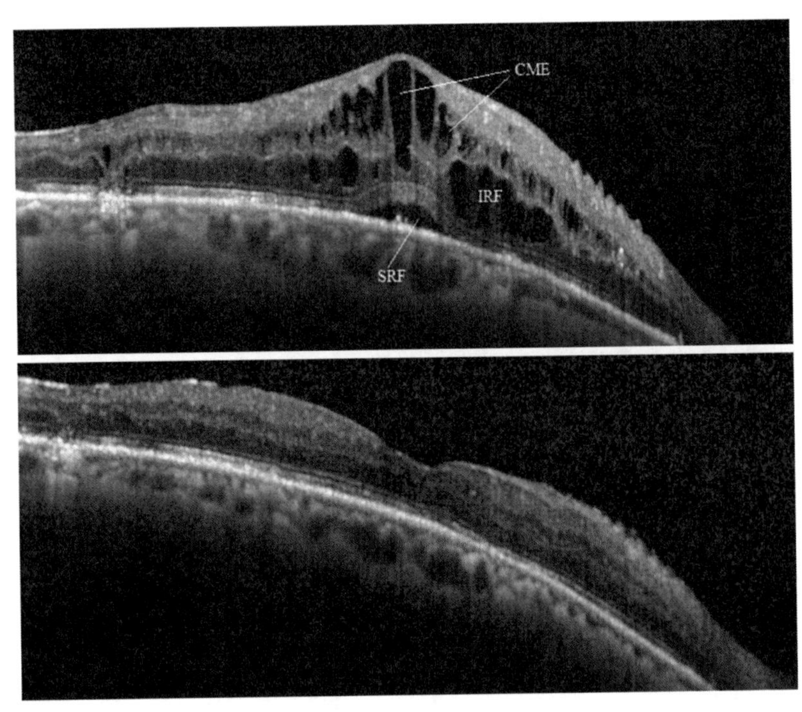

Fig. 3.7 A 61-year-old male presented with CSME. The Spectralis™ HRA + OCT shows cystoid macula edema with intraretinal fluid (IRF), subretinal fluid (SRF) and intact IS/OS junction, and external limiting membrane (ELM). The edema resolved after three consecutive monthly intravitreal ranibizumab injections. BCVA improved from 20/100 to 20/40 after 6 months

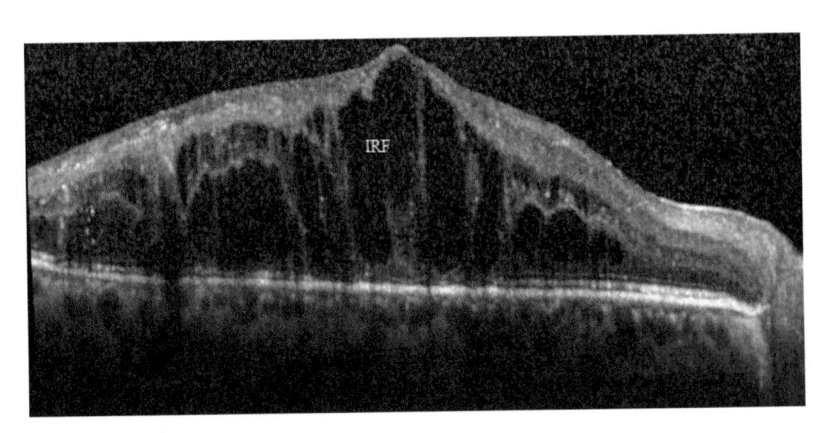

Fig. 3.8 A 54-year-old male presented with chronic CSMO in the left eye and has not responded to conventional focal and grid laser. He was given monthly injection of ranibizumab for 3 months, but there was no improvement of visual acuity. The Spectralis™ HRA + OCT demonstrates presence of diffused cystoid space, intraretinal fluid (IRF) and disturbed IS/OS junction, and external limiting membrane (ELM) which may contribute to poor responds after the treatments

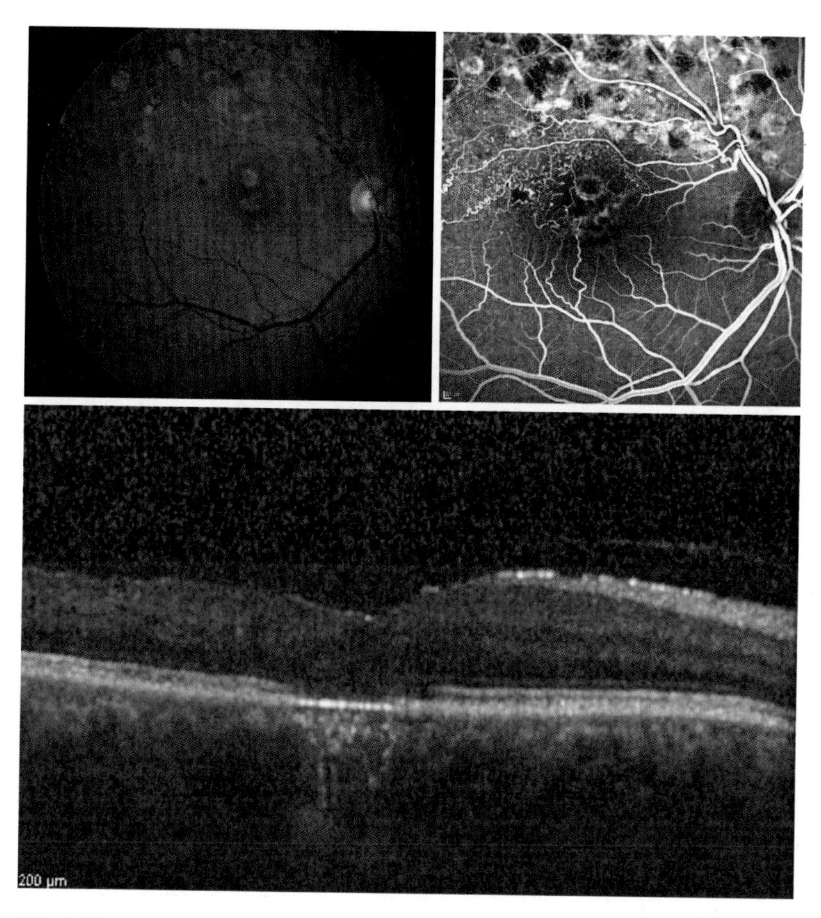

Fig. 3.9 A 55-year-old male presented with ischemic branch retinal vein occlusion of the superior temporal quadrant. The BCVA was 20/50. Color fundus photograph demonstrates laser scars at the superior temporal quadrant and macula edema. FA showed hypofluorescein areas that correspond to laser scars and some hyperfluorescein leaks on the macula. The Spectralis™ HRA+OCT shows diffused macula edema with photoreceptor loss and scar under the foveal area

In 2007, Ota et al. [11] reported a study looking for correlation between integrity of the photoreceptor layer after resolution of macular edema (MO) associated with branch retinal vein occlusion (BRVO) and final visual acuity (VA) to determine prognostic factors for visual outcome. The results show that no statistical differences were found in initial VA or in foveal thickness between eyes with or without complete IS/OS junction at final observation. However, final VA in eyes without a complete IS/OS junction was significantly poorer. Additionally, initial status of the IS/OS junction in the parafoveal area of unaffected retina was associated with final VA. Absence of the IS/OS junction at 500 μm or 1,000 μm from the fovea on initial OCT images was associated with poor visual recovery after resolution of the MO (Fig. 3.9).

In 2008, the same author [46] reported another study using OCT images of eyes with resolved ME associated with CRVO and evaluated the integrity of the foveal photoreceptor layer using the IS/OS line as a hallmark. In addition, they also assessed correlation between visual acuity and integrity of the foveal photoreceptor layer after resolution of the ME in the same eyes. The result shows at the final visit that foveal thickness was decreased to a physiologic level in all eyes. In concordance with resolution of the ME, visual acuity had improved significantly by the final visit. However, final visual acuity in eyes without an IS/OS line was significantly poorer than that in eyes with an IS/OS line. In addition, integrity of the foveal photoreceptor layer after resolution of the ME had a significant correlation with the initial retinal perfusion status and with initial visual acuity.

According to histologic studies of eyes with retinal vein occlusion, ME is detected often in the outer nuclear layer and it can cause liquefaction necrosis, which results in loss of photoreceptor cells and in photoreceptor dysfunction within the fovea [47]. Even though the retinal thickness is successfully reduced to a physiologic level, loss of foveal photoreceptor cells could lead to limited visual recovery. Visualization of an IS/OS line on SD-OCT imaging would indicate better integrity of the photoreceptor layer.

A recently published paper by Wolf-Schnurrbusch et al. [48] looking for predictors for short-term visual outcome after anti-VEGF therapy of ME due to CRVO. The results show that disturbed ELM was accompanied by focal disintegration of IS/OS. Four weeks after treatment with intravitreal bevacizumab (1.25 mg) or ranibizumab (0.5), more than half of the eyes showed clinical relevant increase of BCVA. The mean BCVA increase was statically higher in eyes with intact ELM as compared to eyes with disturbed ELM. In conclusion, intact ELM in SD-OCT imaging is associated with better visual outcome after intravitreal anti-VEGF treatment and has strong predictive value in patients after receiving treatment for ME in BRVO or CRVO (Fig. 3.10).

3.5.4 Retinal Artery Occlusion

Branch retinal artery occlusion (BRAO) and central retinal artery occlusion (CRAO) present as an acute painless loss of visual field in the distribution of the occluded artery. The most common cause of BRAO is secondary to cholesterol embolus originated from aorta-carotid atheromatous plaques followed by platelet-fibrin emboli from thrombotic disease and calcific emboli from cardiac valvular disease. The emboli usually lodge at the bifurcation of the central retinal artery into the branch retinal artery.

In a retrospective study, Yuzurihara et al. [49] reported that both presenting visual acuity and visual final acuity were far worse in patients with CRAO than in patients with BRAO. In a different study, Mason et al. [50] found that visual prognosis in BRAO is good, and 80% of the patients recover final visual acuity greater than 20/40. However, in the subset of patients who had poor final visual acuity, Manson

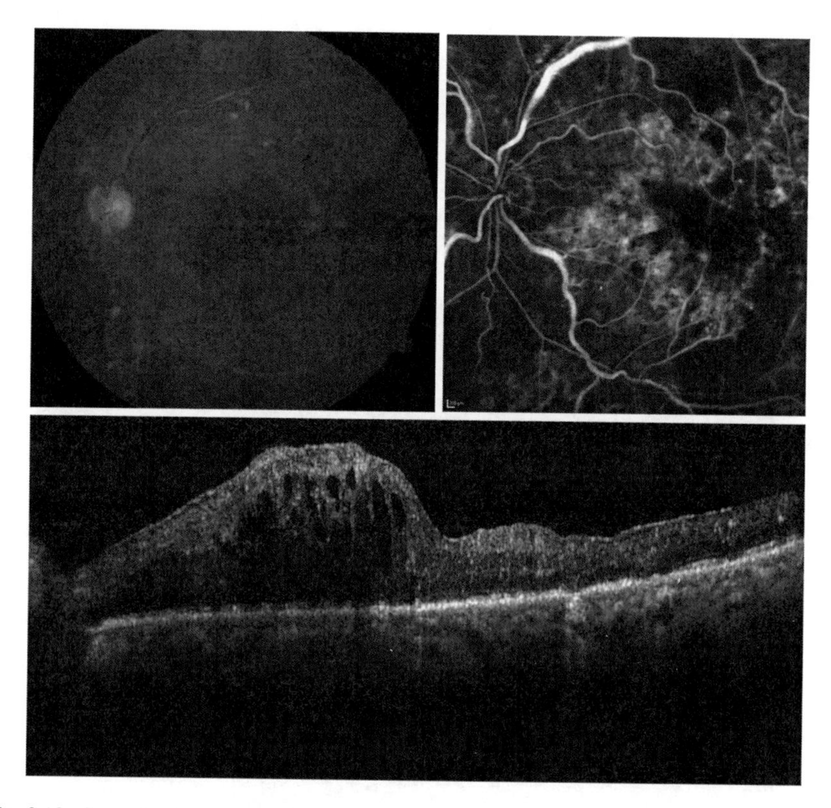

Fig. 3.10 Color fundus photograph demonstrates extensive swelling of the optic disk and sub-retina "flame-shaped" hemorrhage in all four quadrants with dilated tortuous vessels. FA shows hyporeflective area which corresponds to hemorrhagic area on the macula. The Spectralis^{TM} HRA + OCT shows hyporeflective cystic spaces within the retina consistent with macular edema and disturbed outer photoreceptor layers

et al. reported that initial visual acuity of less than 20/100 was the most important negative prognostic factor, and he advocated aggressive treatment in this subset of patients.

SD-OCT findings of BRAO and CRVO have been described in the few case reports [11, 51–53]. In acute CRAO, Falkenberry et al. [52] documented increased reflectivity and thickness of the inner retina, as well as decreased reflectivity of the outer layer of the retina and the retinal pigment epithelium (RPE)/choriocapillaris layer. Beneath the foveal depression, the RPE/choriocapillaris layer demonstrated an area of relatively increased reflectivity compared with other regions of the RPE/choriocapillaris. At the 3-month follow-up examination, the patient showed minimal improvement of visual acuity. However, the OCT images showed decreased reflectivity and thickness of the inner retinal layers and a corresponding increase of reflectivity in the outer retina and RPE/choriocapillaris layer compared with the baseline OCT image (Fig. 3.11).

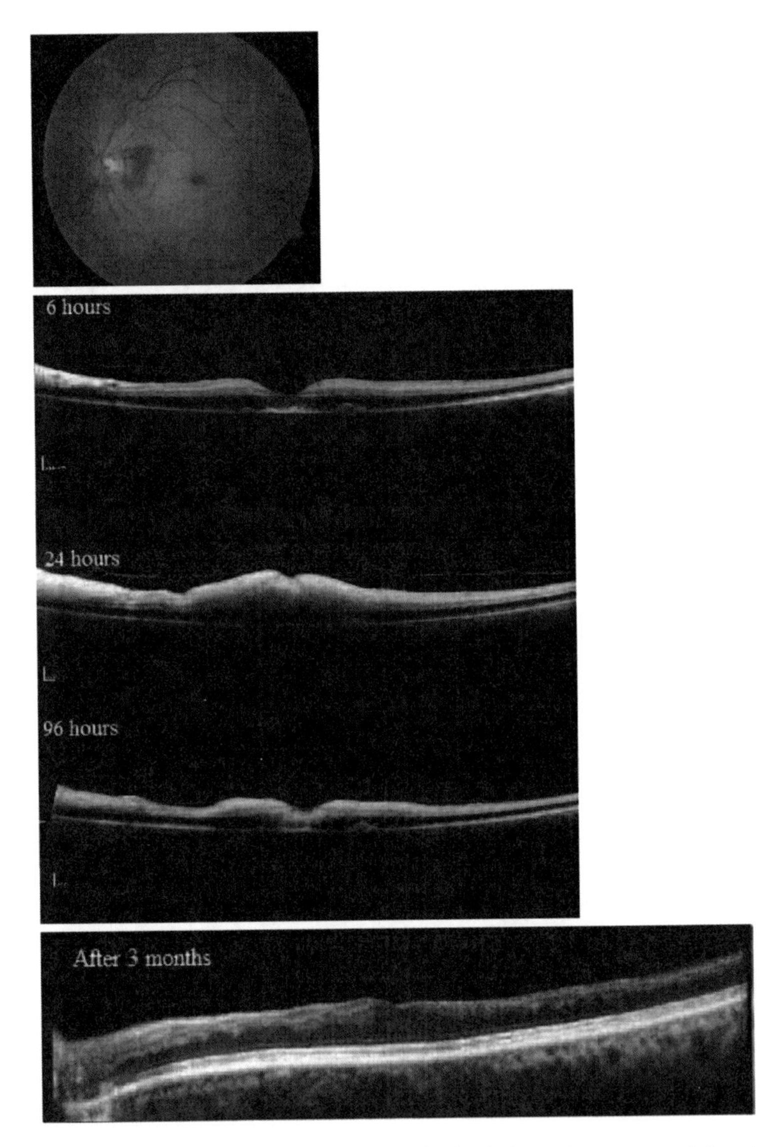

Fig. 3.11 A 34-year-old male with a known case of valvular heart disease presented with sudden onset of reduced vision in the left eye. Color photograph shows pale-looking retina with "cherry red spot" on the macula. The SD-OCT at the early stages shows coagulative necrosis of the inner retinal layers (NFL, GCL, and inner part of INL). After 3 months, SD-OCT image (bottom) shows decreased thickness of the inner retinal layers and a corresponding increased thickness of outer retinal layers as compared to baseline

Histologically, early coagulative necrosis of the inner retina with intracellular edema of neurons is seen after a branch retinal arteriole is obstructed [54]. Prolonged ischemia results in consecutive atrophy of these layers with each layer exhibiting differential sensitivity to the underlying hypoxia. The SD-OCT finding supports

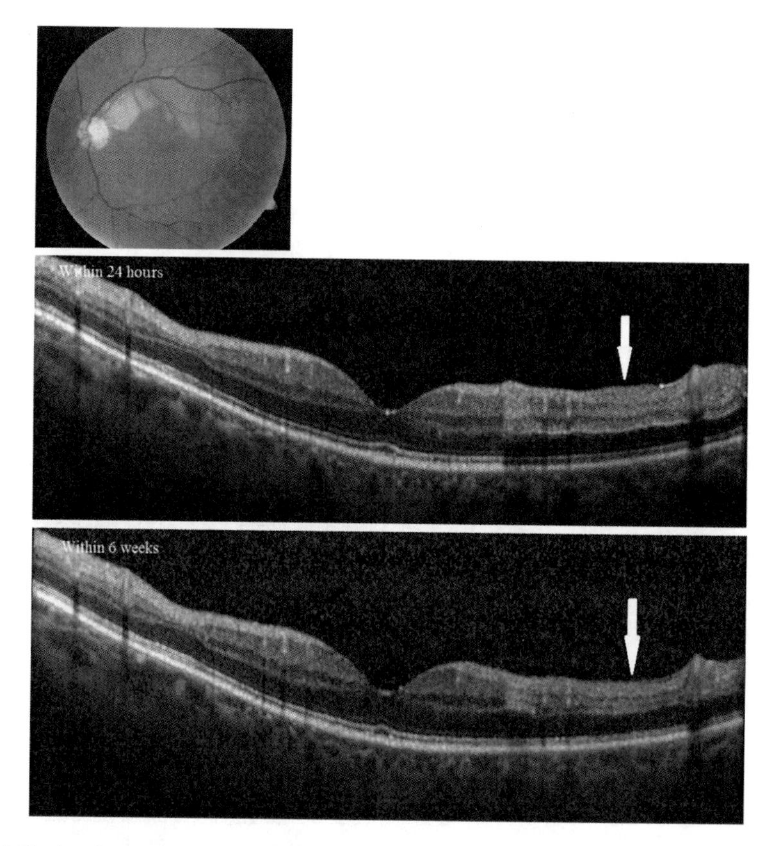

Fig. 3.12 Another patient with BRAO shows SD-OCT changes within 24 h and 6 weeks. Note the inner retinal layer thickness changes (*arrows*) as compared to relatively normal-looking outer retinal layer over the period of time

this description and is useful in monitoring the retinal changes resulting from the embolic events (Fig. 3.12).

3.5.5 Age-Related Macular Degeneration

Age-related macular degeneration (AMD) is the leading cause of visual impairment in patients older than 65 years in the developed world [55, 56]. An estimated eight million persons older than 55 years in the United States have monocular or binocular intermediate AMD or monocular advanced AMD. Drusen, the earliest sign of AMD, are extracellular deposits that accumulate between the retinal pigment epithelium (RPE) and the inner collagenous layer of Bruch's membrane [57, 58] (Fig. 3.13) Curcio et al. [59] reported a correlation between the presence of drusen

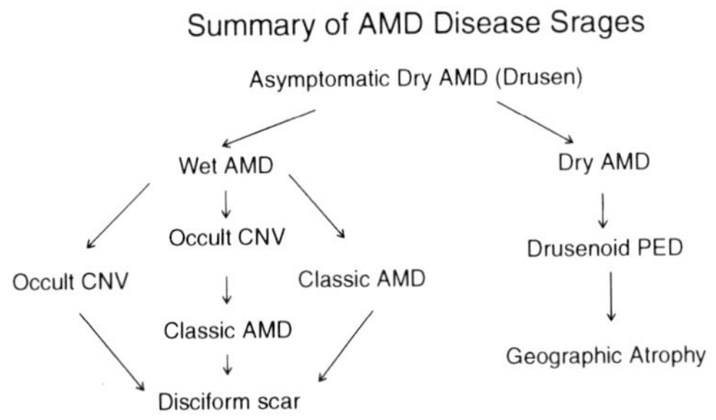

Summary of AMD Disease Srages

Asymptomatic Dry AMD (Drusen)

Wet AMD → Occult CNV → Disciform scar

Wet AMD → Occult CNV → Classic AMD → Disciform scar

Wet AMD → Classic AMD → Disciform scar

Dry AMD → Drusenoid PED → Geographic Atrophy

Fig. 3.13 Summary of AMD disease stages

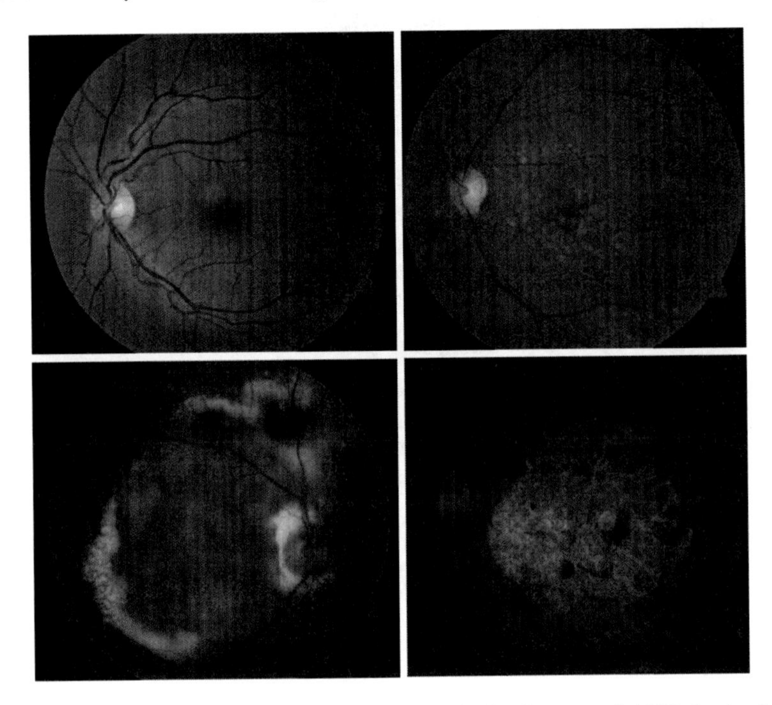

Fig. 3.14 The color photograph shows different stages in development of AMD that begin with normal fundus (*top left*), drusen (*top right*), wet AMD or exudative stage (*bottom left*), and dry AMD or geographic atrophy (*bottom right*)

and photoreceptor cell death in the retina of eyes diagnosed with AMD [60]. Holz et al. [60] and Pauleikhoff et al. [61] have proposed that the accumulation of drusen between the outer retina and the choriocapillaris interferes with the exchange of nutrients and waste products, inducing RPE death. Although, appearance of a few drusen is normal with aging, greater accumulation of extracellular sub-RPE deposits is an indication of AMD, which later may progress to more severe forms such as choroidal neovascularization (CNV) and nonneovascular geographic atrophy (GA) (Fig. 3.14). Interestingly, the Age-Related Eye Disease Study (AREDS) reported

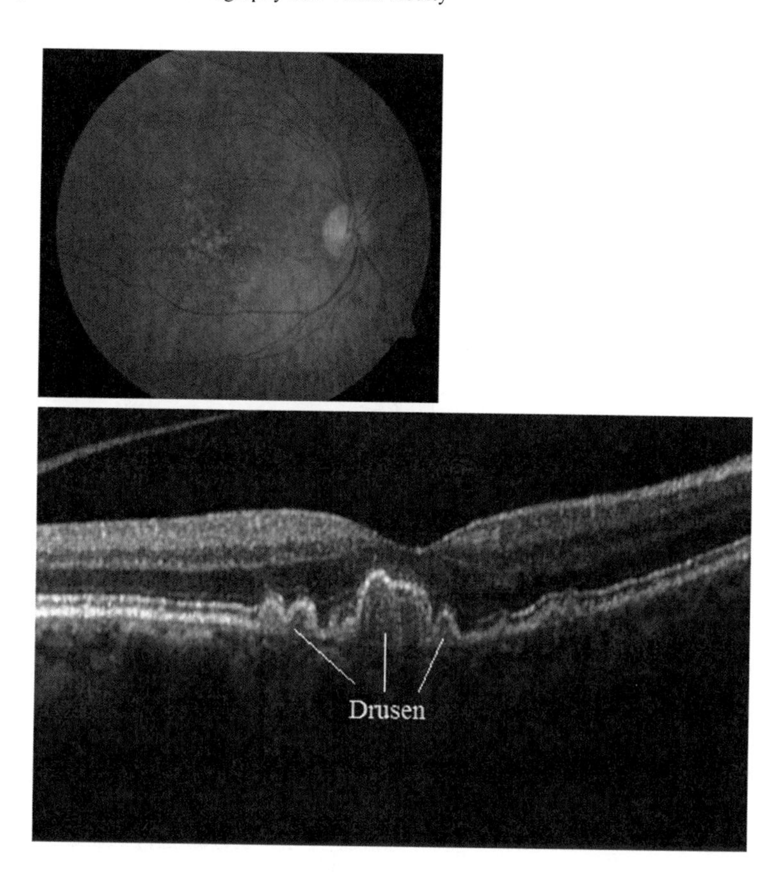

Fig. 3.15 A 75-year-old female presented with visual disturbance over the last 4 years. Color fundus photograph of the right eye shows drusen on the macula. The Spectralis™ HRA + OCT shows basal membrane thickness and prominent elevation under the RPE with disturbed outer nuclear layers at subfoveal region. Note the disturbances of the IS/OS junction in the area of the large drusen

that taking a high-dose vitamin and mineral supplement in the early stage of drusen could delay the onset of severe drusen and vision loss [62].

Schuman et al. [63] used SD-OCT to quantitatively detect the neurosensory retina changes over drusen in AMD. The results show that the photoreceptor layers (PRL) were significantly and profoundly decreased in thickness over drusen compared with age-matched control eyes. Decreased PRL thickness over drusen suggests a degenerative process, with cell loss leading to decreased visual function. This finding is supported by earlier study by Bressler et al. [64] where they found a significant relationship between loss of RPE and photo receptors. In addition, drusen appeared to bulge into the PRL, without corresponding displacement of the overlying inner retinal layers in half of the study eyes; there was definite photoreceptor outer segment loss focally over the drusen. In each case, this was also associated with disruption or loss of the photoreceptor inner segment and outer segment junction (Fig. 3.15). This focal qualitative and quantitative changes in the

PRL over drusen suggest measurable photoreceptor loss and likely dysfunction. In conclusion, photoreceptor loss over drusen detected by SD-OCT may be useful as biomarkers for visual impairment associated with drusen and may predict the subsequent course of disease progression.

Another spectrum of AMD is the geographic atrophy (GA), which is believed to originate from soft drusen. Data indicate that GA also involved not only the retinal pigment epithelial (RPE) but also the outer neurosensory retinal layer and the choriocapillaris. Using the SD-OCT, Wolf-Schnurrbusch et al. [65] have documented morphologic and structural pathogenic mechanisms underlying the atrophic processes. Interestingly, retinal layer alterations were documented, not only in atrophic zones, but also in junctional zones surrounding the geographic atrophy (Fig. 3.16). Disintegration of the retinal layers began in the RPE and adjacent retinal layers, such as the photoreceptor inner and outer segments and external limiting membrane. The same study recorded significant differences between the best corrected VA (BCVA) depending on areas of GA and the configuration of the fovea. The mean VA letter score of all eyes with a normal foveal depression and without atrophic changes at the fovea was significantly higher compared to the eyes without a normal foveal depression and with GA involving the foveal region.

Neovascular AMD or wet AMD is the end spectrum of the disease. It is less common compared to dry AMD, but is typically more damaging. The wet type of macular degeneration is caused by the growth of abnormal blood vessels behind the macula. The abnormal blood vessels tend to bleed, resulting in the formation of scar tissue if left untreated. In some instances, the dry stage of macular degeneration can turn into the wet stage or exudative changes.

Various treatments have been performed on patients with neovascular AMD to resolve the exudative changes resulting from choroidal neovascularization (CNV). Historically, photodynamic therapy (PDT) [66] and angiostatic steroids [67] have been used for treatment of exudative AMD. PDT has been shown to be effective in treating polypoidal choroidal vasculopathy (PCV), another subgroup of occult AMD [68]. The advent of antiangiogenic therapies, such as ranibizumab and bevacizumab, has revolutionized the management of AMD over the last few years [69]. Treatment effects can be evaluated with fluorescein angiography and optical coherence tomography (OCT). However, some eyes with AMD have a poor visual outcome despite complete resolution of the exudative change even after successful treatment.

Hayashia et al. [70] conducted a study to find the association between foveal photoreceptor integrity and visual outcome in neovascular AMD using the SD-OCT. In this study, they divided the patients into typical AMD and PCV groups. Majority of typical AMD groups showed absent of IS/OS line beneath the fovea at the final visit as compared to one third of PVC. Most eyes with typical AMD have an accompanying type 2 CNV (subneural retinal neovascularization), located between the RPE and neurosensory retina, which causes severe macular edema. Severe damage in the neurosensory retina during the active stage of typical AMD may account for the irreversible damage in the foveal photoreceptor layer, with a concomitant poor visual prognosis. In contrast, most components of PCV

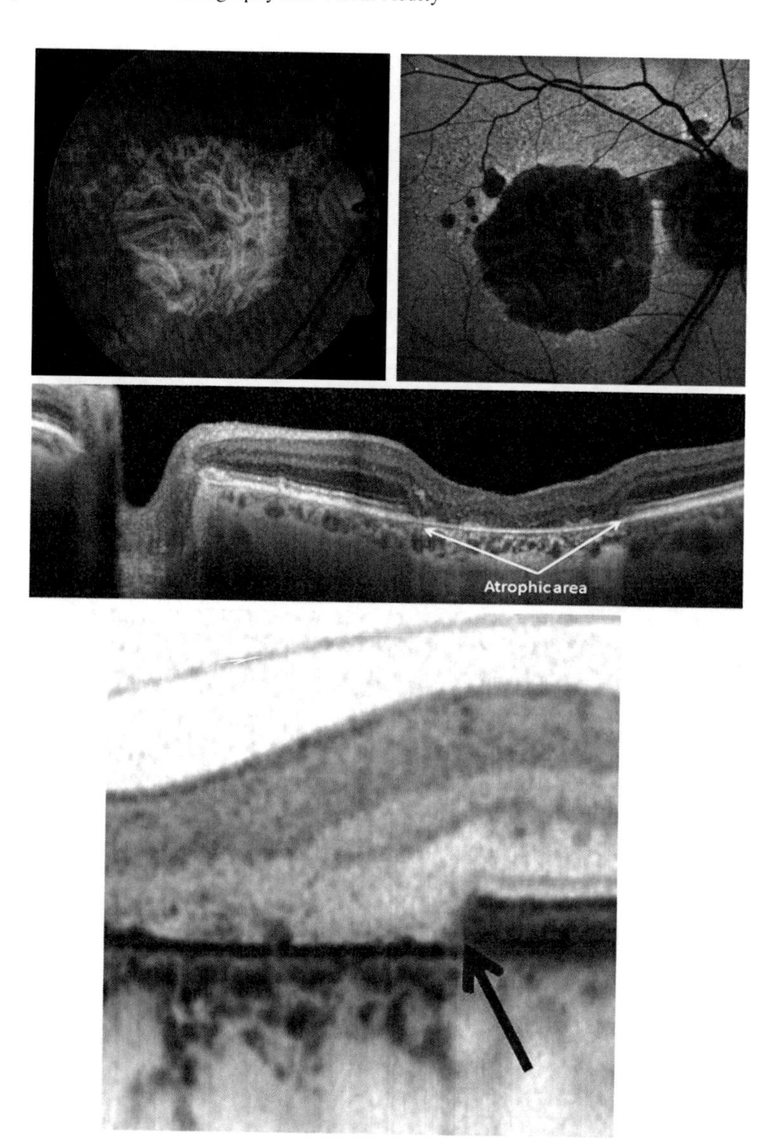

Fig. 3.16 Color fundus photograph shows area of geographic atrophy (GA). Autofluorescence demarcated the GA better (*top right*). The SD-OCT shows loss of photoreceptor layers at the atrophic area over the lesion as well as at the junctional zones (*arrow*) surrounding the geographic atrophy

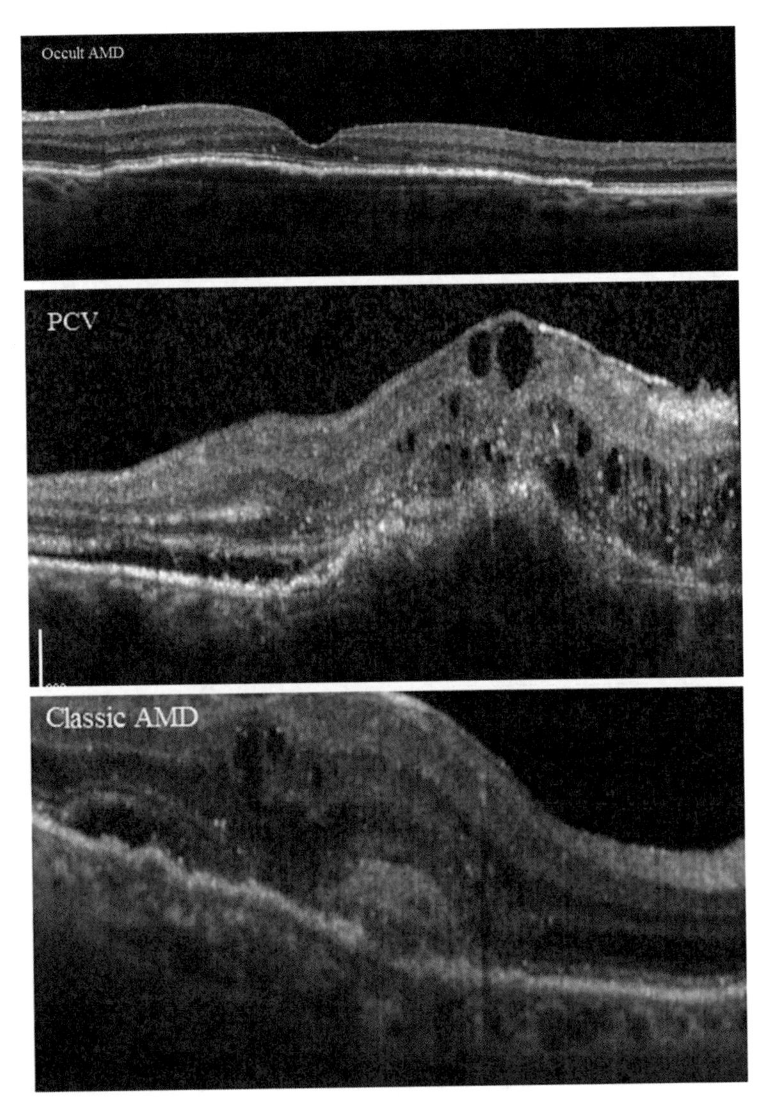

Fig. 3.17 SD-OCT shows type 1 CNV (sub-RPE neovascularization) in occult AMD and PCV and type 2 CNV (subneural retinal neovascularization) in classic AMD

are reported to be located beneath the RPE (Fig. 3.17). The eyes with PCV would have less damage in the neurosensory retina before treatment, even though hemorrhagic complications are seen more frequently. In summary, differences in the pretreatment condition of the foveal photoreceptor layer may account partially for the posttreatment condition and visual prognosis was significantly better in eyes with PCV compared with those with typical AMD.

3.5.6 Central Serous Chorioretinopathy

Central serous chorioretinopathy (CSC) is characterized by serous detachment of the neurosensory retina in the macula often associated with small RPE detachments. These pathologies often cause metamorphopsia and micropsia. Most patients with CSC have good VA despite of macular detachment. Acute CSC is defined as serous macular detachment caused by one or several isolated leaks seen on FA at the level of RPE: Acute CSC may either resolve spontaneously or sometimes reoccur and become chronic. Chronic CSC is defined as a serous retinal detachment associated with area of RPE atrophy and pigment mottling, visible microscopically, where FA displays subtle leaks or ill-defined staining. Acute CSC often resolves rapidly without treatment, whereas chronic CSC usually shows good respond with either laser photocoagulation or PDT [71]. Metamorphopsia and micropsia often persist after resolution of macular detachment, even if the patient has good VA.

SD-OCT has been used to evaluate changes in both acute and chronic CSC. Piccolino et al. [16] observed presence of macular detachment in both acute and chronic CSC. The outer photoreceptor layer (OPL) was intact in patients with symptoms lasting for less than 1 year. Atrophic and granulated OPL was seen in eyes with symptoms of more than 1 year duration. The mean final VA was statically higher in preserved eyes with OPL as compared to eyes with atrophic OPL. In another study, Matsumoto et al. [72] measured outer nuclear layer (ONL) thickness at the central fovea in eyes which had BCVA of less than 1.0 and compared to those with BCVA of 1.0 or better. The same study evaluated the integrity of photoreceptor IS/OS junction using SD-OCT. Discontinuity of the IS/OS junction was prevalent in eye with thinner ONL and lower BCVA.

Long-standing retinal detachment may cause photoreceptor cell apoptosis or loss of outer segment of the foveal photoreceptor cells [73, 74]. Decreased cone cells might contribute to micropsia and loss of central vision which persist after resolved CSC. SD-OCT has contributed to better understanding of the role of foveal photoreceptor layer in visual function in CSC and may predict visual recovery after resolving CSC (Fig. 3.18).

3.5.7 Retinitis Pigmentosa

Retinitis pigmentosa (RP) is a genetic disorder that primarily affects the rod and cone photoreceptors and RPE whereas the inner retina remains relatively intact. RP is a slowly progressive disease, typically affects the rods before the cones and characterized by night blindness (nyctalopia), abnormal electroretinogram (ERG), and visual field constriction. Visual field constriction is typically due to secondary loss of cones, which is highest in density at the center of the fovea. Cystoid macular edema (CME) is a common complication of RP and may affect the VA

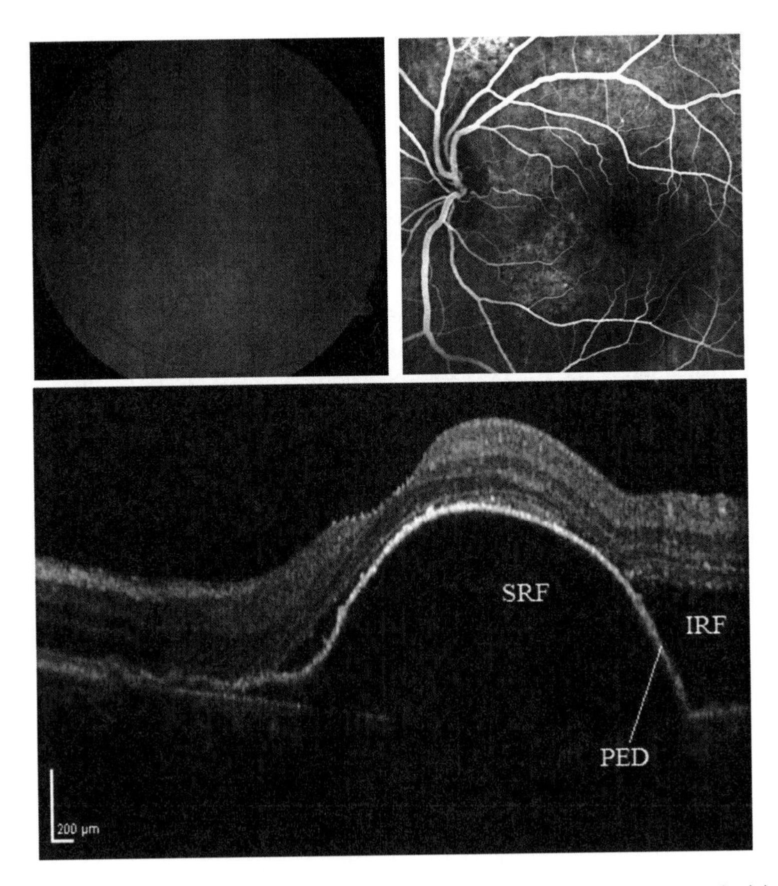

Fig. 3.18 A 46-year-old male presented with metamorphopsia and reduced central vision for 2 months duration. Patient had history of steroid intake for rheumatoid arthritis. Color fundus photograph shows dull-looking foveal reflex. FA demonstrates focal dot-like hyperfluorescein leakage on the macula. The SD-OCT shows pigment epithelial detachment (PED) with SRF and IRF as well as disturbed outer retinal layers

of RP patients even in the less advanced stages [75, 76] (Fig. 3.19). CME in RP has been treated with topical or oral carbonic anhydrase inhibitor [77], periocular or intravitreal injection of steroid [78, 79], intravitreal injection of bevacizumab [80] or ranibizumab [81], and vitrectomy [82]. Although each therapeutic measure provides a benefit for some of the patients, there is still lack of information on the pathophysiology, prevalence, and natural course of CME associated with RP as compared to other retinal vascular diseases. SD-OCT is widely used for the diagnosis and monitoring of RP.

Witkin et al. [83] were among the first who conducted a study to assess the integrity of photoreceptor in patients with RP using SD-OCT and correlate the foveal photoreceptor loss with VA. The result shows macular photoreceptor

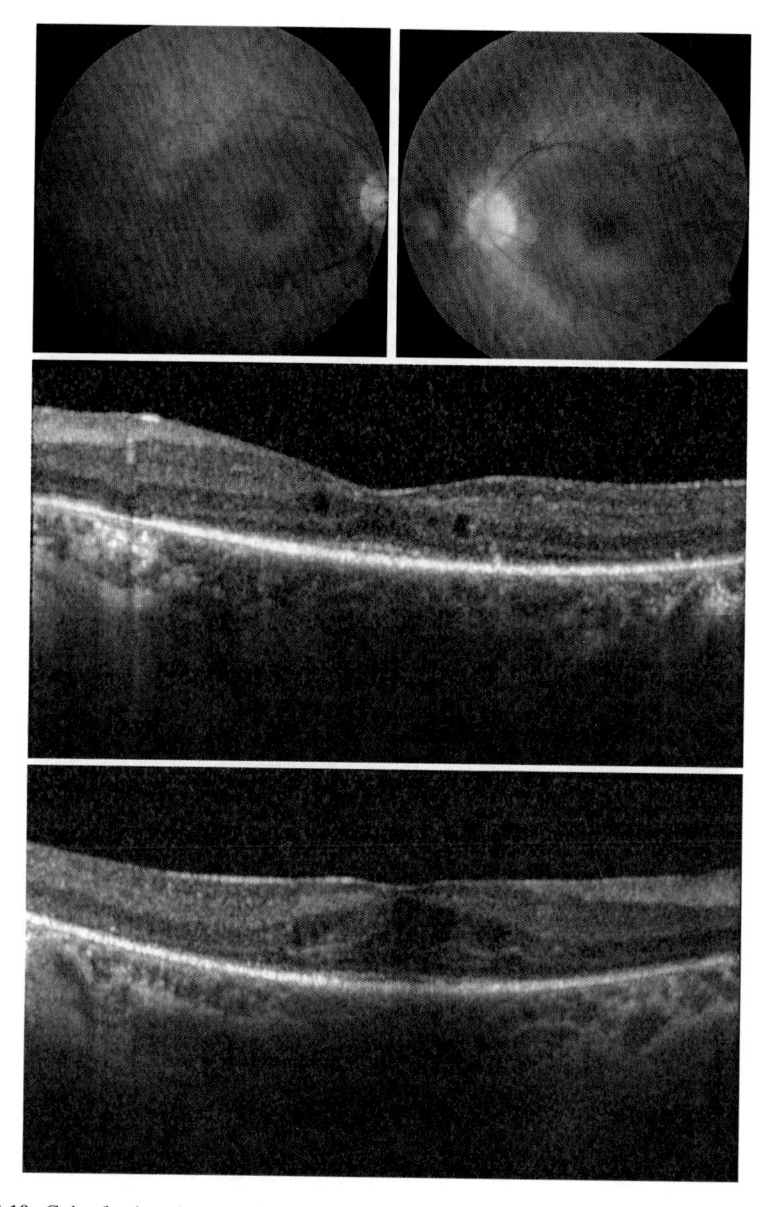

Fig. 3.19 Color fundus photograph shows bilateral retinitis pigmentosa (RP) with macular edema (ME). The SD-OCT shows disturbed photoreceptor layer in both eyes

disruption at the level of the IS/OS junction seen in all patients. A group of patient with good VA had extrafoveal photoreceptor disruption, where photoreceptors near fovea remained relatively normal. In more advanced RP group, both extrafoveal and foveal photoreceptor disruption were documented. In 2007, Aizawa et al. [84] reported almost similar finding where presence of IS/OS junction was significantly

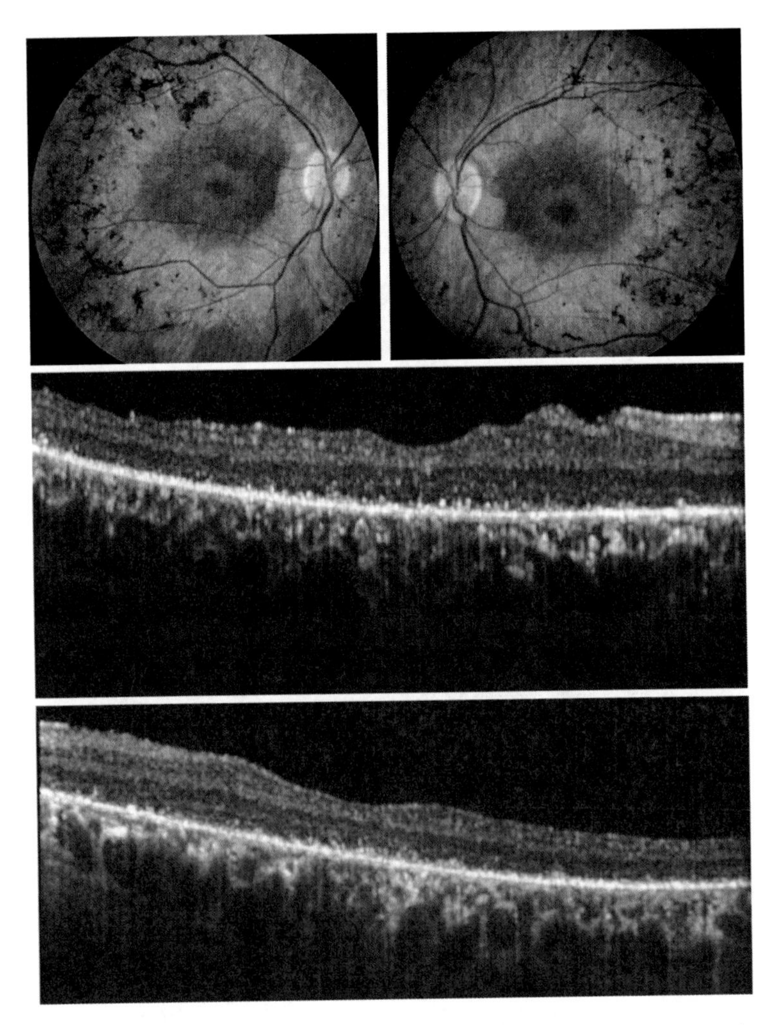

Fig. 3.20 A 35-year-old male presented with history of night blindness and progressive visual field loss over 5 years duration. He had similar family history of night blindness. Color fundus photographs show bilateral waxy pallor optic disk, attenuation of retinal vessels, and peripheral "bone-spicule pigmentation" with cystoid macula edema. The SD-OCT shows photoreceptor disruption at the level of the IS/OS junction

associated with better VA and thicker fovea in RP patients. In 2009, Oishi et al. [85] reported significantly worse VA in the IS/OS junction absent group as compared to intact IS/OS junction in RP patients complicated with CME. Interestingly, foveal thickness, photoreceptor layer thickness, and the cystoid space parameter exhibit no correlation with VA. In conclusion, IS/OS junction is the most significant factor for VA in RP patient with CME (Fig. 3.20).

3.5.8 *Foveal Microstructure and Visual Acuity in Selected Surgical Retina Cases*

3.5.8.1 Retinal Detachment

Retinal detachment (RD) is defined a separation of neurosensory layers to the RPE. Patient with macula-off RD may have poor vision postoperatively, specific color vision defects, or metamorphopsia postoperatively despite of successful anatomical reattachment. Any subtle changes of foveal structure postoperatively, which may cause visual disturbance, can be difficult to identify during standard clinical examination using slit lamp biomicroscopy. Epiretinal membranes (ERMs), CME, pigment migration, macula hole, retina folds, myopic shift, or cataract have been identified as the various factors which may cause reduction in postoperative VA. Decreased final postoperative vision can occur even in the absence of these clinically detectable complications. SD-OCT is useful to demonstrate possible cause of delayed or poor visual recovery after anatomically successful surgical repair (Fig. 3.21).

Histopathology has shown that prolonged RD can cause photoreceptor atrophy and death [73, 74]. The outer retina is deprived of its nutrient supply from the underlying choroid, and thus, photoreceptor atrophy occurs. Photoreceptor atrophy may be responsible for reduced VA seen in patients following anatomically successful RD repair. Schocket et al. [86] observed distortion of the photoreceptor IS/OS junction, CME, pockets of subretinal fluid, and ERM postoperatively among patients who showed incomplete visual recovery after RD surgery. Another interesting finding was that patients with RD not involving the macula (macula-on) also had similar abnormalities seen on SD-OCT. This indicates that healthy-appearing area of attached retina may not be structurally normal in case of RD. Wakabayashi et al. [87] documented disruption of photoreceptor IS/OS junction as well as ELM in preoperative macula-off group of rhegmatogenous retinal detachment (RRD) eyes. The same study reported that the eyes with atrophic changes in both IS/OS junction and ELM did not achieve complete restoration of microstructures in the photoreceptor layer and visual recovery even after anatomically successful RD repair.

In summary, rapid serial SD-OCT is a useful noninvasive tool to evaluate foveal microstructural changes which is not seen clinically. SD-OCT provides a better understanding of the foveal microstructural changes and possible reasons for difference in VA recovery even after anatomically successful RD repair.

3.5.8.2 Macula Hole

Idiopathic macular hole (MH) involves tractional force on the retina, leading to full-thickness retinal defect and visual impairment. In addition to visual impairment, relative scotoma in area of retina surrounding the hole was documented by using microperimetry studies [88]. This indicates absence of photoreceptor cells within

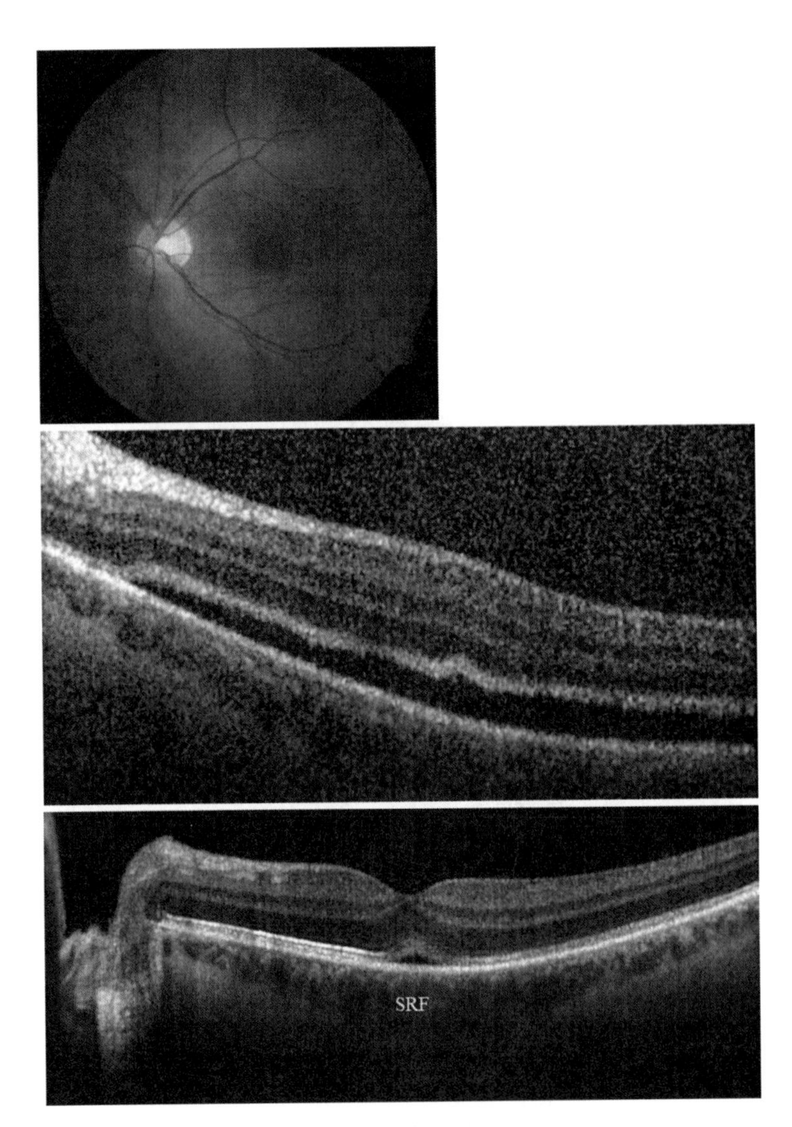

Fig. 3.21 A 22-year-old male presented with recent history of trauma on the left eye followed by reduced visual acuity. Color fundus photograph shows inferior retinal detachment that partially involved the macula. The SD-OCT shows neurosensory detachment from the RPE (*top*). Patient underwent surgery, and postoperatively, the SD-OCT shows attached neurosensory layers with small SRD under the fovea area (*bottom*)

the macular hole itself. Histopathologic study demonstrates intraretinal changes surrounding the hole, including cystic retinal edema and disruption of the photoreceptor layer [89] (Fig. 3.22). OCT has become the gold standard for diagnosis of MH and for confirming anatomic closure after surgery [90, 91]. With the new SD-OCT, alteration of photoreceptor layer in IS/OS junction has been described (Fig. 3.23).

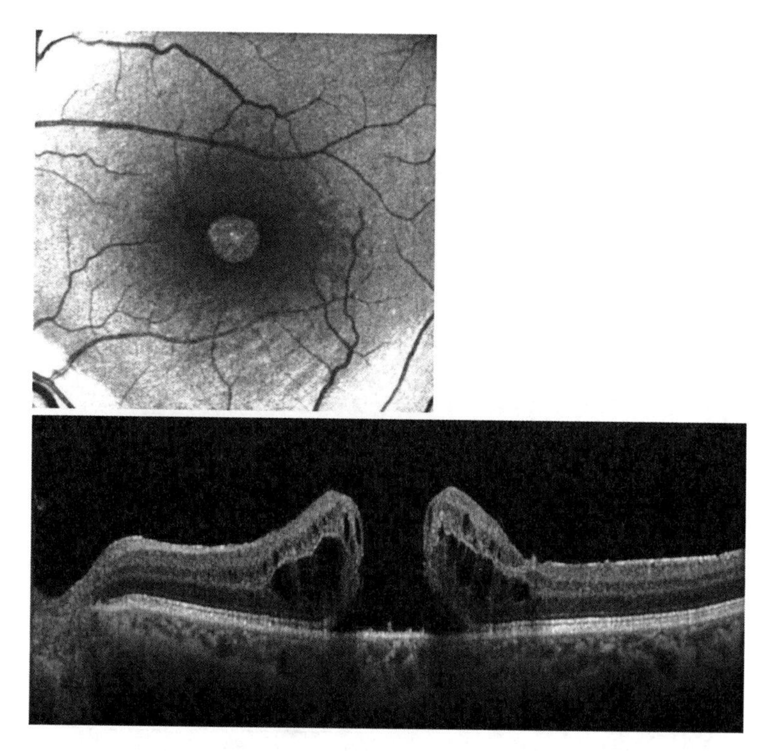

Fig. 3.22 A 71-year-old female presented with macular hole in the left eye. The BCVA was 20/40 in the left eye. SD-OCT shows full-thickness stage 4 macula hole with cystic swelling at the margin and disruption of IS/OS boundary line

Chang et al. [14] observed photoreceptor layers feature in both full-thickness macular hole (FTMH) and closed macula hole (CMH). Disruption of IS/OS junction was seen in all eyes in both FTMH and CMH. The disruption of IS/OS junction persisted after surgery for MH closure but decreased in size. The magnitude of VA changes was not significantly correlated with the change in IS/OS junction. Privat et al. [92] documented persistent defects in the IS/OS junction, and the size of the defect was not correlated with postclosure VA. This finding suggested that photoreceptor abnormalities occur in a larger area than within the macular hole itself and persist after anatomical closure of the retina defect following MH surgery.

3.5.8.3 Epiretinal Membranes

Epiretinal membranes (ERMs) are common macular disorders causing decreased or distorted central vision. ERMs may occur spontaneously (primary ERMs) or following macular diseases such as DME, macular edema following retinal vein occlusion, and uveitis. ERMs are often associated with wrinkling of the retinal surface or

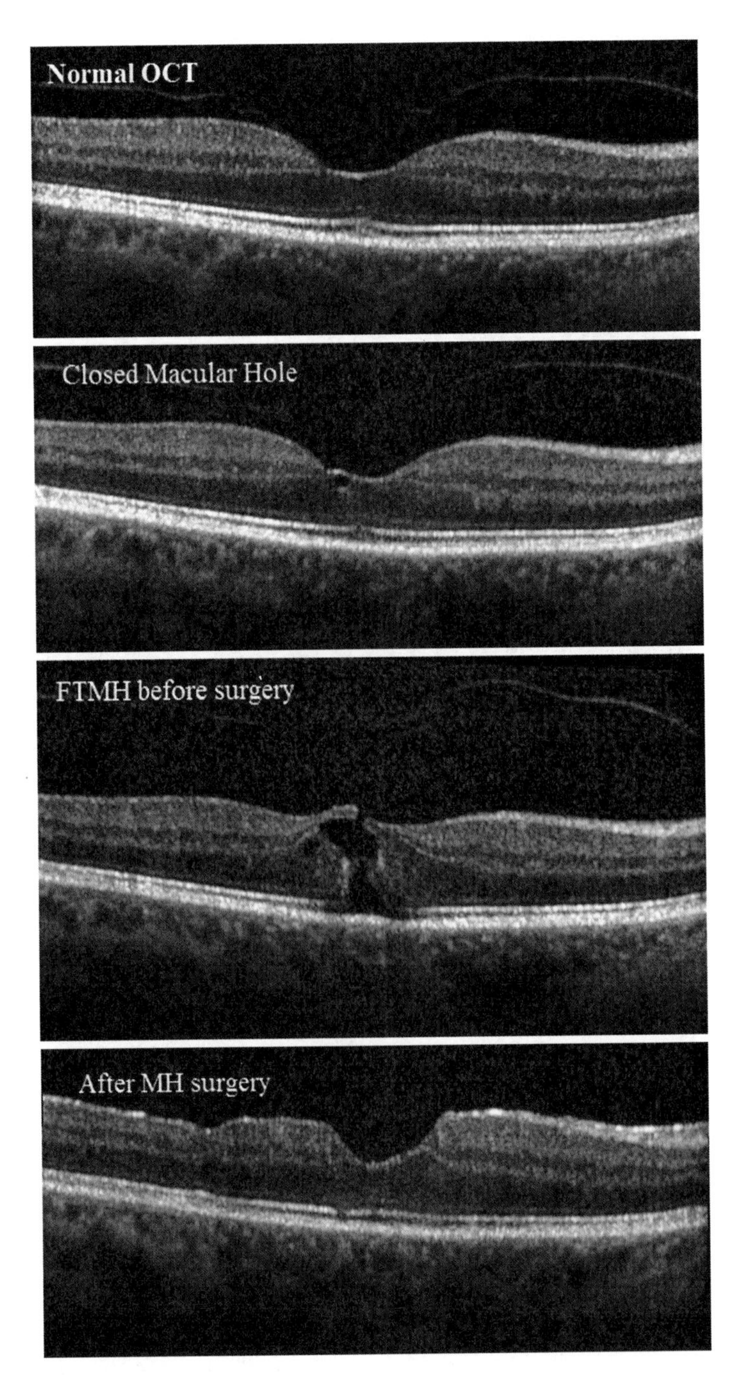

Fig. 3.23 A 75-year-old female presented with progressive reduced vision on the right eye. She had history of macula hole surgery done on the left eye 10 years ago. Series of SD-OCTs of the right eye show progressive development of MH before and after the surgery. Note that the outer retinal layers were disturbed before the surgery and recovered after the surgery

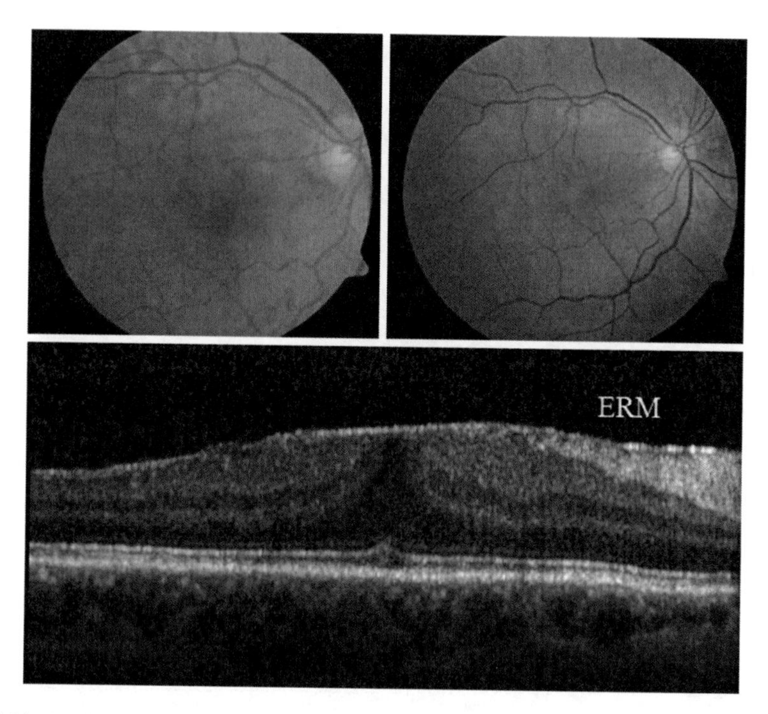

Fig. 3.24 An 83-year-old male presented with idiopathic epiretinal membrane (ERM). Color fundus photograph and red free photograph show wrinkling of the retina surface. The SD-OCT shows both disruption of photoreceptor layers and central retina thickness

macular thickening that can be evaluated clinically or with retinal imaging. PPV and ILM peeling is the standard surgical procedure performed to eliminate the ERMs [93]. Despite of a good success rate of surgery, patient with ERMs may experience persistent poor VA postoperatively. Niwa et al. [94] suggested that an impaired outer retinal structure rarely returns to normal, thus indicating poor visual prognosis in ERMs eyes. SD-OCT is a noninvasive method of examining and management of ERMs. Apart from that, SD-OCT also provides detailed information about the ERMs, retinal structure, and overlying vitreous. Suh et al. [95] found that IS/OS junction correlated with worse VA postoperatively. The eyes with preoperative IS/OS junction disruption, less than a quarter, returned to normal photoreceptor architecture 6 months after surgery. Disruption of photoreceptor layers was a better predictor of postoperative VA than central macular thickness. Oster et al. [13] shared similar finding as Suh et al. In addition, Oster et al. noted that combination of both disruption of photoreceptor layers and central retina thickness were statically significant predictor of poor VA among patients with ERMs.

Prolonged macular traction has been reported as an irreversible cause of photoreceptor loss, and change in the alignments of cones cells thus leads to poor vision in DME and BRVO [96, 97]. Thus, early surgery in ERMs is beneficial to prevent irreversible photoreceptor impairment (Fig. 3.24).

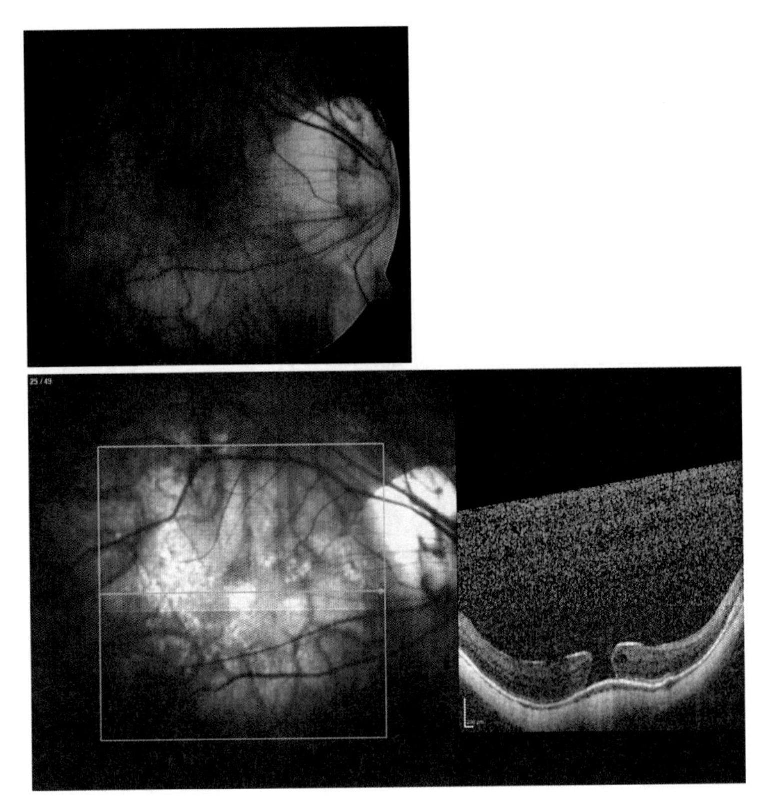

Fig. 3.25 A 45-year-old male presented with reduced VA in the right eye. He is a known case of high myopia with −10 dioptric power on both eyes. Color fundus photograph shows optic disk tilt, peripapillary choroidal atrophy, and posterior staphyloma. SD-OCT shows full-thickness macular hole (FTMH)

3.5.8.4 Myopic Foveoschisis

Myopic foveoschisis is a common complication seen in pathological myopia, and SD-OCT is essential for establishing the diagnosis. The standard treatment for this particular condition is PPV with or without internal limiting membrane (ILM) peeling followed by gas or silicone oil tamponade [98]. Myopic foveoschisis can be divided into two types: foveoschisis type, where the photoreceptor layer is still attached to the RPE, and foveal detachment type, where photoreceptor layer is detached from the RPE. Histological study revealed loss of RPE and photoreceptor layer corresponding to area of degeneration in highly myopic eyes. Sayanagi et al. [15] documented IS/OS defects on SD-OCT were accompanied by myopic chorioretinal atrophy in two third of the eyes. They proposed that IS/OS defects may be specific to, but are not rare, in myopic foveoschisis and may predict the postoperative visual loss (Fig. 3.25).

3.6 Conclusion

Recent development of new-generation SD-OCT has improved the axial image resolution and provides detailed observation of intraretinal microstructure especially the photoreceptor layers IS/OS junction and the ELM. The integrity of IS/OS junction and ELM were found to be significantly correlated with VA in many retinal diseases. Apart from that, photoreceptor loss which is best detected by SD-OC was found to be a strong predictive value for visual recovery following treatment.

3.7 Summary for Doctor

SD-OCT aids in identifying, monitoring, and quantitatively assessing various posterior segment conditions including diabetic macular edema, macular edema due retinal vasculo occlusive diseases, age-related macular degeneration, central serous chorioretinopathy, retinitis pigmentosa, and some surgical retinal conditions which include retinal detachment, macular hole, epiretinal membranes, and myopic foveoschisis.

SD-OCT helps in deciding the treatment protocol of these retinal diseases. The integrity of the photoreceptor was found to be significantly correlated with visual acuity in majority of these diseases. Photoreceptor loss which is best detected by SD-OC was found to be a strong predictive value for visual recovery following treatment.

References

1. D. Huang, E.A. Swanson, C.P. Lin, J.S. Schuman, W.G. Stinson, W. Chang, M.R. Hee, T. Flotte, K. Gregory, C.A. Puliafito, Optical coherence tomography. Science **254**(5035), 1178–1181 (1991)
2. P.E. Stanga, A.C. Bird, Optical Coherence Tomography (OCT): principles of operation, technology, indications in vitreoretinal imaging and interpretation of results. Int. Ophthalmol. **23**(4–6), 191–197 (2001)
3. J.G. Fujimoto, M.E. Brezinski, G.J. Tearney, S.A. Boppart, B. Bouma, M.R. Hee, J.F. Southern, E.A. Swanson, Optical biopsy and imaging using optical coherence tomography. Nat. Med. **1**(9), 970–972 (1995)
4. T.C. Chen, B. Cense, M.C. Pierce, N. Nassif, B. Park, S.H. Yun, B.R. White, B.E. Bouma, G.J. Tearney, J.F. Boer, Spectral domain optical coherence tomography ultra-high-speed, ultra-high resolution ophthalmic imaging. Arch. Ophthalmol. **123**(12), 1715–1720 (2005)
5. U.E.K. Wolf-Schnurrbusch, L. Ceklic, C.K. Brinkmann, M.E. Iliev, M. Frey, S.P. Rothenbuehler, V. Enzmann, S. Wolf, Macular thickness measurement in healthy eyes using six different optical coherence tomography instruments. Invest. Ophthalmol. Vis. Sci. **50**(7), 3432–3437 (2009)
6. G.E. Lang (ed.) *Diabetic retinopathy*. Dev Ophthalmol, vol. 39. (Karger, Basel, 2007), pp. 31–47

7. T. Otani, S. Kishi, Y. Maruyama, Patterns of diabetic macular edema with optical coherence tomography. Am. J. Ophthalmol. **127**(6), 688–693 (1999)
8. G. Panozzo, E. Gusson, B. Parolini, A. Mercanti, Role of optical coherence tomography in the diagnosis and follow up of diabetic macular edema. Semin. Ophthalmol. **18**(2), 74–81 (2003)
9. M. Fleckenstein, I.P. Charbel, H.M. Helb, S. Schmitz-Valckenberg, R.P. Finger, H.P. Scholl, K.U. Loeffler, F.G. Holz, High-resolution spectral domain-OCT imaging in geographic atrophy associated with age-related macular degeneration. Invest. Ophthalmol. Vis. Sci. **49**(9), 4137–4144 (2008)
10. U.E.K. Wolf-Schnurrbusch, V. Enzmann, C.K. Brinkmann, S. Wolf, Morphologic changes in patients with geographic atrophy assessed with a novel spectral OCT-SLO combination. Invest. Ophthalmol. Vis. Sci. **49**(7), 3095–3099 (2008)
11. M. Ota, A. Tsujikawa, T. Murakami, M. Kita, K. Miyamoto, A. Sakamoto, N. Yamaike, N. Yoshimura, Association between integrity of foveal photoreceptor layer and visual acuity in branch retinal vein occlusion. Br. J. Ophthalmol. **91**(12), 1644–1649 (2007)
12. N. Yamaike, A. Tsujikawa, M. Ota, A. Sakamoto, Y. Kotera, M. Kita, K. Miyamoto, N. Yoshimura, M. Hangai, Three-dimensional imaging of cystoid macular edema in retinal vein occlusion. Ophthalmology **115**(2), 355.e2–362.e2 (2008)
13. H.S.F. Oster, F. Mojana, M. Brar, R.M.S. Yuson, L. Cheng, W.R. Freeman, Disruption of the photoreceptor inner segment/outer segment layer on spectral domain-optical coherence tomography is a predictor of poor visual acuity in patients with epiretinal membranes. Retina **30**(5), 713–718 (2010)
14. L.K. Chang, H. Koizumi, R.F. Spaide, Disruption of the photoreceptor inner segment-outer segment junction in eyes with macular holes. Retina **28**(7), 969–975 (2008)
15. K. Sayanagi, Y. Ikuno, K. Soga, Y. Tano, Photoreceptor inner and outer segment defects in myopic foveoschisis. Am. J. Ophthalmol. **145**(5), 902–908 (2008)
16. F.C. Piccolino, R.R. de la Longrais, G. Ravera, C.M. Eandi, L. Ventre, A. Abdollahi, M. Manea, The foveal photoreceptor layer and visual acuity loss in central serous chorioretinopathy. Am. J. Ophthalmol. **139**(1), 87–99 (2005)
17. T. Wakaayashi, Y. Oshima, Restoration of ELM reflection line crucial for visual recovery in surgically closed MH. New SD-OCT imaging insights related to the reconstruction of photoreceptor layer and visual outcomes after macular hole repair. Retina Today 48–51 (2010)
18. E.R. Kandel, J.H. Schwartz, T.M. Jessell, *Principles of Neural Science*, 4th edn (McGraw-Hill, New York, 2000) pp. 507–513. ISBN 0–8385–7701–6
19. R. Klein, L.K.E. Knudtson, R. Gangnon, B.E.K. Klein, The Wisconsin Epidemiologic Study of diabetic retinopathy XXIII: The twenty-five-year incidence of macular edema in persons with type 1 diabetes. Ophthalmology **116**(3), 497–503 (2009)
20. Early Treatment Diabetic Retinopathy Study Research Group, Grading diabetic retinopathy from stereoscopic color fundus photographs: an extension of the modified Airlie House classification: ETDRS Report No 10. Ophthalmology **98**(5 Suppl), 786–806 (1991)
21. Early Treatment of Diabetic Retinopathy Study Research Group, Early photocoagulation for diabetic retinopathy. ETDRS Report Number 9. Ophthalmology **98**(5 Suppl), 766–785 (1991)
22. J.B. Jonas, I. Kreissig, A. Sofker, R.F. Degenring, Intravitreal injection of triamcinolone for diffused diabetic macular oedema. Arch. Ophthalmol. **121**(1), 57–61 (2003)
23. T. AvitabilevA. Lungo, A. Reibaldi, Intravitreal triamcinolone compared with macular laser grid photocoagulation for the treatment of cystoid macular edema. Am. J. Ophthalmol. **140**(4), 695–702 (2005)
24. C.E. Jahn, K. Topfner von Schutz, J. Richter, J. Boller, M. Kron, Improvement of visual acuity in eyes with diabetic macular edema after treatment with pars plana vitrectomy. Ophthalmologica **218**(6), 378–384 (2004)
25. P. Massin, F. Bandello, J.G. Garweg, L.L. Hansen, S.P. Harding, M. Larsen, P. Mitchell, D. Sharp, U.E.K. Wolf-Schnurrbusch, M. Gekkieva, A. Weichselberger, S. Wolf, Safety and efficacy of ranibizumab in diabetic macular edema (RESOLVE Study*): a 12-month, randomized, controlled, double-masked, multicenter phase II study. Diabetes Care **33**(11), 2399–2405 (2010)

26. S. Wolf, Assessing diabetic macular edema with optical coherence tomography, in *Medical Retina Essentials in Ophthalmology*, ed. by R.F. Spaide (Springer, Berlin, 2010), pp. 125–129; DOI: 10.1007/978-3-540-85540-8_11)
27. A. Salam, S. Wolf, C. Framme, U.E.K. Wolf-Schnurrbusch, Association between foveal photoreceptor changes and visual acuity in diabetic macular edema: a Spectral Domain Optical Coherence Tomography Study. 2010; Reference of abstract send for EURETINA 2011
28. T. Otani, Y. Yamaguchi, S. Kishi, Correlation between visual acuity and foveal microstructural changes in diabetic macular edema. Retina **30**(5), 774–780 (2010)
29. A.S. Maheshwary, S.F. Oster, R.M.S. Yuson, L. Cheng, F. Mojan, W.R. Freeman, The association between percent disruption of the photoreceptor inner segment—outer segment junction and visual acuity in diabetic macular edema. Am. J. Ophthalmol. **150**(1), 63.e1–67.e1 (2010)
30. K. Mervin, J. Stone, Developmental death of photoreceptors in the C57BL/6JMouse: association with retinal function and self-protection. Exp. Eye Res. 75(6):703–713 (2002)
31. H. Ozdemir, M. Karacorlu, S. Karacorlu, Serous macular detachment in diabetic cystoid macular edema. Acta Ophthalmol. Scand. **83**(1), 63–66 (2005)
32. Early Treatment Diabetic Retinopathy Study Research Group: Focal photocoagulation treatment of diabetic macular edema. Relationship of treatment effect to fluorescein angiographic and other retinal characteristics at baseline: ETDRS report no. 19. Arch. Ophthalmol. 113(9):1144–1155 (1995)
33. O. Cekic, S. Chang, J.J. Tseng, Y. Akar, G.R. Barile, W.M. Schiff, Cataract progression after intravitreal triamcinolone injection. Am. J. Ophthalmol. **139**(6), 993–998 (2005)
34. D. Thomas, C. Bunce, C. Moorman, D.A. Laidlaw, A randomised controlled feasibility trial of vitrectomy versus laser for diabetic macular oedema. Br. J. Ophthalmol. **89**(1), 81–86 (2005)
35. C. Framme, A. Walter, P. Prahs, R. Regler, D. Theisen-Kunde, C. Alt, R. Brinkmann, Structural changes of the retina after conventional laser photocoagulation and selective retina treatment (SRT) in spectral domain OCT. Curr. Eye Res. **34**(7), 568–579 (2009)
36. L.P. Aiello, S.E. Bursell, A. Clermont, E. Duh, H. Ishii, C. Takagi, F. Mori, T.A. Ciulla, K. Ways, M. Jirousek, L.E. Smith, G.L. King, Vascular endothelial growth factor-induced retinal permeability is mediated by protein kinase C in vivo and suppressed by an orally effective beta-isoform-selective inhibitor. Diabetes **46**(9), 1473–1480 (1997)
37. V. Poulaki, Hypoxia in the pathogenesis of retinal disease, in *Retinal Vascular Disease*, ed. by A.M.Joussen, T.W. Gardner, B. Kirchhof, S.J.Ryan (Springer, Berlin, Germany, 2007), pp. 121–138
S. Wolf, Assessing diabetic macular edema with optical coherence tomography, in *Medical Retina Essentials in Ophthalmology*, ed. by R.F. Spaide (Springer, Berlin, 2010), pp. 125–129; DOI: 10.1007/978-3-540-85540-8_11)
38. W. Einbock, L. Berger, U.E.K. Wolf-Schnurrbusch, J. Fleischhauer, S. Wolf, Predictive factors for visual acuity of patients with diabetic macular edema interpretated form the spectralis HRA + OCT. Invest. Ophthalmol. Vis. Sci. **49**, (2008); ARVO E-Abstract: 3473
39. A. Sakamoto, K. Nishijima, M. Kita, H. Oh, A. Tsujikawa, N. Yoshimura, Association between foveal photoreceptor status and visual acuity after resolution of diabetic macular edema by pars plana vitrectomy. Graefes Arch. Clin. Exp. Ophthalmol. **247**(10), 1325–1330 (2009)
40. The Branch Vein Occlusion Study Group: Argon laser photocoagulation for macular edema in branch vein occlusion. Am. J. Ophthalmol. **98**(3), 271–282 (1984)
41. A. Arnarsson, E. Stefansson, Laser treatment and the mechanism of edema reduction in branch retinal vein occlusion. Invest. Ophthalmol. Vis. Sci. **41**(3), 877–879 (2000)
42. E.M. Opremcak, R.A. Bruce, M.D. Lomeo, C.D. Ridenour,.D. Letson, A.J. Rehmar, Radial optic neurotomy for central retinal vein occlusion: a retrospective pilot study of 11 consecutive cases. Retina **21**(5), 408–415 (2001)
43. J.F. Arevalo, R.A. Garcia, L. Wu, F.J. Rodriguez, J. Dalma-Weiszhausz, H. Quiroz-Mercado, V. Morales-Canton, J.A. Roca, M.H. Berrocal, F. Graue-Wiechers, V. Robledo, Radial optic neurotomy for central retinal vein occlusion: results of the Pan-American Collaborative Retina Study Group (PACORES). Retina **28**(8), 1044–1052 (2008)

44. M. Karacorlu, H. Ozdemir, S.A. Karacorlu, Resolution of serous macular detachment after intravitreal triamcinolone acetonide treatment of patients with branch retinal vein occlusion. Retina **25**(7), 856–860 (2005)
45. J.A. Haller, F. Bandello, R.J. Belfort, M.S. Blumenkranz, M. Gillies, J. Heier, A. Loewenstein, Y.H. Yoon, M.L. Jacques, J. Jiao, X.Y. Li, S.M. Whitecup, OZURDEX GENEVA Study Group, Randomized, sham-controlled trial of dexamethasone intravitreal implant in patients with macular edema due to retinal vein occlusion. Ophthalmology **117**(6), 1134.e3–1146.e3 (2010)
46. M. Ota, A. Tsujikawa, M. Kita, K. Miyamoto, A. Sakamoto, N. Yamaike, Y. Kotera, N. Yoshimura, Integrity of foveal photoreceptor layer in central retinal vein occlusion. Retina **28**(10), 1502–1508 (2008)
47. M.O. Tso, Pathology of cystoid macular edema. Ophthalmology **89**(8), 902–915 (1982)
48. U.E.K. Wolf-Schnurrbusch, R. Ghanem, S.P. Rothenbuehler, V. Enzmann, C. Framme, S. Wolf, Predictors for short term visual outcome after anti-VEGF 1 therapy of macular edema due to central retinal vein occlusion. IOVS Papers in Press. Published on November 18, 2010. doi: 10.1167/iovs.10–6097
49. D. Yuzurihara, H. Iijima, Visual Outcome in Central Retinal and Branch Retinal Artery Occlusion. Jpn. J. Ophthalmol. **48**(5), 490–492 (2004)
50. J.O. Mason 3rd, A.A. Shah, R.S. Vail, P.A. Nixon, E.L. Ready, J.A. Kimble, Branch retinal artery occlusion: visual prognosis. Am. J. Ophthalmol. **146**(3), 455–457 (2008)
51. M. Karacorlu, H. Ozdemir, S. Arf Karacorlu, Optical coherence tomography findings in branch retinal artery occlusion. Eur. J. Ophthalmol. **16**(2), 352–353 (2006)
52. S.M. Falkenberry, M.S. Ip, B.A. Blodi, J.B. Gunther, Optical coherence tomography findings in central retinal artery occlusion. Ophthalmic Surg. Lasers Imaging. **37**(6), 502–505 (2006)
53. R.K. Murthy, S. Grover, K.V. Chalam, Sequential spectral domain OCT documentation of retinal changes after branch retinal artery occlusion. Clin. Ophthalmol. **4**, 327–329 (2010)
54. M. Yanoff, B.S. Fine, *Ocular Pathology, A Text and Atlas*, 2nd ed. Central retinal artery occlusion (Harper & Row, Philadelphia, PA, 1982), pp. 492–494
55. R. Klein, B.E. Klein, S.C. Jensen, S.M. Meuer, The five-year incidence and progression of age-related maculopathy: the Beaver Dam Eye Study. Ophthalmology **104**(1), 7–21 (1997)
56. P. Mitchell, W. Smith, K. Attebo, J.J. Wang, Prevalence of age-related maculopathy in Australia: the Blue Mountains Eye Study. Ophthalmology **102**(10), 1450–1460 (1995)
57. D. Pauleikhoff, M.J. Barondes, D. Minassian, I.H. Chisholm, A.C. Bird, Drusen as a risk factor in age related macular disease. Am. J. Ophthalmol. **109**(1), 38–43 (1990)
58. N.M. Bressler, J.C. Silva, S.B. Bressler, S.L. Fine, W.R. Green, Clinicopathologic correlation of drusen and retinal pigment epithelial abnormalities in age-related macular degeneration. Retina **14**(2), 130–142 (1994)
59. C.A. Curcio, N.E. Medeiros, C.L. Millican, Photoreceptor loss in age-related macular degeneration. Invest. Ophthalmol. Vis. Sci. **37**(7), 1236–1249 (1996)
60. F.G. Holz, D. Pauleikhoff, R. Klein, A.C. Bird, Pathogenesis of lesions in late age-related macular disease. Am. J. Ophthalmol. **137**(3), 504–510 (2004)
61. D. Pauleikhoff, C.A. Harper, J. Marshall, A.C. Bird, Aging changes in Bruch's membrane: a histochemical and morphologic study. Ophthalmology 1990; **97**(2), 171–178 (1990)
62. The Age-Related Eye Disease Research Group, The Age-Related Eye Disease Study (AREDS), A clinical trial of zinc and antioxidants. AREDS Report 2. J. Nutr. **130**(5S Suppl), 1516S–1519S (2000)
63. S.G. Schuman, A.F. Koreishi, S. Farsiu, S. Jung, J.A. Izatt, C.A. Toth, Photoreceptor layer thinning over drusen in eyes with age-related macular degeneration imaged in vivo with spectral-domain optical coherence tomography. Ophthalmology **116**(3), 488.e2–496.e2 (2009)
64. S.B. Bressler, M.G. Maguire, N.M. Bressler, S.L. Fine, Macular Photocoagulation Study Group, Relationship of drusen and abnormalities of the retinal pigment epithelium to the prognosis of neovascular macular degeneration. Arch. Ophthalmol. **108**(10), 1442–1447 (1990)
65. U.E.K. Wolf-Schnurrbusch, V. Enzmann, C.K. Brinkmann, S. Wolf, Morphologic changes in patients with geographic atrophy assessed with a novel spectral OCT–SLO combination. Invest. Ophthalmol. Vis. Sci. **49**(7), 3095–3099 (2008)

66. Treatment of age-related macular degeneration with photodynamic therapy (TAP) Study Group: Photodynamic therapy of subfoveal choroidal neovascularization in age-related macular degeneration with verteporfin: one-year results of 2 randomized clinical trials–TAP report. Arch. Ophthalmol. **117**(10), 1329–1345 (1999)
67. J.B. Jonas, I. Kreissig, P. Hugger, G. Sauder, S. Panda-Jonas, R. Degenring, Intravitreal triamcinolone acetonide for exudative age related macular degeneration. Br. J. Ophthalmol. **87**(4), 462–468 (2003)
68. Y. Kurashige, A. Otani, M. Sasahara, Y. Yodoi, H. Tamura, A. Tsujikawa, N. Yoshimura, Two-year results of photodynamic therapy for polypoidal choroidal vasculopathy. Am. J. Ophthalmol. **146**(4), 513–519 (2008)
69. P. Mitchell, J.F. Korobelnik, P. Lanzetta, F.G. Holz, C. Prünte, U. Schmidt-Erfurth, Y. Tano, S. Wolf, Ranibizumab (Lucentis) in neovascular age-related macular degeneration: evidence from clinical trials. Br. J. Ophthalmol. **94**(1), 2–13 (2010)
70. H. Hayashi, K. Yamashiro, A. Tsujikawa, M. Ota, A. Otani, N. Yoshimura, Association between foveal photoreceptor integrity and visual outcome in neovascular age-related macular degeneration. Am. J. Ophthalmol. 148(1):83.e1–89.e1 (2009)
71. M. Reibaldi, F. Boscia, T. Avitabile, A. Russo, V. Cannemi, M.G. Uva, A. Reibaldi, Low-fluence photodynamic therapy in longstanding chronic central serous chorioretinopathy with foveal and gravitational atrophy. Eur. J. Ophthalmol. **19**(1), 154–158 (2009)
72. H. Matsumoto, T. Sato, S. Kishi, Outer nuclear layer thickness at the fovea determines visual outcomes in resolved central serous chorioretinopathy. Am. J. Ophthalmol. **148**(1), 105.e1–110.e1 (2009)
73. B. Cook, G.P. Lewis, S.K. Fisher, R. Adler, Apoptotic photoreceptor degeneration in experimental retinal detachment. Invest. Ophthalmol. Vis. Sci. **36**(6), 990–996 (1995)
74. L. Berglin, P.V. Algvere, S. Seregard, Photoreceptor decay over time and apoptosis in experimental retinal detachment. Graefes Arch Clin Exp Ophthalmol. **235**(5), 306–312 (1997)
75. M. Hajali, G.A. Fishman, R.J. Anderson, The prevalence of cystoid macular oedema in retinitis pigmentosa patients determined by optical coherence tomography. Br. J. Ophthalmol. **92**(8), 1065–1068 (2008)
76. H. Hirakawa, H. Iijima, T. Gohdo, S. Tsukahara, Optical coherence tomography of cystoid macular edema associated with retinitis pigmentosa. Am. J. Ophthalmol. **128**(2), 185–191 (1999)
77. S. Grover, G.A. Fishman, R.G. Fiscella, A.E. Adelman, Efficacy of dorzolamide hydrochloride in the management of chronic cystoid macular edema in patients with retinitis pigmentosa. Retina **17**(3), 222–231 (1997)
78. V.S. Saraiva, J.M. Sallum, M.E. Farah, Treatment of cystoid macular edema related to retinitis pigmentosa with intravitreal triamcinolone acetonide. Ophthalmic Surg. Lasers Imaging **34**(5), 398–400 (2003)
79. L. Scorolli, M. Morara, A. Meduri, L.B. Reggiani, G. Ferreri, S.Z. Scalinci, R.A. Meduri, Pan-American Collaborative Retina Study Group, Treatment of cystoid macular edema in retinitis pigmentosa with intravitreal triamcinolone. Arch. Ophthalmol. **125**(6), 759–764 (2007)
80. J.F. Arevalo, R.A. Garcia-Amaris, J.A. Roca, J.G. Sanchez, L. Wu, M.H. Berrocal, M. Maia, Primary intravitreal bevacizumab for the management of pseudophakic cystoid macular edema: pilot study of the Pan-American Collaborative Retina Study Group. J. Cataract. Refract. Surg. **33**(12), 2098–2105 (2007)
81. C.R. Shah, M.H. Brent, Treatment of retinitis pigmentosa—related cystoid macular edema with intravitreous ranibizumab. Retinal Cases, Brief Reports. **4**(4), 291–293 (2010)
82. J. García-Arumí, V. Martinez, L. Sararols, B. Corcostegui, Vitreoretinal surgery for cystoid macular edema associated with retinitis pigmentosa. Ophthalmology **110**(6), 1164–1169 (2003)
83. A.J. Witkin, T.H. Ko, J.G. Fujimoto, A. Chan, W. Drexler, J.S. Schuman, E. Reichel, J.S. Duker, Ultra-high resolution optical coherence tomography assessment of photoreceptors in retinitis pigmentosa and related diseases. Am. J. Ophthalmol. **142**(6), 945–952 (2006)

84. S. Aizawa, Y. Mitamura, T. Baba, A. Hagiwara, K. Ogata, S. Yamamoto, Correlation between visual function and photoreceptor inner/outer segment junction in patients with retinitis pigmentosa. Eye (Lond.) **23**(2), 304–308 (2009)
85. A. Oishi, A. Otani, M. Sasahara, H. Kojima, H. Nakamura, M. Kurimoto, N. Yoshimura, Photoreceptor integrity and visual acuity in cystoid macular oedema associated with retinitis pigmentosa. Eye (Lond.) **23**(6), 1411–1416 (2009)
86. L.S. Schocket, A.J. Witkin; J.G. Fujimoto; T.H. Ko; J.S. Schuman; A.H. Rogers; C. Baumal; E. Reichel, J.S. Duker, Ultrahigh-resolution optical. Coherence tomography in patients with decreased visual acuity after retinal detachment repair. Ophthalmology **113**(4), 666–672 (2006)
87. T. Wakabayashi, Y. Oshima, H. Fujimoto, Y. Murakami, H. Sakaguchi, S. Kusaka, Y. Tano, Foveal microstructure and visual acuity after retinal detachment repair: imaging analysis by Fourier-domain optical coherence tomography. Ophthalmology **116**(3), 519–528 (2009)
88. F. Amari, K. Ohta, H. Kojima, N. Yoshimura, Predicting visual outcome after macular hole surgery using scanning laser ophthalmoscope microperimetry. Br. J. Ophthalmol. **85**(1), 96–98 (2001)
89. W.R. Green, The macular hole: histopathologic studies. Arch. Ophthalmol. **124**(3), 317–321 (2006)
90. B. Haouchine, P. Massin, R. Tadayoni, A. Erginay, A. Gaudric, Diagnosis of macular pseudoholes and lamellar macular holes by optical coherence tomography. Am. J. Ophthalmol. **138**(5), 732–739 (2004)
91. J.M. Ruiz-Moreno, C. Staicu, D.P. Piñero, J. Montero, F. Lugo, P. Amat, Optical coherence tomography predictive factors for macular hole surgery outcome. Br. J. Ophthalmol. **92**(5), 640–644 (2008)
92. E. Privat, R. Tadayoni, D. Gaucher, B. Haouchine, P. Massin, A. Gaudric, Residual defect in the foveal photoreceptor layer detected by optical coherence tomography in eyes with spontaneously closed macular holes. Am. J. Ophthalmol. **143**(5), 814–819 (2007)
93. J.W. Lee, I.T. Kim, Outcomes of idiopathic macular epiretinal membrane removal with and without internal limiting membrane peeling: a comparative study. Jpn. J. Ophthalmol. **54**(2), 129–134 (2010)
94. T. Niwa, H. Terasaki, M. Kondo, C.H. Piao, T. Suzuki, Y. Miyake, Function and Morphology of macula before and after removal of idiopathic epiretinal membrane. Invest. Ophthalmol. Vis. Sci. **44**(4), 1652–1656 (2003)
95. M.H. Suh, J.M. Seo, K.H. Park, H.G. Yu, Associations between macular findings by optical coherence tomography and visual outcomes after epiretinal membrane removal. Am. J. Ophthalmol. **147**(3), 473.e3–480.e3 (2009)
96. C.W.T.A. Lardenoye, K. Probst, P.J. DeLint, A. Rothova, Photoreceptor function in eyes with macular edema. Invest. Ophthalmol. Vis. Sci. **41**(12), 4048–4053 (2000)
97. N. Villate, J.E. Lee, A. Venkatraman, W.E. Smiddy, Photoreceptor layer features in eyes with closed macular holes: optical coherence tomography findings and correlation with visual outcomes. Am. J. Ophthalmol. **139**(2), 280–289 (2005)
98. A. Tsujikawa, M. Kikuchi, K. Ishida, A. Nonaka, K. Yamashiro, Y. Kurimoto, Fellow eye of patients with retinal detachment associated with macular hole and bilateral high myopia. Clin. Exp. Ophthalmol. **34**(5), 430–433 (2006)

Chapter 4
Optical Coherence Tomography Updates on Clinical and Technical Developments. Age-Related Macular Degeneration: Drusen and Geographic Atrophy

Monika Fleckenstein, Steffen Schmitz-Valckenberg, and Frank G. Holz

Abstract Age-related macular degeneration (AMD) is a complex disease with both genetic and environmental factors influencing its development. With the advent of high-resolution OCT imaging, the characterization of drusen in AMD has become possible. The in vivo morphologic characteristics imaged with SD-OCT may represent distinct subclasses of drusen variants, may relate closely to ultrastructural drusen elements identified in donor eyes, and may be useful imaging biomarkers for disease severity or risk of progression [Khanifar et al. Ophthalmology 115(11):1883–1890, 2008].

4.1 Introduction

Age-related macular degeneration (AMD) is a complex disease with both genetic and environmental factors influencing its development, and it has become the most common cause of legal blindness in industrialized countries [1–8]. It represents a chronic disease with various phenotypic manifestations, different disease stages, and different rates of progression over time.

The early stage of AMD is characterized by the presence of drusen and pigmentary changes. In advanced AMD, choroidal neovascularization (CNV) is present in the exudative form, and there is atrophy of the retinal pigment epithelium (RPE), the photoreceptors, and the choriocapillaris in the advanced nonexudative form (geographic atrophy, GA). While CNV is the most common cause of severe visual loss in advanced AMD, approximately 20% of AMD patients who are legally blind have lost central vision due to GA [3, 9–12].

F.G. Holz (✉)
Department of Ophthalmology, University of Bonn, Ernst Abbe Strasse 2, 53127 Bonn, Germany
e-mail: frank.holz@ukb.uni-bonn.de

R. Bernardes and J. Cunha-Vaz (eds.), *Optical Coherence Tomography*, Biological and Medical Physics, Biomedical Engineering,
DOI 10.1007/978-3-642-27410-7_4,

Patients with primary GA tend to be older than those with neovascular forms of AMD at the time of initial manifestation. In patients aged 85 years and older, GA occurs four times as often as CNV [4].

4.2 Early AMD

4.2.1 In Vivo Visualization of Drusen and Pigmentary Changes by Spectral-Domain Optical Coherence Tomography

With the advent of high-resolution OCT imaging, the characterization of drusen in AMD has become possible in vivo [13]. Khanifar and coworkers categorized drusen ultrastructure in AMD using spectral-domain (SD)-OCT imaging and correlated the tomographic and photographic drusen appearance [14]: The grading system included evaluation of the internal reflectivity as well as the presence of hyperreflective foci overlying the drusen (Fig. 4.1). SD-OCT imaging obviously gives information over and above conventional imaging techniques and conventional funduscopy. The in vivo morphologic characteristics imaged with SD-OCT may represent distinct subclasses of drusen variants, may relate closely to ultrastructural drusen elements identified in donor eyes, and may be useful imaging biomarkers for disease severity or risk of progression [14].

The same group later reported changes in the neurosensory retina over drusen using SD-OCT imaging [15]. It was concluded that a decreased photoreceptor layer thickness over drusen may represent a degenerative process with cell loss associated with impaired function. Furthermore, several authors speculated that the hyperreflective foci overlying drusen may relate to RPE cell migration into the overlying retina [15–17].

Ho et al. [18] recently described the features of intraretinal RPE migration documented on a prototype spectral-domain, high-speed, ultrahigh-resolution OCT device in a group of patients with early to intermediate dry AMD and correlated intraretinal RPE migration on OCT to hyperpigmentation on fundus photographs. They concluded that the high incidence of intraretinal RPE migration observed above areas of drusen suggests that drusen may play a catalytic role in facilitating intraretinal RPE migration in dry AMD patients [18].

Other observations on SD-OCT images include subretinal spaces associated with large, confluent drusen [19] (Fig. 4.2). It was concluded that specific distribution of the fluid was limited to the depression between adjacent drusen which may indicate that the cluster of coalescent drusen produces mechanical strain to the outer retinal layers that locally pulls the sensory retina away from its normal position. Consequently, the appearance of fluid within subretinal compartment between coalescent drusen in OCT cross-sectional images may not be a reliable marker for the presence of CNV [19].

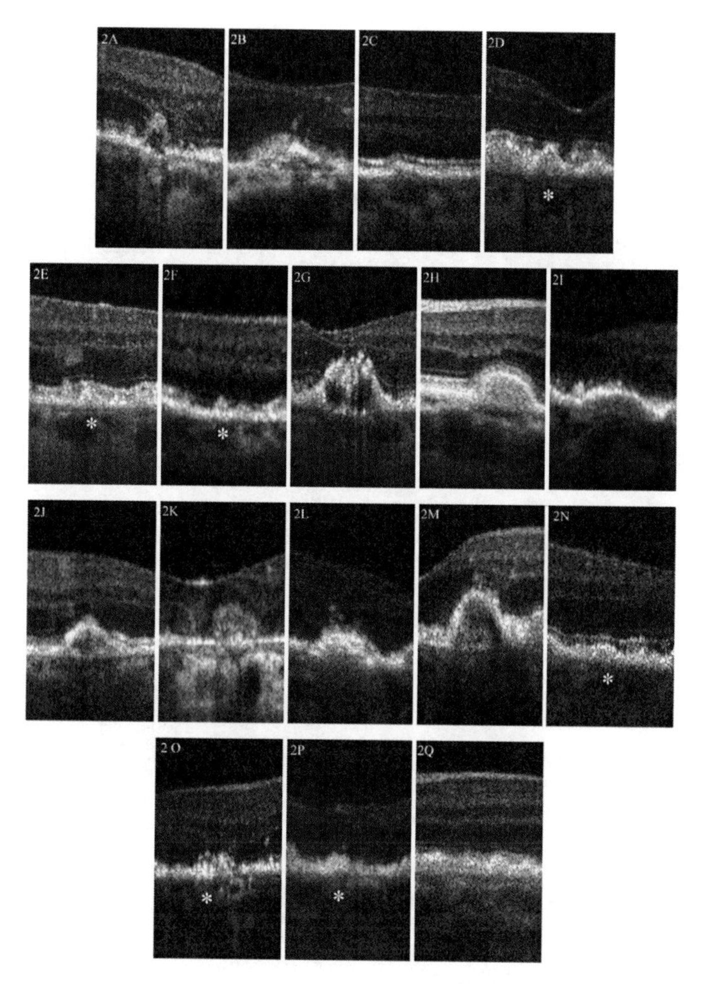

Fig. 4.1 The 17 combined morphologic drusen patterns observed: (**a**) Pattern 1: concave, low reflectivity, nonhomogeneous without core, no overlying foci. (**b**) Pattern 2: concave, low reflectivity, nonhomogeneous with core, overlying foci. (**c**) Pattern 3: concave, medium reflectivity, homogeneous, no overlying foci. (**d**) Pattern 4 (marked with *asterisks*): concave, medium reflectivity, nonhomogeneous without core, no overlying foci. (**e**) Pattern 5 (marked with *asterisks*): concave, medium reflectivity, nonhomogeneous without core, overlying foci. (**f**) Pattern 6 (marked with *asterisks*): concave, high reflectivity, homogeneous, no overlying foci. (**g**) Pattern 7: convex, low reflectivity, nonhomogeneous with core, overlying foci. (**h**) Pattern 8: convex, medium reflectivity, homogeneous, no overlying foci. (**i**) Pattern 9: convex, medium reflectivity, homogeneous, overlying foci. (**j**) Pattern 10: convex, medium reflectivity, nonhomogeneous without core, no overlying foci. (**k**) Pattern 11: convex, medium reflectivity, nonhomogeneous with core, no overlying foci. (**l**) Pattern 12: convex, medium reflectivity, nonhomogeneous without core, overlying foci. (**m**) Pattern 13: convex, medium reflectivity, nonhomogeneous with core, overlying foci. (**n**) Pattern 14 (marked with *asterisks*): convex, high reflectivity, homogeneous, no overlying foci. (**o**) Pattern 15 (marked with *asterisks*): convex, high reflectivity, nonhomogeneous without core, no overlying foci. (**p**) Pattern 16 (marked with *asterisks*): convex, high reflectivity, nonhomogeneous without core, overlying foci. (**q**) Pattern 17: sawtooth, no overlying foci. Copyright ref. [14]

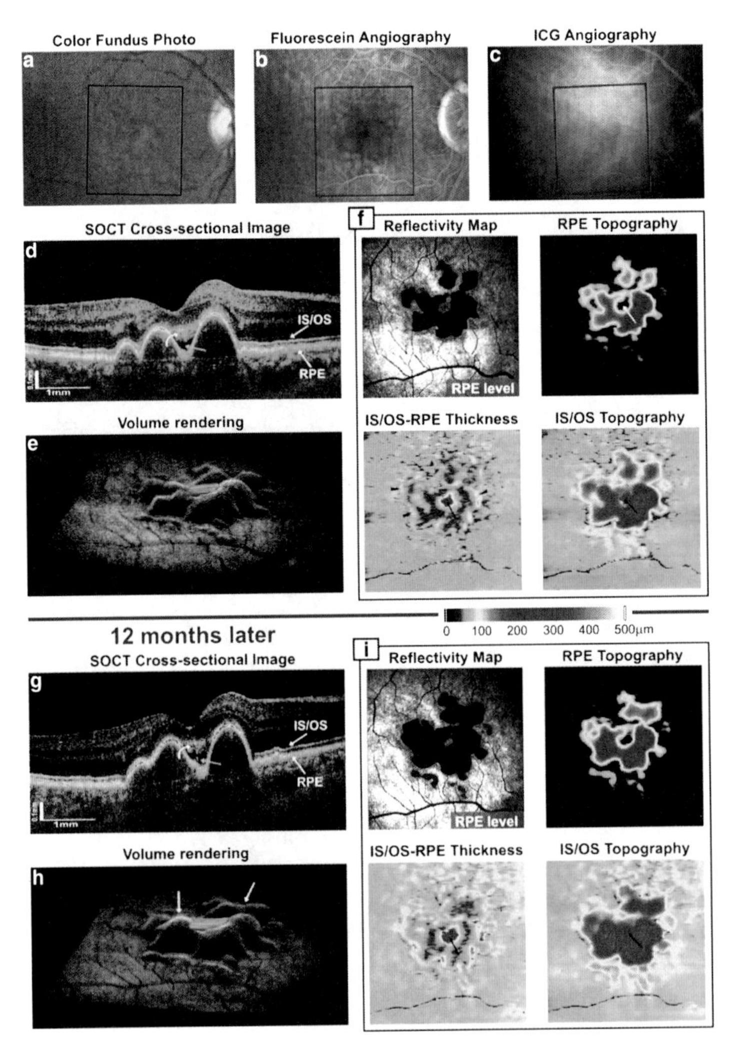

Fig. 4.2 Case 1: right eye of a 71-year-old woman with nonexudative age-related macular degeneration. (**a**) Color fundus photograph. (**b**) Late-phase fluorescein angiogram showing a staining of drusen. (**c**) Indocyanine-green angiography (ICG) with hypofluorescent spots corresponding to drusen. (**d**) Spectral-domain optical coherence tomography (SD-OCT) cross-sectional image. *Gray arrow* points to the subretinal fluid. The thickened detached photoreceptor outer segments are marked with *curved arrow*. (**e**) Volumetric reconstruction of retinal pigment epithelium (RPE) surface showing drusen topography; subretinal fluid is marked in *blue*. (**f**) Group of SD-OCT contour maps. Case 1: at the 12-month follow-up. (**g**) SD-OCT cross-sectional image. *Gray arrow* points to the subretinal fluid. The thickened detached photoreceptor outer segments are marked with *curved arrow*. (**h**) Volumetric reconstruction of RPE surface showing the progression of drusen coalescence (*white arrows*). (**i**) Group of SD-OCT contour maps. The reflectivity and RPE topography maps demonstrate increased horizontal and vertical extent of drusen. The inner segment/outer segment junction (IS/OS) RPE thickness and IS/OS topography maps show no apparent change in fluid volume that is located in the depression between clustering drusen. Copyright ref. [19]

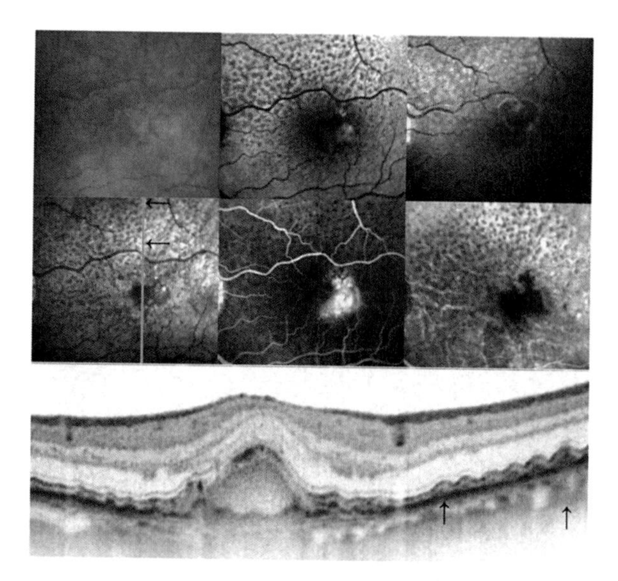

Fig. 4.3 Topographic distribution and the characteristic signal of reticular lesions by different imaging methods in an 88-year-old woman with regressed occult choroidal neovascularization; *top left* color fundus photograph, *top middle* fundus autofluorescence image, *top right* red-free reflectance, *middle left* near-infrared reflectance, *middle middle* late-phase fluorescein angiography (10 min after injection), *middle right* indocyanine-green angiography (10 min), and *bottom* spectral-domain optical coherence tomography (SD-OCT) scan (Spectralis HRA+OCT, Heidelberg Engineering, Heidelberg, Germany). The green line in the near-infrared reflectance images corresponds to the SD-OCT scan; the arrows mark the same location, illustrating the lateral extension of retinal areas involved by reticular drusen superior to the fovea. All en face fundus images have been aligned to each other and therefore have the same scaling, whereas the SD-OCT scan is magnified and thus has a different scaling. Copyright ref. [21]

Besides normal soft drusen, so-called "reticular pseudodrusen" had been noted in eyes with AMD [20]. SD-OCT imaging revealed a signal with undulation of the RPE and photoreceptor segment layer [16], giving the impression of a actual localization anterior to the RPE layer [21–23] (Fig. 4.3). Therefore, some authors suggested that "subretinal drusenoid deposits" [24] are the pathophysiological correlates for reticular pseudodrusen [23].

4.2.2 *Quantification of Drusen Volume by Spectral-Domain Optical Coherence Tomography*

Compared to manual drusen segmentation in SD-OCT imaging [25], a less time-consuming approach for disease monitoring in natural history studies and

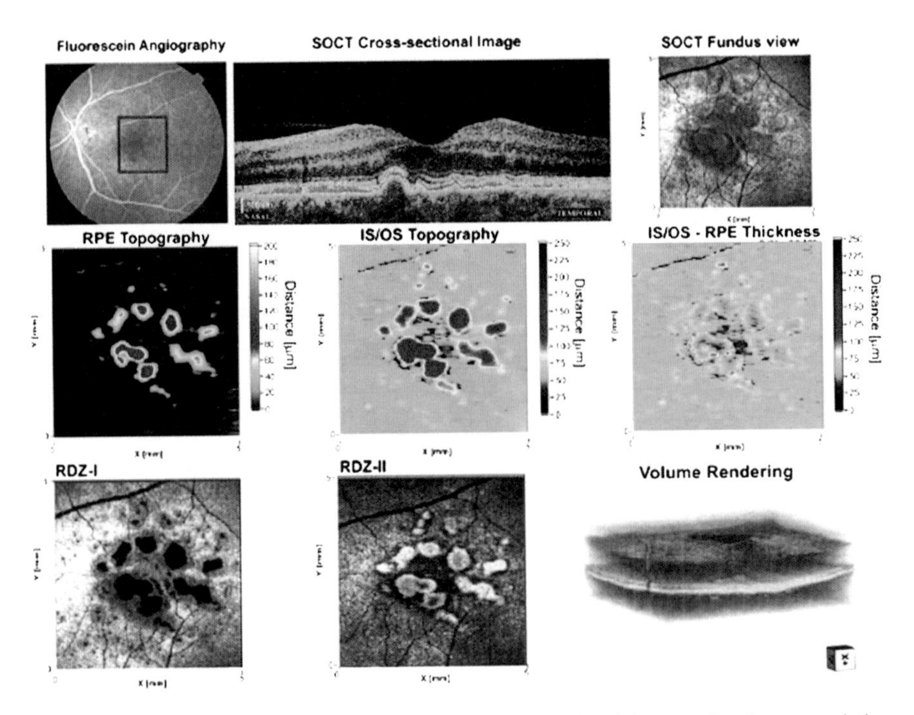

Fig. 4.4 Comparison of retinal topography mapping with reflectivity mapping in nonexudative age-related macular degeneration. Soft drusen in a 76-year-old patient is visualized in the cross-sectional image and five SOCT maps. Full three-dimensional data set is accessible via OSA ISP (view 3). Copyright ref. [27]

interventional clinical trials include automated segmentation and quantification of drusen by OCT imaging. In experimental settings, algorithms to analyze drusen area and volume from SD-OCT images which allows for determination of drusen area, drusen volume, and proportion of drusen SD-OCT systems has been developed [26–28] (Fig. 4.4).

However, Schlanitz and coworkers reported that the commercially available automated segmentation algorithms yet have limitations to reliably quantify number and density of drusen and size [29]. They evaluated the performance of automated analyses integrated in three SD-OCT devices to identify drusen in eyes with early (i.e., nonatrophic and nonneovascular) AMD. The automated segmentation of the RPE using the Cirrus SD-OCT (Carl Zeiss Meditec, Dublin, CA) made significantly fewer errors in detecting drusen than did the 3DOCT-1000 (Topcon, Tokyo, Japan) ($P < 0.001$). The 200×200 scan pattern detected 30% of the drusen with negligible errors.

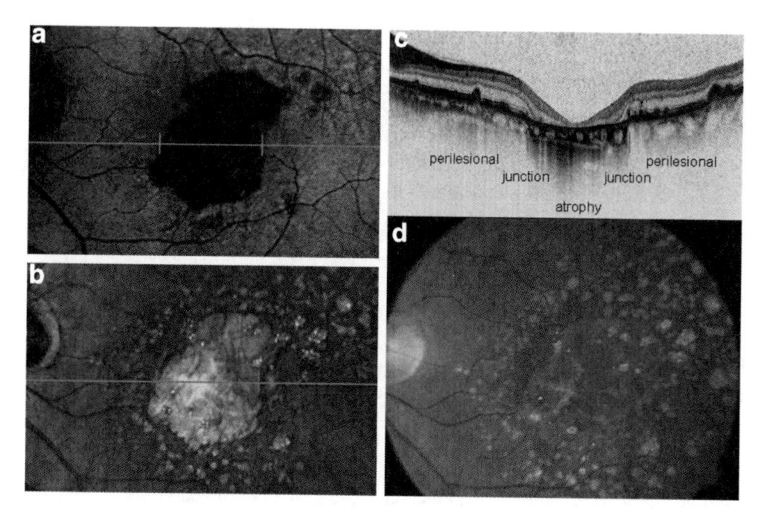

Fig. 4.5 Fundus autofluorescence (**a**) and infrared reflectance (**b**) images with the corresponding SD-OCT scan (**c**) obtained by simultaneous cSLO and SD-OCT (Spectralis HRA+OCT, Heidelberg Engineering, Heidelberg, Germany) imaging in a 70-year-old patient with GA due to AMD. (**a**, **b**) *Horizontal green lines* location of the SD-OCT scan (**c**). (**a**–**c**) *Vertical green lines* junction between the atrophic patch and nonatrophic retina. (**d**) Corresponding fundus picture. Copyright ref. [16]

4.3 Geographic Atrophy

4.3.1 Phenotypic Features of Atrophy and Identification of Predictive Markers by Spectral-Domain Optical Coherence Tomography

SD-OCT has provided new insights into the microstructural alterations associated with GA [16, 30–38] (Fig. 4.5). The atrophic lesion is thereby characterized by loss of the outer nuclear layer, the external limiting membrane (ELM), the inner segment/outer segment of photoreceptors layer (IPRL), and the RPE layer [16, 39] (Figs. 4.6 and 4.7). SD-OCT imaging furthermore revealed a wide spectrum of microstructural abnormalities, both in the surrounding retinal tissue and in the atrophic area (Figs. 4.8 and 4.9). These alterations may reflect different disease stages or, alternatively, heterogeneity on a cellular and molecular level.

Characterization of microstructural changes of the border zone of GA may allow identifying risk characteristics for an increased disease progression. Different types of GA borders have been identified by SD-OCT imaging. Brar and coworkers have recently introduced a dichotomic classification system for the border zone of GA [31]: They differentiated a type 1 border with smooth margins and no alterations of the outer retina and a type 2 border with severe alterations in the outer retinal layers and irregular margins (Fig. 4.8a, b). More recently, Fleckenstein and coworkers have described a third border type, the "splitting" border which is most prevalent in a

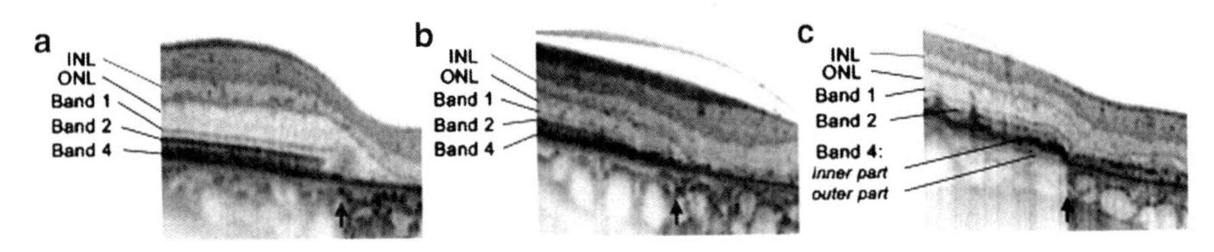

Fig. 4.6 Different types of borders of GA as visualized by spectral-domain optical coherence tomography (SD-OCT) (Spectralis HRA+OCT, Heidelberg Engineering, Heidelberg, Germany). (**a**) and (**b**) (GA other than "diffuse-trickling" phenotype): Band 4 (assumed retinal pigment epithelium (RPE)/Bruch's membrane complex) narrows and an outer layer remains throughout the atrophic area (assumed Bruch's membrane). According to Brar and coworkers [7], the margin depicted in 3A represents a type 1 border with smooth margins and no alterations of the outer retina and that in 3B a type 2 border with severe alterations in the outer retinal layers and irregular margins. (**c**) In contrast to 3A and 3B, the border of GA in the "diffuse-trickling" phenotype is characterized by "splitting" of band 4 in an inner and outer part. Band 1, external limiting membrane; band 2, inner segment/outer segment of photoreceptor layer; band 4, inner part, assumed retinal pigment epithelium; band 4, outer part, assumed Bruch's membrane. Copyright ref. [40]

rapidly progressing GA subtype [40]. This border is characterized by an obvious separation of the RPE/Bruch's membrane complex and may indeed represent a high-risk characteristic for an increased GA progression (Fig. 4.8c).

The eye-tracking software incorporated, e.g., in the Spectralis HRA+OCT (Heidelberg Engineering, Heidelberg, Germany) allows for serial examinations at the same location over time [33, 34]. Using this technology, the dynamic nature of development and progression of atrophy can be longitudinally analyzed in the same eye on a three-dimensional level. Microstructural changes such as the advancing loss of the RPE and photoreceptor bands at the GA border or the subsequent apposition of the outer plexiform layer with Bruch's membrane within the atrophic lesion have been recently described (Fig. 4.9). Of note, these observations can be made within a relatively short period of time [33]. This approach may be helpful for both, monitoring the natural course of the disease and to elucidate pathogenetic mechanisms. In particular, the identification of structural risk factors reflecting disease activity and future GA progression may be important for prognosis and visual function. Furthermore, effects of pharmacological interventions on the border zone may be evaluated in the future.

4.3.2 Correlation with Other Imaging Modalities

The simultaneous recording of confocal scanning laser ophthalmoscopy (cSLO) and SD-OCT images in one instrument with an exact topographic overlay during image acquisition allows for accurate orientation of pathological alteration in different imaging modalities. Herein, the correlation of cSLO fundus autofluorescence (FAF) with SD-OCT imaging has been focused in several studies [31, 35, 37, 41].

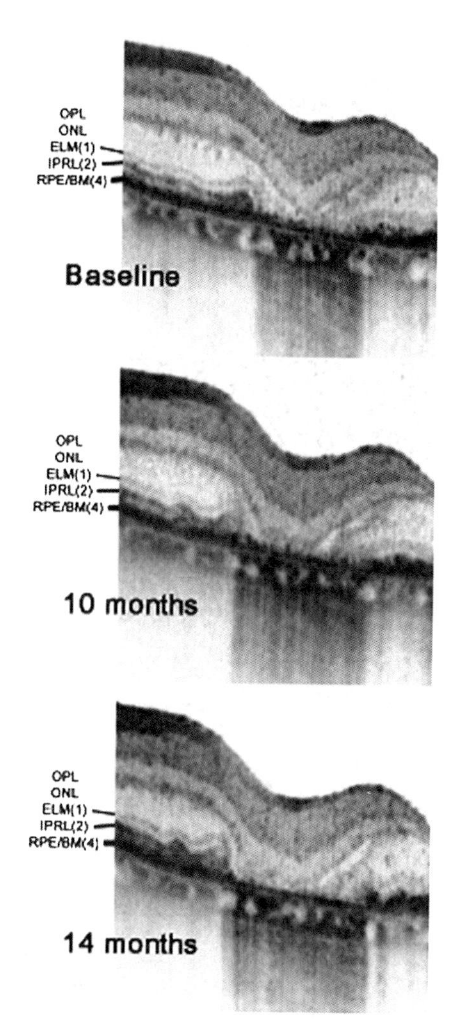

Fig. 4.7 Progression of preexisting geographic atrophy (GA) visualized by serial spectral-domain OCT imaging (Spectralis HRA+OCT, Heidelberg Engineering, Heidelberg, Germany) (14-month follow-up). There is marked enlargement of the atrophic area over time with progressive loss of the RPE layer (inner part of band 4), IPRL (band 2), and ELM (band 1), respectively, and thinning of ONL at the border of atrophy. Within the atrophic lesion, the remaining part of band 4 becomes more homogenous and the ONL progressively thins; subsequently, the OPL approaches the remaining part of band 4 (assumed Bruch's membrane). *OPL* outer plexiform layer; *ONL* outer nuclear layer; *ELM* external limiting membrane (band 1); *IPRL* interface of the inner and outer segments of the photoreceptor layer (band 2); *RPE/BM* retinal pigment epithelium/Bruch's membrane complex (band 4). Copyright ref. [33]

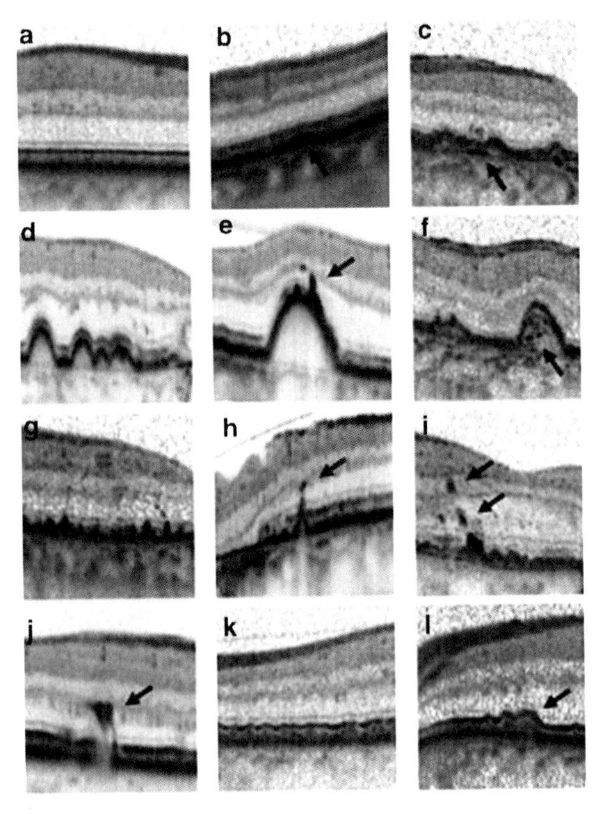

Fig. 4.8 Heterogeneous alterations in the SD-OCT scans (Spectralis HRA+OCT, Heidelberg Engineering, Heidelberg, Germany) in the perilesional zone of atrophy. (**a**) Layers in clinically normal appearing retina; (**b**) thickening of band 4; (**c**) elongated elevation of band 4 with plaques beneath; (**d**) *dome*-shaped elevations with preserved layers above; (**e**) *dome*-shaped elevation with clumps at the *top*; (**f**) *dome*-shaped elevation with backscattering material beneath; (**g**) apical extensions of band 4; (**h**) spike with a clump at the *tip*; (**i**) clumps at different retinal levels; (**j**) clump in the OPL with disrupted bands beneath; (**k**) small elevations of band 2 and mottled band 4; and (**l**) increased distance between band 2 and band 4. Copyright ref. [16]

4.3.2.1 Fundus Autofluorescence Imaging

Several lines of experimental and clinical evidence indicate that the RPE plays in an important role in the pathogenesis of GA associated with AMD. In postmitotic RPE cells, lipofuscin (LF) accumulates in the lysosomal compartment with age and also in various complex and monogenetic retinal diseases including Best's disease, Stargardt's disease, and AMD [42]. LF is thought to be mainly derived from the chemically modified residues of incompletely digested photoreceptor outer segment disks. Experimental findings suggest that certain molecular compounds of LF such as *N*-retinylidene-*N*-retinylethanolamine (A2-E) possess toxic properties and may interfere with normal cell function [43].

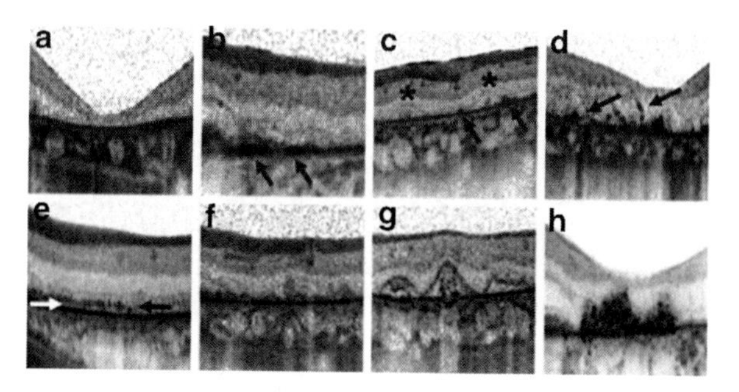

Fig. 4.9 Alterations within the atrophic area visualized by SD-OCT (Spectralis HRA+OCT, Heidelberg Engineering, Heidelberg, Germany). (**a**) Loss of the outer retinal layers and remaining line in extension of the lower part of band 4; (**b**) plaques at the former band 4 level (*arrows*); (**c**) elevations (*asterisks*) and clumps (*arrows*) at the remaining line in extension of the lower part of band 4; (**d**) clumps (*arrows*) at different retinal levels; (**e**) *dotted-line*-appearing formation that remained as an extension of the elevated band 4 (*black arrow*; see Figs. 4.5f, 4.6b); beneath, there was a continuous hyperreflective band (*white arrow*); (**f**) irregular elevations at the remaining extension of the lower part of band 4; (**g**) crown-like elevations with "debris" beneath; (**h**) accumulation of irregular, highly reflective material. Copyright ref. [16]

The accumulation of LF in postmitotic human RPE cells and its harmful effects on normal cell function has been largely studied in vitro with fluorescence microscopic techniques [42]. Delori and coworkers have shown that FAF in vivo is mainly derived from RPE LF [44]. With the advent of cSLO, it is possible to document FAF and its spatial distribution and intensity over large retinal areas in the living human eye [45–48]. FAF imaging gives additional information above and beyond conventional imaging tools such as fundus photography, fluorescein angiography, or OCT.

Due to the lack of RPE cells and, therefore, LF, FAF imaging shows markedly decreased FAF intensity over atrophic patches. Areas of atrophy can be accurately delineated and quantified with image analysis software, and atrophy progression rates can be calculated [49, 50]. Therefore, FAF imaging is an easy feasible, noninvasive method which is a useful tool for following GA patients and progression over time.

Furthermore, it has been shown that areas with increased FAF intensities and, therefore, excessive RPE LF load surrounding atrophy, in the so-called junctional zone of atrophy, can be identified [46, 48]. Areas of increased FAF have been shown to precede the development of new areas of GA or the enlargement of preexisting atrophic patches [51]. In a natural history study—fundus autofluorescence in age-related macular degeneration (FAM) study—it has been shown that eyes with larger areas of increased FAF outside atrophy were associated with higher rates of GA progression over time compared to eyes with smaller areas of increased FAF outside atrophy at baseline [52]. These findings suggest that the area of increased FAF surrounding the atrophy at baseline is positively correlated with the degree of spread of GA over time. A more recent analysis of the FAM study of 195 eyes of 129 patients show that variable rates of progression of GA are dependent on the specific phenotype of abnormal FAF pattern at baseline [53]. Atrophy enlargement was the slowest in eyes with no abnormal FAF pattern (median 0.38 mm^2/year), followed by eyes with the focal FAF pattern (median 0.81 mm^2/year), then by eyes with the diffuse FAF pattern (median 1.77 mm^2/year), and by eyes with the banded FAF pattern (1.81 mm^2/year). The difference in atrophy progression between the groups of no abnormal and focal FAF patterns and the groups of the diffuse and banded FAF patterns was statistically significant ($p < 0.0001$). These results have subsequently been confirmed in another large-scale natural history study (GAP study) [54, 55]. These findings underscore the importance of abnormal FAF intensities around atrophy and the pathophysiological role of increased RPE LF accumulation in patients with GA due to AMD.

4.3.2.2 Simultaneous Fundus Autofluorescence and Spectral-Domain Optical Coherence Tomography Imaging

By applying simultaneous FAF and SD-OCT imaging (Spectralis HRA+OCT, Heidelberg Engineering, Heidelberg, Germany), Schmitz-Valckenberg and coworkers analyzed outer retinal changes within the atrophic lesion in patients with GA [41] (Fig. 4.10). In this study, two readers independently graded the following parameters: width of the atrophic lesion on the FAF image at the site where the SD-OCT scan had been placed and, on the SD-OCT image, widths of the linear disruption of the outer nuclear layer, the ELM, and the IPRL and width of the disruption of choroidal signal enhancement. They found a mean width of the atrophic lesion by FAF imaging of 2.83 mm (95% confidence interval, 2.37–3.29). The linear disruption of choroidal hyperreflectivity showed the closest agreement with 2.83 mm (2.37–3.28), whereas the linear width of disrupted IPRL was larger (3.10 mm; 2.65–3.55). Overall, the width of the atrophic lesion correlated significantly with all five SD-OCT parameters ($P < 0.0001$, $r = 0.96$–0.99). They concluded that these findings demonstrate that the atrophic lesions identified with FAF represent irreversible underlying outer retinal damage [41].

Wolf-Schnurrbusch and coworkers [37] first analyzed SD-OCT alterations underlying specific FAF patterns: They noticed differences in the retinal layer alterations related to certain FAF patterns. When a small, continuous band at the margin of the GA was detected on FAF, they observed extensive, relatively homogenous alterations in the outer retinal layers. These alterations corresponded with the width of the continuous band in the FAF at the margin of the GA. The structural alterations in the group with small focal spots of increased FAF in the junctional zone were very confined. They corresponded pixel to pixel to the focal spots of increased FAF seen in the simultaneous FAF image of the HRA-OCT. In the group with FAF with a diffusely increased FAF at the entire posterior pole, only a few alterations in the retinal layer structure were observed [37].

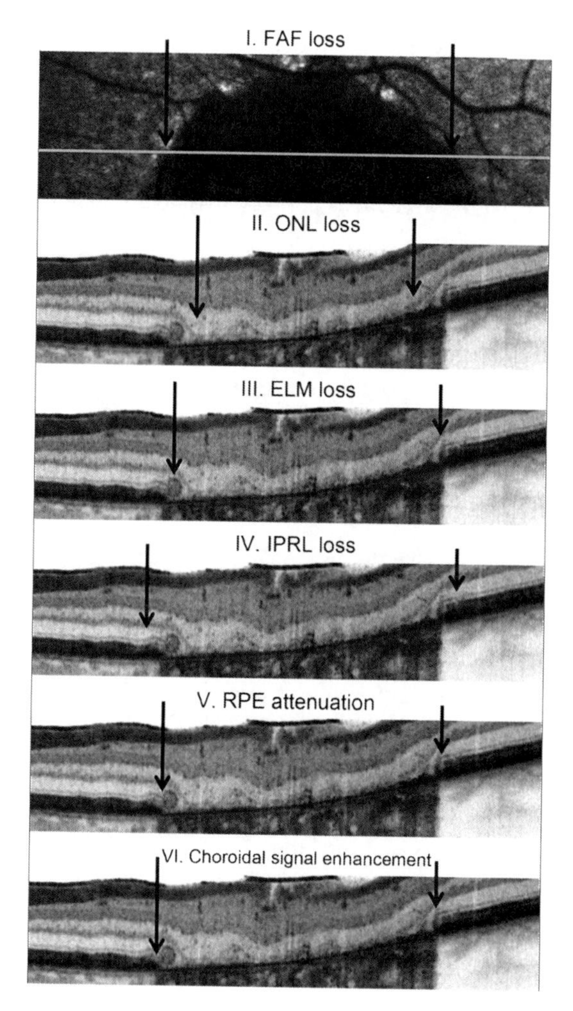

Fig. 4.10 Illustrations of the six linear dimensions that were separately and independently assessed by two graders. The six linear dimensions include the width of the fundus autofluorescence atrophic lesion at the site transected by the SD-OCT scan (Spectralis HRA+OCT, Heidelberg Engineering, Heidelberg, Germany), the width of disruption of the ONL, the ELM, the IRPL, and the width of choroidal hyperreflectivity. The disruption of each parameter was defined at both linear ends of atrophy, and a line was placed horizontally to determine the respective linear dimension. The images in the figure were cropped for illustration purposes and do not show the entire 30° field of the original scans. Copyright ref. [41]

Later, Brar and coworkers studied the appearance of margins of GA in high-resolution OCT images and correlated those changes with FAF imaging [31]. The outer retinal layer alterations were analyzed in the junctional zone between normal retina and atrophic retina and were correlated with corresponding FAF. They found a significant association between OCT findings and the FAF findings

($r = 0.67$; $P < 0.0001$): Severe alterations of the outer retinal layers at margins on spectral-domain OCT corresponded significantly to increased autofluorescence, and smooth margins on OCT corresponded significantly to normal FAF (kappa, 0.7348; $P < 0.0001$). They concluded that visualization of reactive changes in the RPE cells at the junctional zone and correlation with increased FAF—secondary to increased LF—together these methods may serve as determinants of progression of GA [31].

4.3.3 *Quantification of Atrophy by Spectral-Domain Optical Coherence Tomography Imaging*

While the layered structure of the retina is well ordered in healthy eyes, it is sometimes hard to differentiate structures solely based on their reflectivity in eyes with retinal disease [56].

Schütze and coworkers recently investigated the performance of OCT segmentation procedures in commercially available OCT devices to assess morphology and extension of GA [57]. All patients underwent Stratus (model 3000), Cirrus (Carl Zeiss Meditec), Spectralis (Spectralis HRA+OCT; Heidelberg Engineering), and 3D-OCT-1000 (Topcon) imaging. Automated segmentation analyses were compared. An overlay of SLO and three-dimensional retinal thickness (RT) maps were used to investigate whether areas of retinal thinning correspond to areas of RPE atrophy. Herein, GA areas identified in SLO scans were significantly larger than areas of retinal thinning in RT maps. No convincing topographic correlation could be found between areas of retinal thinning and actual GA size as identified in SLO and fundus photography. Spectralis OCT showed significantly more mild and severe segmentation errors than 3D and Cirrus OCT. This study therefore showed substantial limitations in identifying zones of GA reliably when using automatic segmentation procedures in current SD-OCT devices.

However, by using the SD-OCT fundus image to visualize and quantify GA (Cirrus SD-OCT, Carl Zeiss Meditec, Dublin, CA), Yehoshua and coworkers reported an excellent reproducibility. They further emphasized the advantages by using SD-OCT technology in GA assessment that include the convenience and assurance of using a single imaging technique that permits simultaneous visualization of GA along with the loss of photoreceptors and the RPE that should correlate with the loss of visual function [38].

Recently, segmentation of the RPE was demonstrated by polarization-sensitive (PS) OCT based on that layer's—in contrast to most of the other layers—depolarizing character [58]. Using this tissue-specific contrast, the same group showed the potential of PS-OCT to image the RPE in patients with AMD [59, 60] and recently presented PS-OCT for quantitative assessment of retinal pathologies in AMD [56] (Fig. 4.11). On the basis of the polarization scrambling characteristics of the RPE, novel segmentation algorithms were developed that allow one to segment

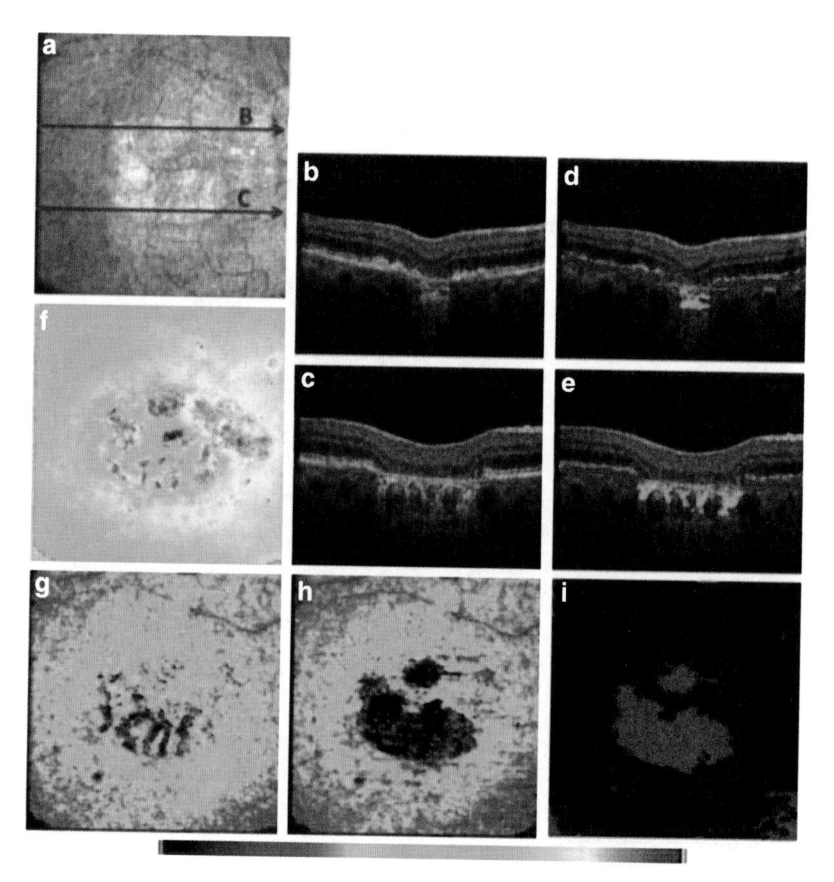

Fig. 4.11 Geographic atrophy: (**a**) En face fundus image reconstructed from three-dimensional OCT data set; (**b**, **c**) reflectivity B-scan images; (**d**, **e**) overlay of depolarizing structures within and outside the evaluation band shown in *red* and *green color*, respectively; and (**f**) retinal thickness map computed as axial distance between ILM and $DOPU_{min}$ position. The $DOPU_{min}$ position is usually located in the RPE. In the case of atrophy, this position was located in the choroid [green pixels in (**d**) and (**e**)]. (**g**) Map of overall number of depolarizing pixels per A-line. Zones of RPE atrophy are masked by choroidal depolarization. (**h**) Map displaying the thickness of the depolarizing layer inside the evaluation band. (**i**) Binary map of atrophic zones. The color map scales from 70 to 325 μm for (**f**) and from 0 to 39 pixels for (**g**) and (**h**). Copyright ref. [56]

pathologic features such as drusen and atrophic zones in dry AMD as well as to determine their dimensions. PS-OCT therefore appears as a promising imaging modality for three-dimensional retinal imaging and ranging with additional contrast based on the structures' tissue-inherent polarization properties [56].

Using the Spectralis HRA+OCT (Heidelberg Engineering, Heidelberg, Germany), the lateral spread of GA and the reduction in retinal thickness were confirmed as surrogate markers for disease progression at the GA border that were both quantifiable and corresponded to loss of outer retinal layers (Fig. 4.12). However, an increase in retinal thickness was also observed that was related to

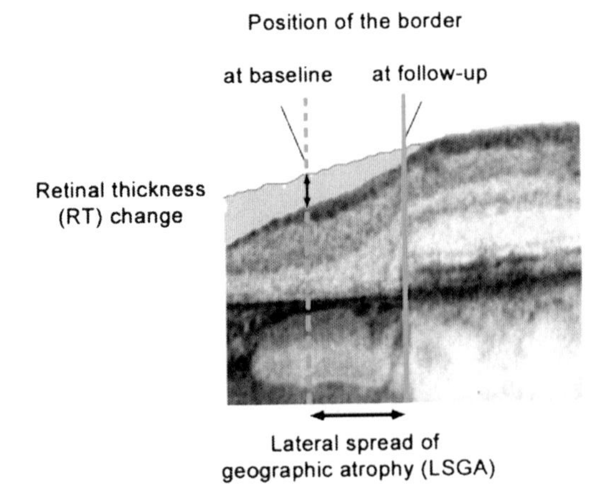

Fig. 4.12 Quantification of morphological changes at the border of geographic atrophy (GA) in spectral-domain OCT imaging. The position of the border at follow-up is marked by the *solid green vertical line*; the position of the border at baseline is indicated by the *dotted green line*. The change in retinal thickness (RT) and the lateral spread of GA (LSGA), respectively, are indicated by *black arrows*. Copyright ref. [33]

confounders such as epiretinal membrane formation or marked collateral changes such as development of RPE elevations or sub-RPE deposits. Furthermore, it was shown in patients with Stargardt's macular dystrophy that reduced retinal thickness (i.e., atrophy of retinal layers) did not correlate with the transverse extent of photoreceptor loss [33, 61]. These observations indicate that tracking of GA progression by changes in retinal thickness measurements alone should be interpreted with caution.

4.4 Future Perspectives

Simultaneous SD-OCT imaging with microperimetric (MP) assessment is a promising approach to gain further information on structure–function correlation in retinal diseases [62] (Fig. 4.13).

Iwama and coworkers studied the relationship between retinal sensitivity and morphologic changes in eyes with confluent soft drusen [63]. Although not applying simultaneous assessment of SD-OCT and MP, they found that eyes with confluent soft drusen often show focal areas with reduced retinal function, areas that are consistent with irregularity of the RPE or of the IPRL [63].

A further trend-setting approach is the assessment of the photoreceptor mosaic over drusen using adaptive optics and SD-OCT. The purpose of the study by Godara and coworkers [64] was to illustrate how SD-OCT and adaptive optics fundus imaging can be used to quantitatively analyze the integrity of the overlying photoreceptors in a single subject with macular drusen. This imaging approach and

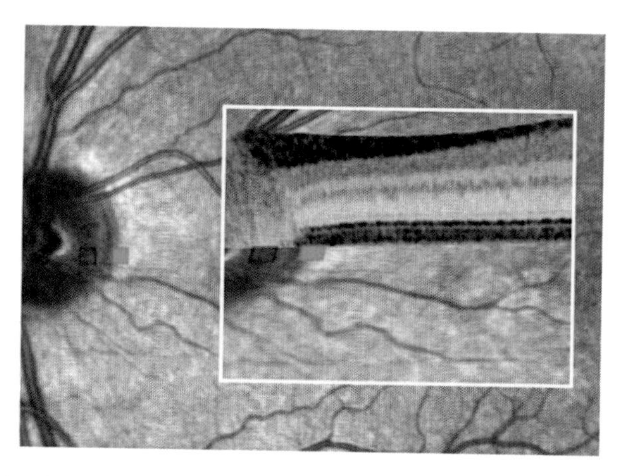

Fig. 4.13 Ability of coregistration using the MultiModalMapper software to exactly map functional testing on anatomical recordings in a normal subject. The optic nerve head is the anatomical correlate for the physiological blind spot in the visual field. The hollow *red square* represents the brightest stimulus using microperimetry and was not detected by the subject (absolute scotoma) when projected onto the optic nerve head. In contrast, a much dimmer testing point (*green square*, 18 dB decreased light intensity compared to the brightest stimulus) projected just outside the optic nerve area head was detected by the subject and shows normal light sensitivity. An overlay with the SD-OCT scan (*inset*, framed in *white*) shows the normal retinal layers within the area bordering the optic nerve head. Copyright ref. [62]

the image analysis metrics introduced may serve as the foundation for valuable imaging-based biomarkers for detecting the earliest stages of disease, tracking progression, and monitoring treatment response [64].

References

1. J. Ambati, B.K. Ambati, S.H. Yoo, S. Ianchulev, A.P. Adamis, Age-related macular degeneration: etiology, pathogenesis, and therapeutic strategies. Surv. Ophthalmol. **48**(3), 257–293 (2003)
2. U. Chakravarthy, C. Augood, G.C. Bentham, P.T. de Jong, M. Rahu, J. Seland, G. Soubrane, L. Tomazzoli, F. Topouzis, J.R. Vingerling, J. Vioque, I.S. Young, A.E. Fletcher, Cigarette smoking and age-related macular degeneration in the EUREYE Study. Ophthalmology **114**(6), 1157–1163 (2007)
3. C.C. Klaver, R.C. Wolfs, J.R. Vingerling, A. Hofman, P.T. de Jong, Age-specific prevalence and causes of blindness and visual impairment in an older population: the Rotterdam Study. Arch. Ophthalmol. **116**(5), 653–658 (1998)
4. R. Klein, B.E. Klein, M.D. Knudtson, S.M. Meuer, M. Swift, R.E. Gangnon, Fifteen-year cumulative incidence of age-related macular degeneration: the Beaver Dam Eye Study. Ophthalmology **114**(2), 253–262 (2007)
5. S. Schmitz-Valckenberg, M. Fleckenstein, H.P. Scholl, F.G. Holz, Fundus autofluorescence and progression of age-related macular degeneration. Surv. Ophthalmol **54**(1), 96–117 (2009)

6. S.C. Tomany, J.J. Wang, R. Van Leeuwen, R. Klein, P. Mitchell, J.R. Vingerling, B.E. Klein, W. Smith, P.T. De Jong, Risk factors for incident age-related macular degeneration: pooled findings from 3 continents. Ophthalmology **111**(7), 1280–1287 (2004)
7. R. van Leeuwen, C.C. Klaver, J.R. Vingerling, A. Hofman, P.T. de Jong, Epidemiology of age-related maculopathy: a review. Eur. J. Epidemiol. **18**(9), 845–854 (2003)
8. J.J. Wang, E. Rochtchina, A.J. Lee, E.M. Chia, W. Smith, R.G. Cumming, P. Mitchell, Ten-year incidence and progression of age-related maculopathy: the blue Mountains Eye Study. Ophthalmology. **114**(1), 92–98 (2007)
9. D.S. Friedman, B.J. O'Colmain, B. Muñoz, S.C. Tomany, C. McCarty, P.T. de Jong, B. Nemesure, P. Mitchell, J. Kempen, Eye Diseases Prevalence Research Group, Prevalence of age-related macular degeneration in the United States. Arch. Ophthalmol. **122**(4), 564–572 (2004)
10. R. Klein, B.E. Klein, S.C. Tomany, S.M. Meuer, G.H. Huang, Ten-year incidence and progression of age-related maculopathy: the Beaver Dam eye study. Ophthalmology **109**(10), 1767–1779 (2002)
11. P. Mitchell, W. Smith, K. Attebo, J.J. Wang, Prevalence of age-related maculopathy in Australia. The Blue Mountains Eye Study. Ophthalmology **102**(10), 1450–1460 (1995)
12. J.S. Sunness, J. Gonzalez-Baron, C.A. Applegate, N.M. Bressler, Y. Tian, B. Hawkins, Y. Barron, A. Bergman, Enlargement of atrophy and visual acuity loss in the geographic atrophy form of age-related macular degeneration. Ophthalmology **106**(9), 1768–1779 (1999)
13. S. Wolf, U. Wolf-Schnurrbusch, Spectral-domain optical coherence tomography use in macular diseases: a review. Ophthalmologica **224**(6), 333–340 (2010)
14. A.A. Khanifar, A.F. Koreishi, J.A. Izatt, C.A. Toth, Drusen ultrastructure imaging with spectral domain optical coherence tomography in age-related macular degeneration. Ophthalmology **115**(11), 1883–1890 (2008)
15. S.G. Schuman, A.F. Koreishi, S. Farsiu, S.H. Jung, J.A. Izatt, C.A. Toth, Photoreceptor layer thinning over drusen in eyes with age-related macular degeneration imaged in vivo with spectral-domain optical coherence tomography. Ophthalmology **116**(3), 488.e2–496.e2 (2009)
16. M. Fleckenstein, P. Charbel Issa, H.M. Helb, S. Schmitz-Valckenberg, R.P. Finger, H.P. Scholl, K.U. Loeffler, F.G. Holz, High-resolution spectral domain-OCT imaging in geographic atrophy associated with age-related macular degeneration. Invest. Ophthalmol. Vis. Sci. **49**(9), 4137–4144 (2008)
17. C.G. Pieroni, A.J. Witkin, T.H. Ko, J.G. Fujimoto, A. Chan, J.S. Schuman, H. Ishikawa, E. Reichel, J.S. Duker, Ultrahigh resolution optical coherence tomography in non-exudative age related macular degeneration. Br. J. Ophthalmol. **90**(2), 191–197 (2006)
18. J. Ho, A.J. Witkin, J. Liu, Y. Chen, J.G. Fujimoto, J.S. Schuman, J.S. Duker, Documentation of intraretinal retinal pigment epithelium migration via high-speed ultrahigh-resolution optical coherence tomography. Ophthalmology **118**(4), 687–693 (2011)
19. B.L. Sikorski, D. Bukowska, J.J. Kaluzny, M. Szkulmowski, A. Kowalczyk, M. Wojtkowski, Drusen with accompanying fluid underneath the sensory retina. Ophthalmology **118**(1), 82–92 (2011)
20. G. Mimoun, G. Soubrane, G. Coscas, [Macular drusen]. J. Fr. Ophtalmol. **13**(10), 511–530 (1990)
21. S. Schmitz-Valckenberg, J.S. Steinberg, M. Fleckenstein, S. Visvalingam, C.K. Brinkmann, F.G. Holz, Combined confocal scanning laser ophthalmoscopy and spectral-domain optical coherence tomography imaging of reticular drusen associated with age-related macular degeneration. Ophthalmology **117**(6), 1169–1176 (2010)
22. R.F. Spaide, C.A. Curcio, Drusen characterization with multimodal imaging. Retina **30**(9), 1441–1454 (2010)
23. S.A. Zweifel, R.F. Spaide, C.A. Curcio, G. Malek, Y. Imamura, Reticular pseudodrusen are subretinal drusenoid deposits. Ophthalmology **117**(2), 303.e1–312.e1 (2010)
24. M. Rudolf, G. Malek, J.D. Messinger, M.E. Clark, L. Wang, C.A. Curcio, Sub-retinal drusenoid deposits in human retina: organization and composition. Exp. Eye Res. **87**(5), 402–408 (2008)

25. S.R. Freeman, I. Kozak, L. Cheng, D.U. Bartsch, F. Mojana, N. Nigam, M. Brar, R. Yuson, W.R. Freeman, Optical coherence tomography-raster scanning and manual segmentation in determining drusen volume in age-related macular degeneration. Retina **30**(3), 431–435 (2010)
26. N. Jain, S. Farsiu, A.A. Khanifar, S. Bearelly, R.T. Smith, J.A. Izatt, C.A. Toth, Quantitative comparison of drusen segmented on SD OCT versus drusen delineated on color fundus photographs. Invest. Ophthalmol. Vis. Sci. **51**(10), 4875–4883 (2010)
27. M. Wojtkowski, B.L. Sikorski, I. Gorczynska, M. Gora, M. Szkulmowski, D. Bukowska, J. Kaluzny, J.G. Fujimoto, A. Kowalczyk, Comparison of reflectivity maps and outer retinal topography in retinal disease by 3-D Fourier domain optical coherence tomography. Opt. Express. **17**(5), 4189–4207 (2009)
28. K. Yi, M. Mujat, B.H. Park, W. Sun, J.W. Miller, J.M. Seddon, L.H. Young, J.F. de Boer, T.C. Chen, Spectral domain optical coherence tomography for quantitative evaluation of drusen and associated structural changes in non-neovascular age-related macular degeneration. Br. J. Ophthalmol. **93**(2), 176–181 (2009)
29. F.G. Schlanitz, C. Ahlers, S. Sacu, C. Schütze, M. Rodriguez, S. Schriefl, I. Golbaz, T. Spalek, G. Stock, U. Schmidt-Erfurth, Performance of drusen detection by spectral-domain optical coherence tomography. Invest. Ophthalmol. Vis. Sci. **51**(12), 6715–6721 (2010)
30. S. Bearelly, F.Y. Chau, A. Koreishi, S.S. Stinnett, J.A. Izatt, C.A. Toth, Spectral domain optical coherence tomography imaging of geographic atrophy margins. Ophthalmology **116**(9), 1762–1769 (2009)
31. M. Brar, I. Kozak, L. Cheng, D.U. Bartsch, R. Yuson, N. Nigam, S.F. Oster, F. Mojana, W.R. Freeman, Correlation between spectral-domain optical coherence tomography and fundus autofluorescence at the margins of geographic atrophy. Am. J. Ophthalmol. **148**(3), 439–444 (2009)
32. S.Y. Cohen, L. Dubois, S. Nghiem-Buffet, S. Ayrault, F. Fajnkuchen, B. Guiberteau, C. Delahaye-Mazza, G. Quentel, R. Tadayoni, Retinal pseudocysts in age-related geographic atrophy. Am. J. Ophthalmol. **150**(2), 211.e1–217.e1 (2010)
33. M. Fleckenstein, S. Schmitz-Valckenberg, C. Adrion, I. Krämer, N. Eter, H.M. Helb, C.K. Brinkmann, P. Charbel Issa, U. Mansmann, F.G. Holz, Tracking progression with spectral-domain optical coherence tomography in geographic atrophy caused by age-related macular degeneration. Invest. Ophthalmol. Vis. Sci. **51**(8), 3846–3852 (2010)
34. H.M. Helb, P. Charbel Issa, M. Fleckenstein, S. Schmitz-Valckenberg, H.P. Scholl, C.H. Meyer, N. Eter, F.G. Holz, Clinical evaluation of simultaneous confocal scanning laser ophthalmoscopy imaging combined with high-resolution, spectral-domain optical coherence tomography. Acta Ophthalmol. **88**(8), 842–849 (2010)
35. B.J. Lujan, P.J. Rosenfeld, G. Gregori, F. Wang, R.W. Knighton, W.J. Feuer, C.A. Puliafito, Spectral domain optical coherence tomographic imaging of geographic atrophy. Ophthalmic Surg. Lasers Imaging **40**(2), 96–101 (2009)
36. S. Schmitz-Valckenberg, M. Fleckenstein, H.M. Helb, P. Charbel Issa, H.P. Scholl, F.G. Holz, In vivo imaging of foveal sparing in geographic atrophy secondary to age-related macular degeneration. Invest. Ophthalmol. Vis. Sci. **50**(8), 3915–3921 (2009)
37. U.E. Wolf-Schnurrbusch, V. Enzmann, C.K. Brinkmann, S. Wolf, Morphologic changes in patients with geographic atrophy assessed with a novel spectral OCT-SLO combination. Invest. Ophthalmol. Vis. Sci. **49**(7), 3095–3099 (2008)
38. Z. Yehoshua, P.J. Rosenfeld, G. Gregori, W.J. Feuer, M. Falcão, B.J. Lujan, C. Puliafito, Progression of geographic atrophy in age-related macular degeneration imaged with spectral domain optical coherence tomography. Ophthalmology **118**(4), 679–686 (2011)
39. M. Fleckenstein, U. Wolf-Schnurrbusch, S. Wolf, C. von Strachwitz, F.G. Holz, S. Schmitz-Valckenberg, [Imaging diagnostics of geographic atrophy]. Ophthalmologe **107**(11), 1007–1015 (2010)
40. M. Fleckenstein, S. Schmitz-Valckenberg, C. Martens, S. Kosanetzky, C.K. Brinkmann, G.S. Hageman, F.G. Holz, Fundus autofluorescence and spectral domain optical coherence tomography characteristics in a rapidly progressing form of geographic atrophy. Invest. Ophthalmol. Vis. Sci. **52**(6):3761–3766 (2011)

41. S. Schmitz-Valckenberg, M. Fleckenstein, A.P. Göbel, T.C. Hohman, F.G. Holz, Optical coherence tomography and autofluorescence findings in areas with geographic atrophy due to age-related macular degeneration. Invest. Ophthalmol. Vis. Sci. **52**(1), 1–6 (2011)
42. M. Boulton, P. Dayhaw-Barker, The role of the retinal pigment epithelium: topographical variation and ageing changes. Eye (Lond.) **15**(Pt 3), 384–389 (2001)
43. F. Schütt, S. Davies, J. Kopitz, F.G. Holz, M.E. Boulton, Photodamage to human RPE cells by A2-E, a retinoid component of lipofuscin. Invest. Ophthalmol. Vis. Sci. **41**(8), 2303–2308 (2000)
44. F.C. Delori, C.K. Dorey, G. Staurenghi, O. Arend, D.G. Goger, J.J. Weiter, In vivo fluorescence of the ocular fundus exhibits retinal pigment epithelium lipofuscin characteristics. Invest. Ophthalmol. Vis. Sci. **36**(3), 718–729 (1995)
45. C. Bellmann, F.G. Holz, O. Schapp, H.E. Völcker, T.P. Otto, [Topography of fundus autofluorescence with a new confocal scanning laser ophthalmoscope] Ophthalmologe **94**(6), 385–391 (1997)
46. U. Solbach, C. Keilhauer, H. Knabben, S. Wolf, Imaging of retinal autofluorescence in patients with age-related macular degeneration. Retina **17**(5), 385–389 (1997)
47. A. von Rückmann, F.W. Fitzke, A.C. Bird, Distribution of fundus autofluorescence with a scanning laser ophthalmoscope. Br. J. Ophthalmol. **79**(5), 407–412 (1995)
48. A. von Rückmann, F.W. Fitzke, A.C. Bird, Fundus autofluorescence in age-related macular disease imaged with a laser scanning ophthalmoscope. Invest. Ophthalmol. Vis. Sci. **38**(2), 478–486 (1997)
49. A. Deckert, S. Schmitz-Valckenberg, J. Jorzik, A. Bindewald, F.G. Holz, U. Mansmann, Automated analysis of digital fundus autofluorescence images of geographic atrophy in advanced age-related macular degeneration using confocal scanning laser ophthalmoscopy (cSLO). BMC Ophthalmol. **5**, 8 (2005)
50. S. Schmitz-Valckenberg, J. Jorzik, K. Unnebrink, F.G. Holz, FAM Study Group, Analysis of digital scanning laser ophthalmoscopy fundus autofluorescence images of geographic atrophy in advanced age-related macular degeneration. Graefes Arch. Clin. Exp. Ophthalmol. **240**(2), 73–78 (2002)
51. F.G. Holz, C. Bellman, S. Staudt, F. Schütt, H.E. Völcker, Fundus autofluorescence and development of geographic atrophy in age-related macular degeneration. Invest. Ophthalmol. Vis. Sci. **42**(5), 1051–1056 (2001)
52. S. Schmitz-Valckenberg, A. Bindewald-Wittich, J. Dolar-Szczasny, J. Dreyhaupt, S. Wolf, H.P. Scholl, F.G. Holz, Fundus Autofluorescence in Age-Related Macular Degeneration Study Group, Correlation between the area of increased autofluorescence surrounding geographic atrophy and disease progression in patients with AMD. Invest. Ophthalmol. Vis. Sci. **47**(6), 2648–2654 (2006)
53. F.G. Holz, A. Bindewald-Wittich, M. Fleckenstein, J. Dreyhaupt, H.P. Scholl, S. Schmitz-Valckenberg, FAM-Study Group, Progression of geographic atrophy and impact of fundus autofluorescence patterns in age-related macular degeneration. Am. J. Ophthalmol. **143**(3), 463–472 (2007)
54. S. Schmitz-Valckenberg, G.J. Jaffe, M. Fleckenstein, P. Kozma, T. Hohman, F.G. Holz, GAP-Study Group. *Lesion Characteristics and Progression in the Natural History of Geographic Atrophy (GAP)-Study.* ARVO Meeting Abstracts April 11 2009; 50:3914
55. F.G. Holz, S. Schmitz-Valckenberg, M. Fleckenstein, G.J. Jaffe, T. Hohman. *Lesion Characteristics and Progression in the Natural History of Geographic Atrophy (GAP)-Study.* ARVO Meeting Abstracts April 11 2010; 51:94
56. B. Baumann, E. Gotzinger, M. Pircher, H. Sattmann, C. Schuutze, F. Schlanitz, C. Ahlers, U. Schmidt-Erfurth, C.K. Hitzenberger, Segmentation and quantification of retinal lesions in age-related macular degeneration using polarization-sensitive optical coherence tomography. J. Biomed. Opt. **15**(6), 061704 (2010)
57. C. Schütze, C. Ahlers, S. Sacu, G. Mylonas, R. Sayegh, I. Golbaz, G. Matt, G. Stock, U. Schmidt-Erfurth, Performance of OCT segmentation procedures to assess morphology and extension in geographic atrophy. Acta Ophthalmol. **89**(3), 235–240 (2011)

58. E. Götzinger, M. Pircher, W. Geitzenauer, C. Ahlers, B. Baumann, S. Michels, U. Schmidt-Erfurth, C.K. Hitzenberger, Retinal pigment epithelium segmentation by polarization sensitive optical coherence tomography. Opt. Express. **16**(21), 16410–16422 (2008)
59. C. Ahlers, E. Götzinger, M. Pircher, I. Golbaz, F. Prager, C. Schütze, B. Baumann, C.K. Hitzenberger, U. Schmidt-Erfurth, Imaging of the retinal pigment epithelium in age-related macular degeneration using polarization-sensitive optical coherence tomography. Invest. Ophthalmol. Vis. Sci. **51**(4), 2149–2157 (2010)
60. S. Michels, M. Pircher, W. Geitzenauer, C. Simader, E. Götzinger, O. Findl, U. Schmidt-Erfurth, C.K. Hitzenberger, Value of polarisation-sensitive optical coherence tomography in diseases affecting the retinal pigment epithelium. Br. J. Ophthalmol. **92**(2), 204–209 (2008)
61. E. Ergun, B. Hermann, M. Wirtitsch, A. Unterhuber, T.H. Ko, H. Sattmann, C. Scholda, J.G. Fujimoto, M. Stur, W. Drexler, Assessment of central visual function in Stargardt's disease/fundus flavimaculatus with ultrahigh-resolution optical coherence tomography. Invest. Ophthalmol. Vis. Sci. **46**(1), 310–316 (2005)
62. P. Charbel Issa, E. Troeger, R. Finger, F.G. Holz, R. Wilke, H.P. Scholl, Structure-function correlation of the human central retina. PLoS One **5**(9), e12864 (2010)
63. D. Iwama, A. Tsujikawa, Y. Ojima, H. Nakanishi, K. Yamashiro, H. Tamura, S. Ooto, N. Yoshimura, Relationship between retinal sensitivity and morphologic changes in eyes with confluent soft drusen. Clin. Exp. Ophthalmol. **38**(5), 483–488 (2010)
64. P. Godara, C. Siebe, J. Rha, M. Michaelides, J. Carroll, Assessing the photoreceptor mosaic over drusen using adaptive optics and SD-OCT. Ophthalmic Surg. Lasers Imaging **41**(Suppl), S104–S108 (2010)

Chapter 5
Optical Coherence Tomography in Glaucoma

Fatmire Berisha, Esther M. Hoffmann, and Norbert Pfeiffer

Abstract Retinal nerve fiber layer (RNFL) thinning and optic nerve head cupping are key diagnostic features of glaucomatous optic neuropathy. The higher resolution of the recently introduced SD-OCT offers enhanced visualization and improved segmentation of the retinal layers, providing a higher accuracy in identification of subtle changes of the optic disc and RNFL thinning associated with glaucoma.

5.1 Introduction

Retinal nerve fiber layer (RNFL) thinning and optic nerve head cupping are key diagnostic features of glaucomatous optic neuropathy. Optical coherence tomography (OCT), which provides high-resolution objective and quantitative measurements of the optic disc parameters and RNFL thickness, has been widely used for detection of glaucomatous damage and disease progression. Recent introduction of spectral domain (SD)-OCT technology, also known as Fourier-domain (FD)-OCT, offers significant advantages over the previous time-domain (TD)-OCT, allowing three-dimensional (3D) imaging of the retina and optic disc with ultrahigh acquisition speed and ultrahigh resolution. The higher resolution of SD-OCT offers enhanced visualization and improved segmentation of the retinal layers, providing a higher accuracy in identification of subtle changes of the optic disc and RNFL thinning associated with glaucoma. While SD-OCT is the current commercially available state-of-the-art technology, future developments of new OCT instruments such as polarization-sensitive (PS)-OCT and swept-source (SS)-OCT as well as adaptive optics (AO)-OCT reveal further details and promise additional visualization of retinal features such as axons and cellular structures.

F. Berisha (✉)
Johannes Gutenberg University, University Medical Center Mainz, Lagenbeckstr. 1, 55101 Mainz, Germany
e-mail: fatmire.berisha@gmx.de

R. Bernardes and J. Cunha-Vaz (eds.), *Optical Coherence Tomography*, Biological and Medical Physics, Biomedical Engineering,
DOI 10.1007/978-3-642-27410-7_5, © Springer-Verlag Berlin Heidelberg 2012

Doppler OCT is currently under investigation and may be useful for the optic nerve and retinal blood flow measurements.

5.2 Imaging in Glaucoma

Glaucoma is a progressive neurodegenerative disease characterized by pathologic loss of ganglion cells, optic nerve damage, and visual field defects. Worldwide, it is the second leading cause of irreversible blindness [1, 2]. A major risk factor for blindness is often the late detection of the disease [1–4]. Glaucomatous optic neuropathy is biologically identified by the death of retinal ganglion cells. Ganglion cell axons are slowly lost, leading to thinning of the RNFL, thinning of the neuroretinal rim, and the typical appearance of the glaucomatous cupping of the optic nerve head. Glaucomatous changes in the optic nerve and RNFL are irreversible and cause progressive visual field defects. Clinical examination and stereophotography of the optic disc are essential for diagnosis and monitoring of glaucoma. However, the role of imaging devices in glaucoma detection has been increasing in recent years because they can provide useful objective and quantitative measurement of structural damage. Furthermore, newly emerging imaging technologies enable accurate detection of glaucoma progression, especially in the early stages. Several types of imaging devices such as confocal scanning laser ophthalmoscopy (CSLO; HRT), scanning laser polarimetry (SLP; GDx), and OCT are currently used for glaucoma management.

HRT uses a scanning laser to create images of serial sections of the optic nerve head providing a 3D map of the optic disc topography, and the software calculates various quantitative measures that are used to identify glaucomatous optic disc damage [5, 6]. The strength of HRT is detection of glaucoma progression using topographic change analysis, standardized and absolute stereometric parameter change analysis, and trend of contour line height variation profile.

Scanning laser polarimetry (GDx) is based on the birefringence properties of ganglion cell axon microtubules that alter the laser scanning beam polarization proportional to the RNFL thickness [7, 8]. Several studies have shown that using the GDx Nerve Fiber Analyzer with enhanced corneal compensation (ECC) significantly improves the ability for glaucoma detection and structure–function relationship [9–11].

OCT provides quantitative and objective measurement of RNFL and optic nerve head with high resolution and enables the detection of glaucomatous optic neuropathy [12]. Although time-domain OCT (Stratus OCT) has been the prevailing instrument for glaucoma detection and management [13, 14], the incorporation of spectral domain OCT (SD-OCT) offers significant advantages for identifying glaucomatous changes [15].

Imaging techniques (OCT, SLP, and CSLO) are available for clinical use and continue to be developed and investigated intensively for improved ability to identify glaucomatous damage and to monitor progression of the disease. The information obtained from imaging devices is useful in clinical practice when analyzed in combination with other relevant clinical parameters that define glaucoma diagnosis and progression.

5.3 Time-Domain Optical Coherence Tomography in Glaucoma

OCT uses low-coherence interferometry to provide high-resolution objective and quantitative measurement of the retinal layers and topographical measurements of the optic nerve head. The first demonstrations of the OCT images performed in the retina ex vivo were published by Huang and colleagues in 1991 [16]. In vivo retinal imaging with OCT was first demonstrated in 1993 [17, 18], and early studies showing normal retinal structure [19], macular pathology [20], and glaucomatous optic neuropathy [21, 22] were published in 1995.

RNFL profile created by retinal segmentation is of particular importance in glaucoma. Peripapillary RNFL thickness measurement using a 3.4-mm-diameter circle scan centered on the optic nerve head is the most important TD-OCT (Stratus OCT) scan option for diagnosing and following glaucomatous damage. Schuman et al. [23] defined the 3.4-mm circumpapillary RNFL thickness scan as the standard for TD-OCT for glaucoma assessment. Segmentation algorithms allow measurement of the RNFL, which is the anterior most layer of the retina and appears as the highly backscattering layer, due to the structure of the fibers being perpendicular to the direction of the light beam. Image analysis using automated segmentation and measurement of the RNFL with the Stratus OCT allows the comparison with the normative database (available since 2003) as well as evaluation over time. Standard RNFL and fast RNFL scan protocols are available for the measurement of the peripapillary RNFL thickness using Stratus OCT, and both scan options are equally reproducible in normal and glaucomatous eyes [24, 25]. On the other hand, it has been shown that the fast protocol tends to overestimate RNFL thickness; therefore, higher-resolution RNFL scan (Standard RNFL) provides better diagnostic sensitivity in glaucoma detection [26]. Standard RNFL measurement consists of 512 scans taken in a circle three times around the optic disc, and the fast RNFL scan protocol consists of three consecutive circles with 256 A-scans in each circle. RNFL circle scans are positioned around the optic disc by the operator, and the patient has to fixate on an internal fixation target. A segmentation algorithm from the OCT software is used to determine the RNFL thickness for each of the three scans, which are averaged to determine the final RNFL thickness profile. Parameters including mean overall RNFL thickness, mean RNFL thickness in four quadrants (temporal, superior, nasal, and inferior), and 12 clock hours are generated automatically in the analysis report. RNFL thickness measurement has been reported to have good reproducibility, with a standard deviation ~2.5 μm for mean overall RNFL thickness [27, 28]. Several studies have demonstrated good sensitivity and specificity of the Stratus OCT for diagnosing glaucoma [29–31]. The average RNFL thickness as well as the inferior quadrant RNFL have been reported as the best parameters for differentiating between normal and glaucomatous eyes. Furthermore, it has been reported that Stratus OCT has the potential to detect localized RNFL defects [32] as well as glaucoma progression [33, 34].

Optic nerve head (ONH) measurements are performed using the Stratus OCT "fast optic nerve head" scan protocol that consists of six 4-mm radial intersecting line scans centered on the ONH. Stratus OCT software analysis allows determination of topographic data derived from each individual linear scan into an integrated analysis characterizing the entire optic disc topography. Optic disc parameters include cup-disc area ratio, cup area, and neuroretinal rim area. Medeiros et al. [35] reported the cup-disc area ratio as the best Stratus OCT ONH parameter to differentiate between healthy and glaucomatous eyes ($AUROC = 0.88$), and the combination of the RNFL thickness and ONH parameters improved the diagnostic ability for glaucoma detection (AUROC of 0.97).

Macular thickness measured using TD-OCT has been demonstrated as an indicator of retinal ganglion cell loss in glaucoma [36]. A significant reduction in macular thickness associated with peripapillary RNFL loss and visual field defects was reported.

However, due to the moderate sensitivity and specificity of macular parameters (thickness and volume) measured, using Stratus OCT is of limited diagnostic value in patients with glaucoma [35].

It has been concluded that the sensitivity and specificity of imaging instruments for detection of glaucoma are comparable with that of expert interpretation of stereophotography (First AIGS Consensus Meeting on "Structure and Function in the Management of Glaucoma"). A study based on subjective assessment of progressive optic disc changes using stereoscopic photographs showed only a slight to fair interobserver agreement among glaucoma specialists [37]. Another study reported that the diagnostic ability of imaging techniques including Stratus OCT showed better performance than subjective assessment of the optic disc by general ophthalmologists, but not by a glaucoma expert [38].

5.4 Spectral Domain Optical Coherence Tomography in Glaucoma

Recently, SD-OCT technology, also known as Fourier-domain (FD)-OCT, was introduced, which offers significant advantages over the traditional TD-OCT. The SD-OCT technology enables 3D imaging of the retina and optic nerve head with ultrahigh acquisition speed and ultrahigh resolution. The basic principles of SD-OCT have been described elsewhere [39, 40]. Currently, several companies manufacture SD-OCT instruments: Cirrus HD-OCT (Carl Zeiss Meditec), RTVue-100 (Optovue), Spectralis OCT (Heidelberg), 3D OCT-1000 (Topcon), etc.

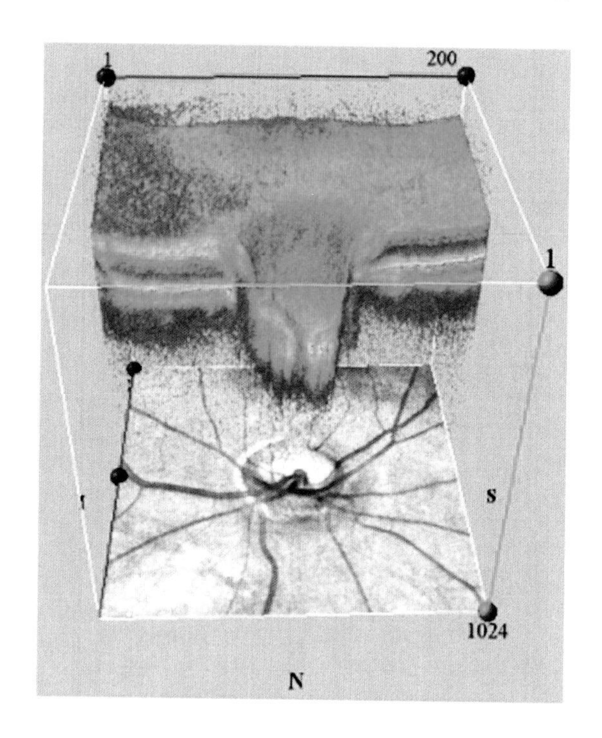

Fig. 5.1 Cirrus OCT 3D visualization of the optic nerve head

Compared with TD-OCT, all SD-OCT techniques have the advantage of better visualization of retinal layers and optic nerve head. However, some differences exist between the SD-OCT instruments produced by different manufacturer, and different segmentation algorithms are used.

Cirrus SD-OCT provides the optic disc cube scan protocol for imaging the optic disc and the peripapillary retinal region covering an area of $6 \times 6\,\text{mm}^2$ and enables 3D visualization of the optic disc (Fig. 5.1). The RNFL thickness at each scan point is analyzed, and an RNFL thickness map is generated. The built-in algorithm identifies the disc center from the edges of the retinal pigment epithelium and automatically extracts a peripapillary scan circle of 3.46 mm diameter. The average quadrant and clock-hour peripapillary RNFL thickness values are reported in the analysis printout. The RNFL thickness measurement at the $6 \times 6\,\text{mm}^2$ parapapillary area is analyzed and displayed in the RNFL thickness deviation map. The Cirrus normative database uses the same percentile cutoffs as the Stratus normative database, with yellow band borderline and red band outside normal limits representing abnormal thinning of the RNFL. Furthermore, the new software (version 5.0) allows automatic evaluation of several optic nerve head parameters: rim area, disc area, cup volume, average cup-disc ratio, and vertical cup-disc ratio (Fig. 5.2).

RTVue SD-OCT offers the glaucoma scan protocol including two scan options for peripapillary RNFL thickness measurements: RNFL 3.45 and the optic nerve head map (NHM4). The RNFL 3.45 mode provides an average of four peripapillary scans

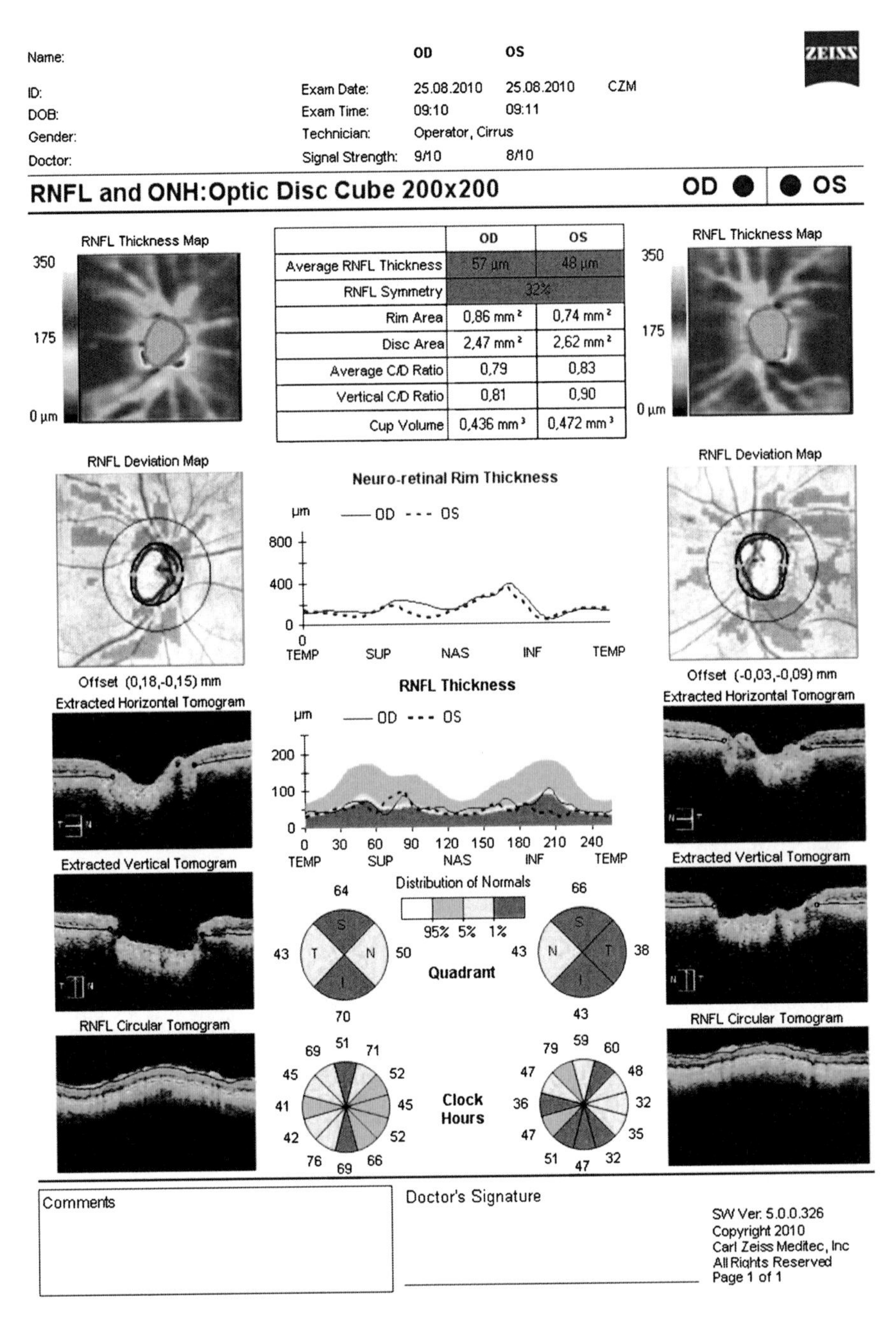

Fig. 5.2 Cirrus OCT RNFL thickness map, optic disc parameters, circumpapillary RNFL thickness at 3.46 mm (displayed in sectors and clock hours), and RNFL thickness deviation map (comparison with the normative database)

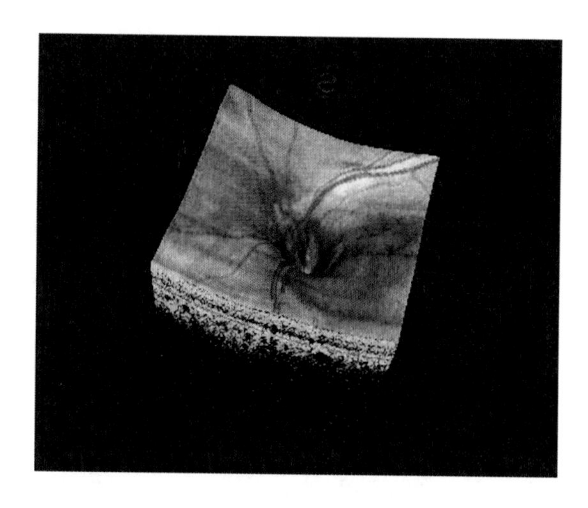

Fig. 5.3 RTVue OCT 3D visualization of the optic nerve head

measuring RNFL thickness along a circle 3.45 mm in diameter centered at the optic disc. The optic nerve head map 4-mm-diameter (NHM4) scan protocol is composed of 12 radial scans 3.4 mm in length and six concentric circle scans ranging from 2.5 to 4.0 mm in diameter all centered on the optic disc. A polar RNFL thickness map is provided, and the RNFL thickness is measured by recalculating data along a circle 3.45 mm in diameter around the optic disc. RTVue OCT provides 16 regional RNFL thickness maps using the RNFL 3.45 or the NHM4 mode. The optic disc parameters are measured using the 3D disc and the NHM4 scan protocols. The 3D disc protocol is a 4 × 4 mm raster scan centered on the optic disc and provides a 3D image of the optic disc and peripapillary area (Fig. 5.3). The "en face" image generated by the 3D disc scanning is used to draw the contour line describing the disc margin by using the retinal pigment epithelium (RPE) visualized in the B-scan that is required to generate optic disc parameters from the NHM4 scan protocol. The optic cup is defined automatically by RTVue software, and the optic disc parameters obtained are optic disc area, cup area, neuroretinal rim area, optic disc volume, cup volume, rim volume, cup-disc area ratio, and cup-disc horizontal and vertical ratios (Fig. 5.4). The RTVue OCT has a new, expanded ethnic specific normative databases of RNFL and optic disc parameters in software version 4.0. Furthermore, RTVue includes a ganglion cells complex (GCC) scan protocol for measuring macular inner retinal layer thickness, which comprises the ganglion cell layer, RNFL, and inner plexiform layer. The GCC thickness values are analyzed and compared to the normative database, and the GCC map is provided (Fig. 5.5).

Spectralis SD-OCT uses a circle scan with a diameter 3.45 mm manually positioned at the center of the optic disc for imaging the peripapillary RNFL. This device simultaneously captures infrared fundus and SD-OCT images, and an integrated real-time tracking system compensates for eye movements. The eye tracking automatically positions follow-up scans (AutoRescan), allowing for accurate monitoring of disease progression. Images at the same location are captured and averaged automatically by the built-in software, and the RNFL in each image

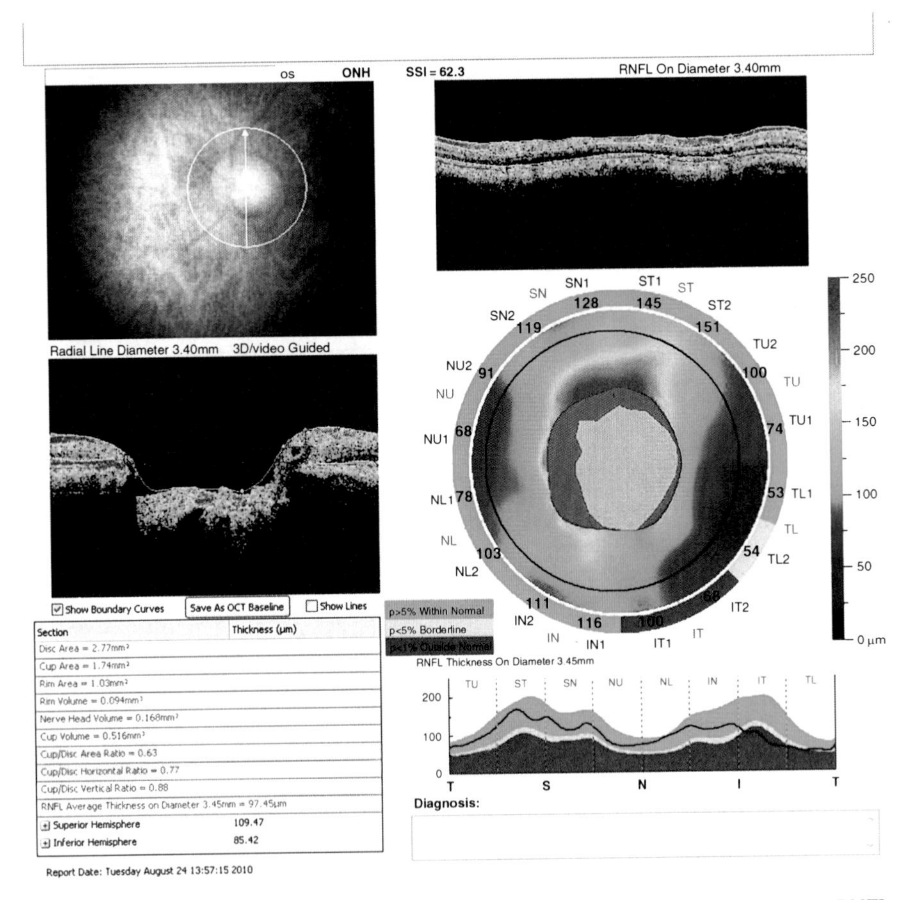

Fig. 5.4 RTVue OCT optic nerve head scan, morphometric parameters, circumpapillary RNFL thickness at 3.45 mm, RNFL thickness map, and comparison with the normative database

is automatically segmented. The analysis printout including average and regional (temporal, superotemporal, superonasal, nasal, inferonasal, and inferotemporal) RNFL thickness is provided. The comparison with normative database indicates the classification in green (normal), yellow (borderline), or red (outside normal limits). Optic nerve head scanning is performed using the 4 × 4 mm raster scan protocol providing 3D volumetric data.

Topcon SD-OCT provides 3D volumetric imaging of the optic disc and peripapillary area. RNFL thickness measurement is performed using a circular scan 3.4 mm in diameter centered on the optic disc. The 3D-OCT algorithm identifies the center of the disc, and peripapillary RNFL thickness is measured using the segmentation algorithm. The average sectorial (temporal, inferior, nasal, and superior) and clock-hour RNFL thickness parameters are calculated and can be monitored via RNFL trend analysis and comparison with normative data. Disc

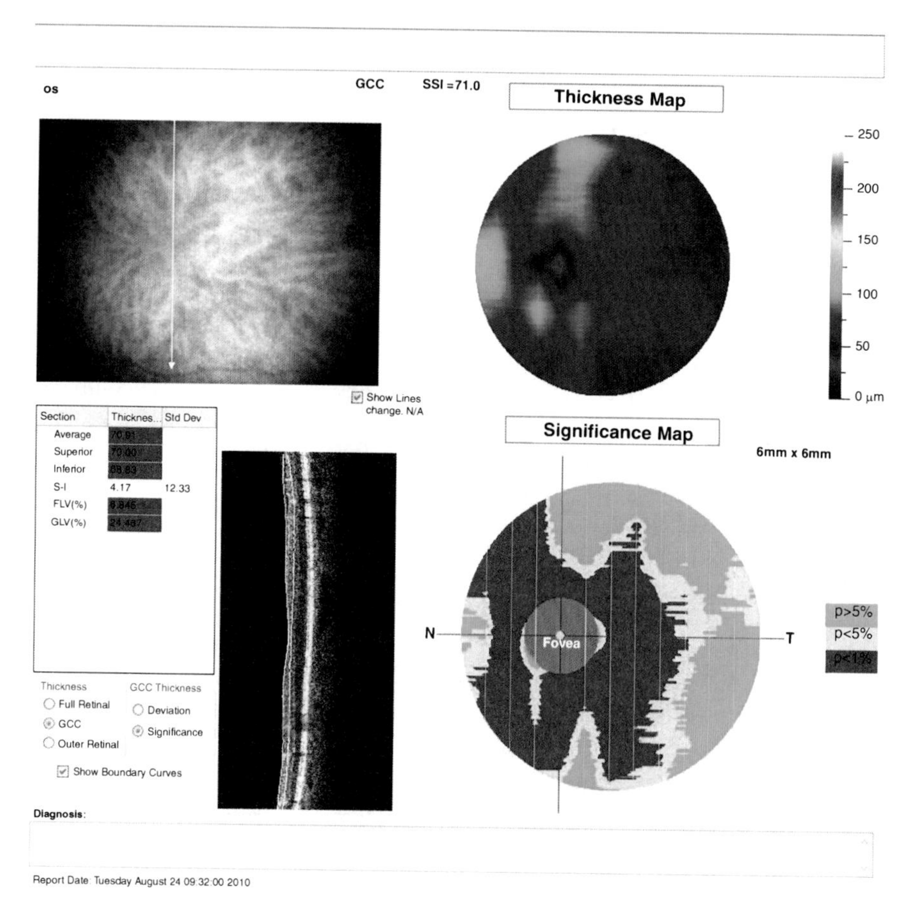

Fig. 5.5 RTVue OCT ganglion cell complex thickness map (macular ganglion cell layer, RNFL, and inner plexiform layer) compared to the normative database

topography can be displayed and analyzed including the comparison with normative database. A 3D-tracking system compensates for microsaccades, enabling to obtain consistent quality images.

5.5 Diagnostic Ability of Spectral Domain OCT in Glaucoma

SD-OCT imaging improves the discriminating power and increases the diagnostic value of OCT technology for glaucoma. Several studies have shown high reproducibility of the optic disc parameters and RNFL thickness measurements using SD-OCT in healthy and glaucomatous eyes [41–45]. We have also shown that RNFL thickness measurements obtained with the RTVue are highly reproducible in normal and glaucomatous eyes [46]. A study by Hoffmann et al. [47] found that the measurements of the optic disc parameters obtained with RTVue OCT in glaucoma and healthy eyes are not influenced by pupil size; however, the measured RNFL in glaucoma patients was thicker after dilation.

Using RTVue SD-OCT, it has been shown that the average RNFL thickness is the best OCT parameter to detect glaucoma. Additional measurements of macular inner retinal layer (ganglion cell complex) and optic disc parameters, however, improve the discriminating ability for glaucoma diagnosis [48]. Similar results have been reported by other investigators [49, 50]. A study from our group found a significant thinning of the ganglion cell complex (GCC) in glaucoma patients compared to healthy subjects [51]. It has been demonstrated that RNFL thickness measurements using RTVue OCT have similar discriminating power compared with Straus OCT [52]. However, measurements obtained using SD-OCT provide significantly different values of the RNFL thickness than those obtained with TD-OCT [52–55]. This difference should be taken into account during the follow-up of a patient using different OCT instruments.

Leung et al. [56] reported that both TD-OCT (Stratus OCT) and SD-OCT (Cirrus OCT) have high sensitivity and specificity for glaucoma detection. However, additional analysis of RNFL thickness deviation map using Cirrus OCT may improve the diagnostic performance and minimize the measurement variability based on the algorithms for automatic placement of the scan circle. Other investigators evaluating the diagnostic power of the RNFL thickness measurements demonstrated high sensitivity and specificity of results by Cirrus OCT, which were comparable with Stratus OCT in distinguishing glaucoma patients from healthy control subjects [55, 57]. Recent studies have shown that RNFL measurements obtained with Cirrus OCT are not influenced by pupil size [58, 59].

Nakatani et al. [42] reported that macular parameters measured with Topcon SD-OCT have comparably high discriminating power for early glaucoma and high reproducibility comparable with peripapillary RNFL parameters.

Horn et al. [60] evaluated the correlation between localized glaucomatous visual field defects and corresponding RNFL loss measured with SLO (GDx) and Spectralis SD-OCT. The authors concluded that the structural-functional relationship is stronger for SD-OCT. There is evidence that RNFL measurement with Spectralis OCT has a higher diagnostic sensitivity for glaucoma detection than optic disc measurement with HRT [61]. Another study showed that the RTVue OCT measurement of the optic disc topographic parameters had better diagnostic performance for detection of glaucomatous damage than the HRT3 [62].

A recent study evaluating the agreement between SD-OCT instruments (Cirrus, RTVue, and Spectralis) found that the OCT systems provide different values for RNFL thickness [63]. These discrepancies are most probably due to differences in RNFL detection algorithms. The results of this study indicate that different OCT systems cannot be used interchangeably for the measurement of the RNFL thickness. Table 5.1 shows several spectral domain OCT instruments and the differences among them.

Table 5.1 Spectral domain OCT instruments

Device	Axial resolution	A-scans/s	Eye tracking	3D-OCT
Carl Zeiss Meditec Cirrus	5 μm	27,000	No	Yes
Optovue RTVue-100	5 μm	26,000	No	Yes
Heidelberg Spectralis	7 μm optical 3.5 μm digital	40,000	Yes	Yes
Topcon 3D OCT-2000	6 μm	27,000	No	Yes

5.6 Future Directions of Optical Coherence Tomography in Glaucoma

While SD-OCT system is the current commercially available state-of-the-art technology, further improvement in image quality and advanced visualization of retinal layers is possible with research OCT instruments. Swept laser source (SS)-SD-OCT for imaging of the retina, optic disc, and choroid with high scan density and improved transverse resolution has been demonstrated [64, 65]. A femtosecond titanium/sapphire laser light has been used for research OCT instruments, allowing retinal imaging and detection of changes in cellular structures with ultrahigh resolution of ~3 μm [66, 67]. However, femtosecond laser sources are expensive for commercial instrument manufacturers. Applications of adaptive optics to OCT systems (AO-SD-OCT) allow ultrahigh resolution of 2–3 μm, providing cellular imaging of the retina and optic nerve head [68, 69]. Future applications of polarization-sensitive (PS)-SD-OCT techniques provide quantitative information on RNFL birefringence and potentially enable more accurate detection of RNFL damage, possibly improving glaucoma diagnosis [70, 71].

Doppler-SD-OCT enables measurements of optic nerve head and retinal blood flow in addition to obtaining morphological images [72, 73]. A study by Berisha et al. [74] investigated the correlation between RNFL thickness and retinal blood flow using Stratus OCT and Canon laser Doppler blood flow instrument in patients with early glaucoma and healthy subjects. An inverse association between retinal blood flow and RNFL thickness was found in patients with early glaucomatous optic neuropathy. Another study using color Doppler imaging, however, demonstrated a positive correlation between central retinal artery blood speed and neuroretinal rim area and volume in patients with more advanced glaucomatous optic neuropathy [75]. These studies indicate that the relationship between blood flow and progressive glaucomatous damage may be bimodal. Retinal blood flow appears to increase with increasing RNFL loss until a critical level is reached and then decreases with further RNFL loss. It is likely that abnormalities in the retinal circulation also occur in the pre-glaucoma stage of the disease. Ocular blood flow measurement and imaging using Doppler OCT may become clinically useful for detection and monitoring of glaucoma, allowing an evaluation of the relationship between morphological changes and abnormalities in the ocular blood flow in glaucomatous optic neuropathy. Further developments of this technology toward measurements of the aqueous outflow or blood oxygen saturation provide possible future applications of OCT for glaucoma diagnosis and management.

In summary, 3D spectral domain OCT imaging has clinical value for the quantitative assessment of optic nerve damage and shows promise for detection of early and progressing glaucoma.

References

1. H.A. Quigley, A.T. Broman, The number of people with glaucoma worldwide in 2010 and 2020. Br. J. Ophthalmol. **90**(3), 262–267 (2006)
2. Resnikoff, D. Pascolini, D. Etya'ale, I. Kocur, R. Pararajasegaram, G.P. Pokharel, S.P. Mariotti, Global data on visual impairment in the year 2002. Bull. World Health Organ. **82**(11), 844–851 (2004)
3. W.M. Grant, J.F. Burke Jr., Why do some people go blind from glaucoma? Ophthalmology **89**(9), 991–998 (1982)
4. M.C. Leske, A. Heijl, M. Hussein, B. Bengtsson, L. Hyman, E. Komaroff, Early Manifest Glaucoma Trial Group, Factors for glaucoma progression and the effect of treatment: the Early Manifest Glaucoma Trial. Arch. Ophthalmol. **121**(1), 48–56 (2003)
5. L. Zangwill, F. Medeiros, C. Bowd, R. Weinreb, Optic nerve imaging: recent advances, in *Essentials in Ophthalmology* ed. by F. Grehn, R. Stamper Glaucoma. (Springer, Berlin, 2004), pp. 63–91
6. N. Strouthidis, D. Garway-Heath, New developments in Heidelberg retina tomograph for glaucoma. Curr. Opin. Ophthalmol. **19**(2), 141–148 (2008)
7. R.W. Knighton, X.R. Huang, D.S. Greenfield, Analytical model of scanning laser polarimetry for retinal nerve fiber layer assessment. Invest. Ophthalmol. Vis. Sci. **43**(2), 383–392 (2002)
8. X.R. Huang, R.W. Knighton, Microtubules contribute to the birefringence of the retinal nerve fiber layer. Invest. Ophthalmol. Vis. Sci. **46**(12), 4588–4593 (2005)
9. C. Bowd, I.M. Tavares, F.A. Medeiros, L.M. Zangwill, P.A. Sample, R.N. Weinreb, Retinal nerve fiber layer thickness and visual sensitivity using scanning laser polarimetry with variable and enhanced corneal compensation. Ophthalmology **114**(7), 1259–1265 (2007)
10. F.A. Medeiros, C. Bowd, L.M. Zangwill, C. Patel, R.N. Weinreb, Detection of glaucoma using scanning laser polarimetry with enhanced corneal compensation. Invest. Ophthalmol. Vis. Sci. **48**(7), 3146–3153 (2007)
11. T.A. Mai, N.J. Reus, H.G. Lemij, Structure-function relationship is stronger with enhanced corneal compensation than with variable corneal compensation in scanning laser polarimetry. Invest. Ophthalmol. Vis. Sci. **48**(4), 1651–1658 (2007)
12. D.S. Greenfield, R.N. Weinreb, Role of optic nerve imaging in glaucoma clinical practice and clinical trials. Am. J. Ophthalmol. **145**(4), 598–603 (2008)
13. G. Wollstein, J.S. Schuman, L.L. Price, A. Aydin, P.C. Stark, E. Hertzmark, E. Lai, H. Ishikawa, C. Mattox, J.G. Fujimoto, L.A. Paunescu, Optical coherence tomography longitudinal evaluation of retinal nerve fiber layer thickness in glaucoma. Arch. Ophthalmol. **123**(4), 464–470 (2005)
14. C.K. Leung, C.Y. Cheung, R.N. Weinreb, K. Qiu, S. Liu, H. Li, G. Xu, N. Fan, C.P. Pang, K.K. Tse, D.S. Lam, Evaluation of retinal nerve fiber layer progression in glaucoma: a study on optical coherence tomography guided progression analysis. Invest. Ophthalmol. Vis. Sci. **51**(1), 217–222 (2010)
15. J.S. Schuman, Spectral domain optical coherence tomography for glaucoma (an AOS thesis). Trans. Am. Ophthalmol. Soc. **106**, 426–458 (2008)
16. D. Huang, E.A. Swanson, C.P. Lin, J.S. Schuman, W.G. Stinson, W. Chang, M.R. Hee, T. Flotte, K. Gregory, C.A. Puliafito, Optical coherence tomography. Science **254**(5035), 1178–1181 (1991)

17. E.A. Swanson, J.A. Izatt, M.R. Hee, D. Huang, C.P. Lin, J.S. Schuman, C.A. Puliafito, J.G. Fujimoto, In vivo retinal imaging by optical coherence tomography. Opt. Lett. **18**(21), 1864–1866 (1993)
18. A.F. Fercher, C.K. Hitzenberger, W. Drexler, G. Kamp, H. Sattmann, In vivo optical coherence tomography. Am. J. Ophthalmol. **116**(1), 113–114 (1993)
19. M.R. Hee, J.A. Izatt, E.A. Swanson, D. Huang, J.S. Schuman, C.P. Lin, C.A. Puliafito, J.G. Fujimoto, Optical coherence tomography of the human retina. Arch. Ophthalmol. **113**(3), 325–332 (1995)
20. C.A. Puliafito, M.R. Hee, C.P. LinvE. Reichel, J.S. Schuman, J.S. Duker, J.A. Izatt, E.A. Swanson, J.G. Fujimoto, Imaging of macular diseases with optical coherence tomography. Ophthalmology **102**(2), 217–229 (1995)
21. J.S. Schuman, M.R. Hee, A.V. Arya, T. Pedut-Kloizman, C.A. Puliafito, J.G. Fujimoto, E.A. Swanson, Optical coherence tomography: a new tool for glaucoma diagnosis. Curr. Opin. Ophthalmol. **6**(2), 89–95 (1995)
22. J.S. Schuman, M.R. Hee, C.A. Puliafito, C. Wong, T. Pedut-Kloizman, C.P. Lin, E. Hertzmark, J.A. Izatt, E.A. Swanson, J.G. Fujimoto, Quantification of nerve fiber layer thickness in normal and glaucomatous eyes using optical coherence tomography. Arch. Ophthalmol. **113**(5), 586–596 (1995)
23. J.S. Schuman, T. Pedut-Kloizman, E. Hertzmark, M.R. Hee, J.R. Wilkins, J.G. Coker, C.A. Puliafito, J.G. Fujimoto, E.A. Swanson, Reproducibility of nerve fiber layer thickness measurements using optical coherence tomography. Ophthalmology **103**(11), 1889–1898 (1996)
24. D.L. Budenz, R.T. Chang, X. Huang, R.W. Knighton, J.M. Tielsch, Reproducibility of retinal nerve fiber thickness measurements using the Stratus OCT in normal and glaucomatous eyes. Invest. Ophthalmol. Vis. Sci. **46**(7), 2440–2443 (2005)
25. S. Zafar, R. Gurses-Ozden, M. Makornwattana, R. Vessani, J.M. Liebmann, C. Tello, R. Ritch, Scanning protocol choice affects optical coherence tomography (OCT-3) measurements. J. Glaucoma. **13**(2), 142–144 (2004)
26. C.K. Leung, W.H. Yung, A.C. Ng, J. Woo, M.K. Tsang, K.K. Tse, Evaluation of scanning resolution on retinal nerve fiber layer measurement using optical coherence tomography in normal and glaucomatous eyes. J. Glaucoma. **13**(6), 479–485 (2004)
27. L.A. Paunescu, J.S. Schuman, L.L. Price, P.C. Stark, S. Beaton, H. Ishikawa, G. Wollstein, J.G. Fujimoto, Reproducibility of nerve fiber layer thickness and optic nerve head measurements using Stratus OCT. Invest Ophthalmol Vis Sci. **45**(6), 1716–1724 (2004)
28. D.L. Budenz, M.J. Fredette, W.J. Feuer, D.R. Anderson, Reproducibility of peripapillary retinal nerve fiber thickness measurements with Stratus OCT in glaucomatous eyes. Ophthalmology **115**(4), 661.e4–666.e4 (2008)
29. D.L. Budenz, A. Michael, R.T. Chang, J. McSoley, J. Katz, Sensitivity and specificity of the StratusOCT for perimetric glaucoma. Ophthalmology **112**(1), 3–9 (2005)
30. F.A. Medeiros, L.M. Zangwill, C. Bowd, R.M. Vessani, R. Susanna Jr., R.N. Weinreb, Evaluation of retinal nerve fiber layer, optic nerve head, and macular thickness measurements for glaucoma detection using optical coherence tomography. Am. J. Ophthalmol. **139**(1), 44–55 (2005)
31. J.L. Hougaard, A. Heijl, B. Bengtsson, Glaucoma detection by Stratus OCT. J. Glaucoma. **16**(3), 302–306 (2007)
32. T.W. Kim, U.C. Park, K.H. Park, D.M. Kim, Ability of Stratus OCT to identify localized retinal nerve fiber layer defects in patients with normal standard automated perimetry results. Invest. Ophthalmol. Vis. Sci. **48**(4), 1635–1641 (2007)
33. E.J. Lee, T.W. Kim, K.H. Park, M. Seong, H. Kim, D.M. Kim, Ability of Stratus OCT to detect progressive retinal nerve fiber layer atrophy in glaucoma. Invest. Ophthalmol. Vis. Sci. **50**(2), 662–668 (2009)
34. F.A. Medeiros, L.M. Zangwill, L.M. Alencar, C. Bowd, P.A. Sample, R. Susanna Jr., R.N. Weinreb, Detection of glaucoma progression with Stratus OCT retinal nerve fiber layer, optic nerve head, and macular thickness measurements. Invest. Ophthalmol. Vis. Sci. **50**(12), 5741–5748 (2009)

35. F.A. Medeiros, L.M. Zangwill, C. Bowd, R.M. Vessani, R. Susanna Jr., R.N. Weinreb, Evaluation of retinal nerve fiber layer, optic nerve head, and macular thickness measurements for glaucoma detection using optical coherence tomography. Am. J. Ophthalmol. **139**(1), 44–55 (2005)
36. D.S. Greenfield, H. Bagga, R.W. Knighton, Macular thickness changes in glaucomatous optic neuropathy detected using optical coherence tomography. Arch. Ophthalmol. **121**(1), 41–46 (2003)
37. H.D. Jampel, D. Friedman, H. Quigley, S. Vitale, R. Miller, F. Knezevich, Y. Ding, Agreement among glaucoma specialists in assessing progressive disc changes from photographs in open-angle glaucoma patients. Am. J. Ophthalmol. **147**(1), 39.e1–44.e1 (2009)
38. R.M. Vessani, R. Moritz, L. Batis, R.B. Zagui, S. Bernardoni, R. Susanna, Comparison of quantitative imaging devices and subjective optic nerve head assessment by general ophthalmologists to differentiate normal from glaucomatous eyes. J. Glaucoma. **18**(3), 253–261 (2009)
39. N. Nassif, B. Cense, B.H. Park, S.H. Yun, T.C. Chen, B.E. Bouma, G.J. Tearney, J.F. de Boer, In vivo human retinal imaging by ultrahigh-speed spectral domain optical coherence tomography. Opt. Lett. **29**(5), 480–482 (2004)
40. M. Wojtkowski, V. Srinivasan, J.G. Fujimoto, T. Ko, J.S. Schuman, A. Kowalczyk, J.S. Duker, Three-dimensional retinal imaging with high-speed ultrahigh-resolution optical coherence tomography. Ophthalmology **112**(10), 1734–1746 (2005)
41. A.O. González-García, G. Vizzeri, C. Bowd, F.A. Medeiros, L.M. Zangwill, R.N. Weinreb, Reproducibility of RTVue retinal nerve fiber layer thickness and optic disc measurements and agreement with Stratus optical coherence tomography measurements. Am. J. Ophthalmol. **147**(6), 1067–1074, 1074.e1 (2009)
42. Y. Nakatani, T. Higashide, S. Ohkubo, H. Takeda, K. Sugiyama, Evaluation of macular thickness and peripapillary retinal nerve fiber layer thickness for detection of early glaucoma using spectral domain optical coherence tomography. J. Glaucoma. **20**(4), 252–259 (2011)
43. M.N. Menke, P. Knecht, V. Sturm, S. Dabov, J. Funk, Reproducibility of nerve fiber layer thickness measurements using 3D Fourier-domain OCT. Invest. Ophthalmol. Vis. Sci. **49**(12), 5386–5391 (2008)
44. J.C. Mwanza, R.T. Chang, D.L. Budenz, M.K. Durbin, M.G. Gendy, W. Shi, W.J. Feuer, Reproducibility of peripapillary retinal nerve fiber layer thickness and optic nerve head parameters measured with Cirrus TM HD-OCT in glaucomatous eyes. Invest. Ophthalmol. Vis. Sci. **51**(11), 5724–5730 (2010)
45. S.H. Lee, S.H. Kim, T.W. Kim, K.H. Park, D.M. Kim, Reproducibility of retinal nerve fiber thickness measurements using the test-retest function of Spectral OCT/SLO in normal and glaucomatous eyes. J. Glaucoma. **19**(9), 637–642 (2010)
46. F. Berisha, S. Aliyeva, N. Pfeiffer, E.M. Hoffmann, Intraobserver reproducibility of retinal nerve fiber layer thickness measurements using the RTVue OCT in normal and glaucomatous eyes. Invest. Ophthalmol. Vis. Sci. (2009). ARVO E-Abstract#2261
47. E.M. Hoffmann, A. Schulze, J. Lamparter, N. Pfeiffer, Comparison of Fourier Domain Optical Coherence Tomography (FD-OCT) measurements in subjects with and without pupil dilation. Invest. Ophthalmol. Vis. Sci. (2009). ARVO E-Abstract#239
48. J.Y. Huang, M. Pekmezci, N. Mesiwala, A. Kao, S. Lin, Diagnostic power of optic disc morphology, peripapillary retinal nerve fiber layer thickness, and macular inner retinal layer thickness in glaucoma diagnosis with Fourier-domain optical coherence tomography. J. Glaucoma. **20**(2), 87–94 (2011)
49. O. Tan, V. Chopra, A.T. Lu, J.S. Schuman, H. Ishikawa, G. Wollstein, R. Varma, D. Huang, Detection of macular ganglion cell loss in glaucoma by Fourier-domain optical coherence tomography. Ophthalmology **116**(12), 2305–2314.e1–2 (2009)
50. S. Li, X. Wang, S. Li, G. Wu, N. Wang, Evaluation of optic nerve head and retinal nerve fiber layer in early and advance glaucoma using frequency-domain optical coherence tomography. Graefes Arch. Clin. Exp. Ophthalmol. **248**(3), 429–434 (2010)

51. A. Schulze, N. Pfeiffer, S. Günther, E.M. Hoffmann, Measurement of retinal ganglion cell complex in glaucoma, ocular hypertension and healthy subjects with Fourier domain optical coherence tomographic (Rtvue 100, Optovue). Invest. Ophthalmol. Vis. Sci. (2009). ARVO E-Abstract#3323
52. M. Sehi, D.S. Grewal, C.W. Sheets, D.S. Greenfield, Diagnostic ability of Fourier-domain vs time-domain optical coherence tomography for glaucoma detection. Am. J. Ophthalmol. **148**(4), 597–605 (2009)
53. C.J. Shin, K.R. Sung, T.W. Um, Y.J. Kim, S.Y. Kang, J.W. Cho, S.B. Park, J.R. Park, M.S. Kook, Comparison of retinal nerve fibre layer thickness measurements calculated by the optic nerve head map (NHM4) and RNFL3.45 modes of spectral-domain optical coherence tomography (RTVue-100). Br. J. Ophthalmol. **94**(6), 763–767 (2010)
54. O.J. Knight, R.T. Chang, W.J. Feuer, D.L. Budenz, Comparison of retinal nerve fiber layer measurements using time domain and spectral domain optical coherent tomography. Ophthalmology **116**(7), 1271–1277 (2009)
55. C.K. Leung, C.Y. Cheung, R.N. Weinreb, Q. Qiu, S. Liu, H. Li, G. Xu, N. Fan, L. Huang, C.P. Pang, D.S. Lam, Retinal nerve fiber layer imaging with spectral-domain optical coherence tomography: a variability and diagnostic performance study. Ophthalmology **116**(7), 1257–1263, 1263.e1–e2 (2009)
56. C.K. Leung, S. Lam, R.N. Weinreb, S. Liu, C. Ye, L. Liu, J. He, G.W. Lai, T. Li, D.S. Lam, Retinal nerve fiber layer imaging with spectral-domain optical coherence tomography analysis of the retinal nerve fiber layer map for glaucoma detection. Ophthalmology **117**(9), 1684–1691 (2010)
57. R.T. Chang, O.J. Knight, W.J. Feuer, D.L. Budenz, Sensitivity and specificity of time-domain versus spectral-domain optical coherence tomography in diagnosing early to moderate glaucoma. Ophthalmology **116**(12), 2294–2299 (2009)
58. G.C. Massa, V.G. Vidotti, F. Cremasco, A.P. Lupinacci, V.P. Costa, Influence of pupil dilation on retinal nerve fibre layer measurements with spectral domain OCT. Eye (Lond.) **24**(9), 1498–1502 (2010)
59. G. Savini, M. Carbonelli, V. Parisi, P. Barboni, Effect of pupil dilation on retinal nerve fibre layer thickness measurements and their repeatability with Cirrus HD-OCT. Eye (Lond.) **24**(9), 1503–1508 (2010)
60. F.K. Horn, C.Y. Mardin, R. Laemmer, D. Baleanu, A.M. Juenemann, F.E. Kruse, R.P. Tornow, Correlation between local glaucomatous visual field defects and loss of nerve fiber layer thickness measured with polarimetry and spectral domain OCT. Invest. Ophthalmol. Vis. Sci. **5 0**(5), 1971–1977 (2009)
61. C.K. Leung, C. Ye, R.N. Weinreb, C.Y. Cheung, Q. Qiu, S. Liu, G. Xu, D.S. Lam, Retinal nerve fiber layer imaging with spectral-domain optical coherence tomography a study on diagnostic agreement with Heidelberg retinal tomograph. Ophthalmology **117**(2), 267–274 (2010)
62. A. Schulze, J. Lamparter, N. Pfeiffer, E.M. Hoffmann, Comparison of optic disc topographic measurements using Fourier-domain optical coherence tomography and confocal scanning laser ophthalmoscopy in glaucoma patients and normal subjects. Invest. Ophthalmol. Vis. Sci. (2010). ARVO E-Abstract#4895
63. M.T. Leite, H.L. Rao, R.N. Weinreb, L.M. Zangwill, C. Bowd, P.A. Sample, A. Tafreshi, F.A. Medeiros, Agreement among spectral-domain optical coherence tomography instruments for assessing retinal nerve fiber layer thickness. Am. J. Ophthalmol. **151**(1), 85.e1–92.e1 (2011)
64. M. Choma, M. Sarunic, C. Yang, J. Izatt, Sensitivity advantage of swept source and Fourier domain optical coherence tomography. Opt. Express **11**(18), 2183–2189 (2003)
65. E.C. Lee, J.F. de Boer, M. Mujat, H. Lim, S.H. Yun, In vivo optical frequency domain imaging of human retina and choroid. Opt. Express **14**(10), 4403–4411 (2006)
66. T.H. Ko, J.G. Fujimoto, J.S. Schuman, L.A. Paunescu, A.M. Kowalevicz, I. Hartl, W. Drexler, G. Wollstein, H. Ishikawa, J.S. Duker, Comparison of ultrahigh- and standard-resolution optical coherence tomography for imaging macular pathology. Ophthalmology **112**(11), 1992.e1–1992.15 (2005)

67. U. Schmidt-Erfurth, R.A. Leitgeb, S. Michels, B. Povazay, S. Sacu, B. Hermann, C. Ahlers, H. Sattmann, C. Scholda, A.F. Fercher, W. Drexler, Three-dimensional ultrahigh-resolution optical coherence tomography of macular diseases. Invest. Ophthalmol. Vis. Sci. **46**(9), 3393–3402 (2005)
68. E.J. Fernández, B. Povazay, B. Hermann, A. Unterhuber, H. Sattmann, P.M. Prieto, R. Leitgeb, P. Ahnelt, P. Artal, W. Drexler, Three-dimensional adaptive optics ultrahigh-resolution optical coherence tomography using a liquid crystal spatial light modulator. Vision Res. **45**(28), 3432–3444 (2005)
69. R.J. Zawadzki, S.S. Choi, A.R. Fuller, J.W. Evans, B. Hamann, J.S. Werner, Cellular resolution volumetric in vivo retinal imaging with adaptive optics-optical coherence tomography. Opt. Express **17**(5), 4084–4094 (2009)
70. E. Götzinger, M. Pircher, B. Baumann, C. Hirn, C. Vass, C.K. Hitzenberger, Retinal nerve fiber layer birefringence evaluated with polarization sensitive spectral domain OCT and scanning laser polarimetry: a comparison. J. Biophotonics. **1**(2), 129–139 (2008)
71. M. Yamanari, M. Miura, S. Makita, T. Yatagai, Y. Yasuno, Phase retardation measurement of retinal nerve fiber layer by polarization-sensitive spectral-domain optical coherence tomography and scanning laser polarimetry. J. Biomed. Opt. **13**(1), 014013 (2008)
72. Y. Wang, B.A. Bower, J.A. Izatt, O. Tan, D. Huang, Retinal blood flow measurement by circumpapillary Fourier domain Doppler optical coherence tomography. J. Biomed. Opt. **13**(6), 064003 (2008)
73. R.M. Werkmeister, N. Dragostinoff, M. Pircher, E. Götzinger, C.K. Hitzenberger, R.A. Leitgeb, L. Schmetterer, Bidirectional Doppler Fourier-domain optical coherence tomography for measurement of absolute flow velocities in human retinal vessels. Opt. Lett. **33**(24), 2967–2969 (2008)
74. F. Berisha, G.T. Feke, T. Hirose, J.W. McMeel, L.R. Pasquale, Retinal blood flow and nerve fiber layer measurements in early-stage open-angle glaucoma. Am. J. Ophthalmol. **146**(3), 466–472 (2008)
75. N. Plange, M. Kaup, A. Weber, K.O. Arend, A. Remky, Retrobulbar haemodynamics and morphometric optic disc analysis in primary open-angle glaucoma. Br. J. Ophthalmol. **90**(12), 1501–1504 (2006)

Chapter 6
Anterior Segment OCT Imaging

Alexandre Denoyer, Antoine Labbé, and Christophe Baudouin

Abstract Imaging techniques have been developed to overcome limitations of light biomicroscopy as well as to provide new static and dynamic 3D measurement of the anterior segment of the eye.

The goal of this chapter is to present a large variety of OCT applications in the anterior eye in order to render this new imaging device more familiar.

6.1 Introduction

In the past decade, new imaging techniques have been developed to overcome limitations of light biomicroscopy as well as to provide new static and dynamic 3D measurement of the anterior segment of the eye. Since optical coherence tomography (OCT) was first developed for analysis of the posterior part of the eye, modifications have been made in the technique such that good resolution images of the anterior segment could be obtained. More recently, the development of Fourier-domain OCT has dramatically increased the imaging resolution. As a result, many AS-OCT applications have been proposed, notably in the field of refractive surgery, corneal graft, and glaucoma.

Like for any new imaging technique, analysis of the images is only possible after a thorough grounding in the semiology. The goal of this chapter is to present a large variety of OCT applications in the anterior eye in order to render this new imaging device more familiar, making it possible to use such a new routinely useful technique with optimum efficiency in pathological states.

C. Baudouin (✉)
Department of Ophthalmology, Versailles-Saint-Quentin-en-Yvelines Medical School,
Centre National d'Ophtalmologie des Quinze-Vingts,
Paris, France
e-mail: chrbaudouin@aol.com

R. Bernardes and J. Cunha-Vaz (eds.), *Optical Coherence Tomography*,
Biological and Medical Physics, Biomedical Engineering,
DOI 10.1007/978-3-642-27410-7_6,

6.2 Principles of Anterior Segment OCT

Many clinical situations require a precise assessment of both spatial relationships and dimensions of the structures constitutive of the front of the eye. OCT is a noninvasive *in vivo* technique based on low-coherence interferometry principle cross sections of tissue structures. Delay and intensity of the light reflected from the structure being analyzed are compared with the light reflected by a reference mirror, and the combination of these two signals results in a phenomenon termed interference. The signal intensity depends on the optical properties of the tissues, and the device uses this record to construct axial cross section of the structure being examined. In fact, the operating principle behind OCT is similar to that of the ultrasound scan except that a light is used instead of ultrasound. Since the speed of light is a million times faster than that of ultrasound, this type of imaging allows axial resolution levels of the order of a few microns to be achieved.

Marketed for the first time in 1995 by Carl Zeiss Meditec (Dublin, USA), OCT has until now mostly been used in ophthalmology to produce images of the posterior segment of the eye. The first optical coherence images of the anterior segment were made in 1994 using an OCT with a wavelength of 820 nm designed for examination of the posterior segment. This new imaging technique was then used to assess anatomical changes during refractive surgery and, adapted for use with the slit lamp, for routine clinical examination of the anterior segment. However, the anterior segment images obtained with this system were not always very good in quality. This is because the time taken for image acquisition was between 1 and 5 s, meaning that distortion due to micromovements of the eye caused blurred images. In addition, the image calculation algorithm designed for the (concave) retina could produce distortions in the images of the (convex) cornea and anterior segment. Finally, the wavelength of 820 nm gave only limited penetration of the signal through the sclera, limbus, or iris, making it impossible to view more posterior structures. It was only in 2001 that a high-speed OCT (eight images per second) became available using a wavelength of 1,310 nm allowing optimum visualization of the anterior segment. This wavelength gives better penetration through light-retaining tissues such as the sclera or limbus, making it possible to analyze the iridocorneal angle. Since then, new spectral domain OCTs have become commercially available for the analysis of the anterior segment and the cornea in particular. As a new imaging technique providing the means of obtaining cross sections of the anterior segment in vivo, anterior segment OCT (AS-OCT) has already found many applications.

6.3 Clinical Applications

6.3.1 *Corneal Thickness Assessment*

Biometric assessment of the anterior segment has drawn great benefit from this reproducible and totally noninvasive technique. New AS-OCT devices are able to measure the whole corneal thickness as well as the thickness of each corneal layer from the epithelium to the endothelium, allowing, for example, to determine the exact thickness of both the post-LASIK corneal flap and posterior stroma in the whole cornea (Fig. 6.1). AS-OCT-measured corneal pachymetry is reliable, reproducible, and closely correlated with ultrasound pachymetry especially using Stratus or Fourier-domain OCT [1, 2]. 3D mapping of the epithelial thickness may also be greatly improved, even though these new data have still to be validated.

6.3.2 *Corneal Grafts*

Imaging of the ocular anterior structures is one of the key points in the analysis and follow-up of corneal grafts. AS-OCT provides today useful qualitative and quantitative data, especially by imaging corneal architecture in new procedures for corneal graft such as deep anterior lamellar keratoplasty and endothelial graft. Moreover, AS-OCT contributes to the diagnosis of postoperative complications by investigating the graft neighboring and imaging the other anterior structures,

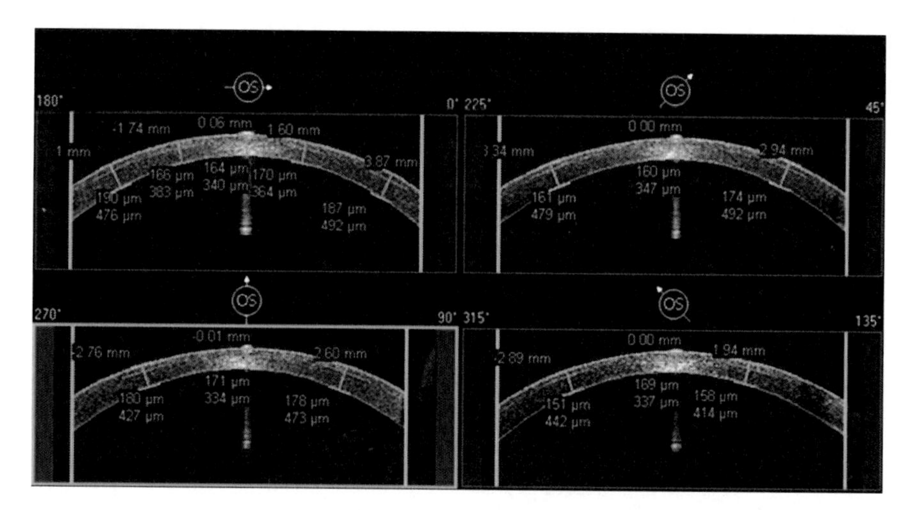

Fig. 6.1 AS-OCT (time domain OCT-3, Zeiss) for pachymetric measurement of the cornea. Corneal flap as well as posterior corneal thickness can be measured in order to create a pachymetric map

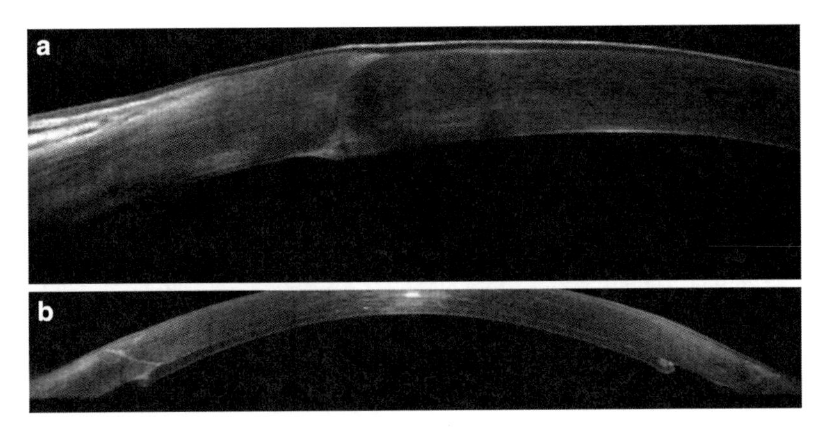

Fig. 6.2 Anterior segment OCT in corneal grafts. (**a**) AS-OCT (spectral domain, Optovue OCT®, EBC) image after penetrating keratoplasty showing a perfect interface between the cornea and the corneal graft. (**b**) AS-OCT (spectral domain, Spectralis®, Heidelberg) image of a Descemet stripping automated endothelial keratoplasty (DSAEK)

offering special usefulness when pathological corneal changes make standard clinical examination difficult.

After penetrating keratoplasty, AS-OCT allows to analyze the entire graft position as well as its macroscopic features such as thickness and curvature. Since laser-assisted penetrating keratoplasty was developed, AS-OCT has become a useful tool to assess the interface between the graft and the cornea (Fig. 6.2a), contributing to the improvement of surgical techniques. Moreover, long-term complications such as Descemet's membrane rupture or secondary glaucoma can be better diagnosed using AS-OCT even when important edema exists. In new graft procedures, AS-OCT helps to analyze preoperatively the corneal disease along with the anterior segment dimensions in order to better define the right procedure.

Postoperative management of new developments in endothelial grafts, i.e., Descemet stripping automated endothelial keratoplasty (DSAEK), has been dramatically improved thanks to AS-OCT (Fig. 6.2b) by following the corneal thickness related to endothelial function [3] as well as by diagnosing early and late complications including Descemet detachment or late pupillary blockade due to iris-graft synechiae. Other unexpected complications such as epithelial ingrowth in the interface can be imaged using AS-OCT.

Last, AS-OCT can substitute for clinical examination after amniotic membrane graft by following in vivo the corneal wound healing above the graft as well as amniotic membrane integration into the superficial cornea.

6.3.3 Corneal Refractive Surgery

Corneal refractive surgery, including surface ablation, laser in situ keratomileusis, femtosecond laser-assisted flaps (Fig. 6.3), and intracorneal rings (Fig. 6.4), induces

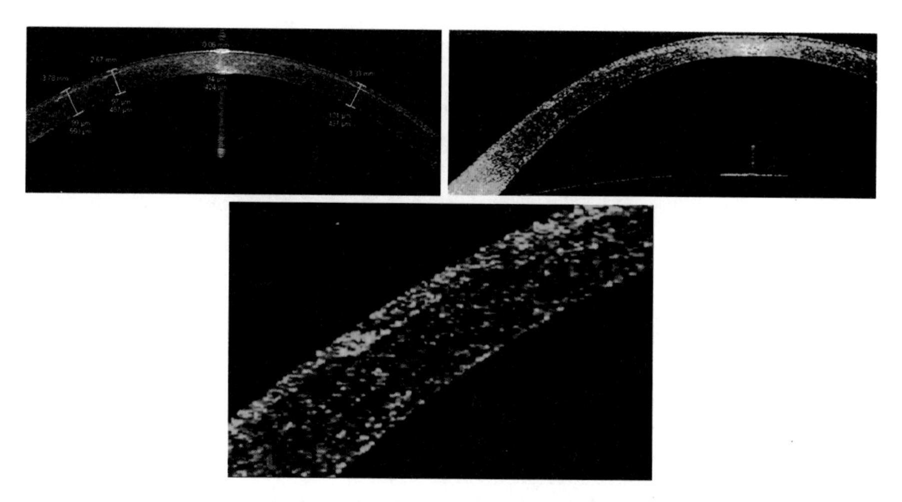

Fig. 6.3 LASIK flap imaging. (**a**) 2D mapping of the flap thickness using AS-OCT (Visante OCT®). (**b**, **c**) Imaging of the flap edge after femtosecond corneal disruption

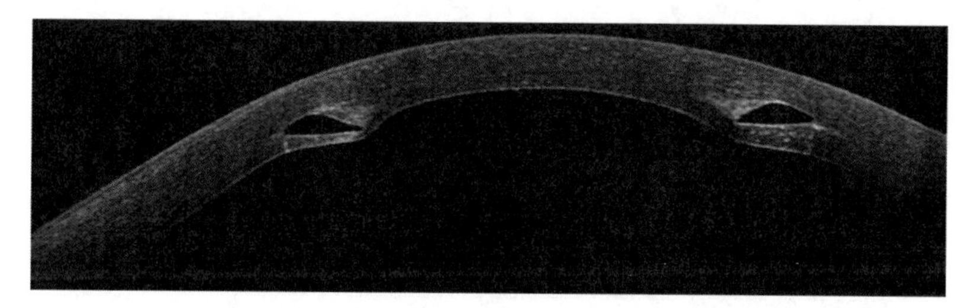

Fig. 6.4 AS-OCT (Spectralis®) images of corneal rings in keratoconus

deep changes in corneal morphology and optical properties along with cell structure remodeling. Great improvements in corneal imaging allow today microscopic in vivo follow-up of mechanisms involved in corneal wound healing, becoming helpful tools for optimizing therapeutic decision as well as surgical procedure.

Preoperatively, high-resolution AS-OCTs, i.e., Stratus and Fourier-domain OCTs, allow precise measurement of corneal thickness, intracorneal scare localization and volume, or previous LASIK interface depth using qualitative imaging but also quantitative A-mode. Moreover, corneal hydration as well as corneal refractive index can nowadays be assessed thanks to new interferometry-based imaging procedures [4]. After corneal refractive photoablation, wound healing can be followed in vivo with 3D mapping of epithelial or entire corneal thickness using high-resolution AS-OCT [5]. LASIK flaps can be precisely imaged from the center to the edges, and light diffraction at the interface can be measured in order to investigate its influence on contrast perception. Indeed, femtosecond laser-assisted LASIK has been demonstrated to perform more precise and homogenous flap cutting thanks to AS-OCT imaging [6]. Moreover, measuring changes in corneal curvature and refractive index allows today a better understanding of the relation between corneal ablation and refractive changes, becoming an essential tool before secondary corneal/intraocular surgery.

Complicated evolution of corneal remodeling and/or refraction pattern after corneal surgery can be analyzed using high-resolution AS-OCT to better identify causal mechanisms, from slight epithelial remodeling to true ectasia. Normal or pathological epithelial wound healing can be precisely followed up thanks to recently developed Fourier-domain AS-OCT (Fig. 6.5). Epithelial ingrowth, flap interface deposits, or microfolds can be easily characterized combining AS-OCT and confocal microscopy in order to better define the therapeutic strategy (Fig. 6.6). Indeed, AS-OCT is able to show undiagnosed microfolds, which might be responsible for postoperative visual impairments [7]. Secondary structural pathologies including basal membrane dystrophy and Salzmann's-like nodular degeneration can be complementarily diagnosed using high-resolution AS-OCT.

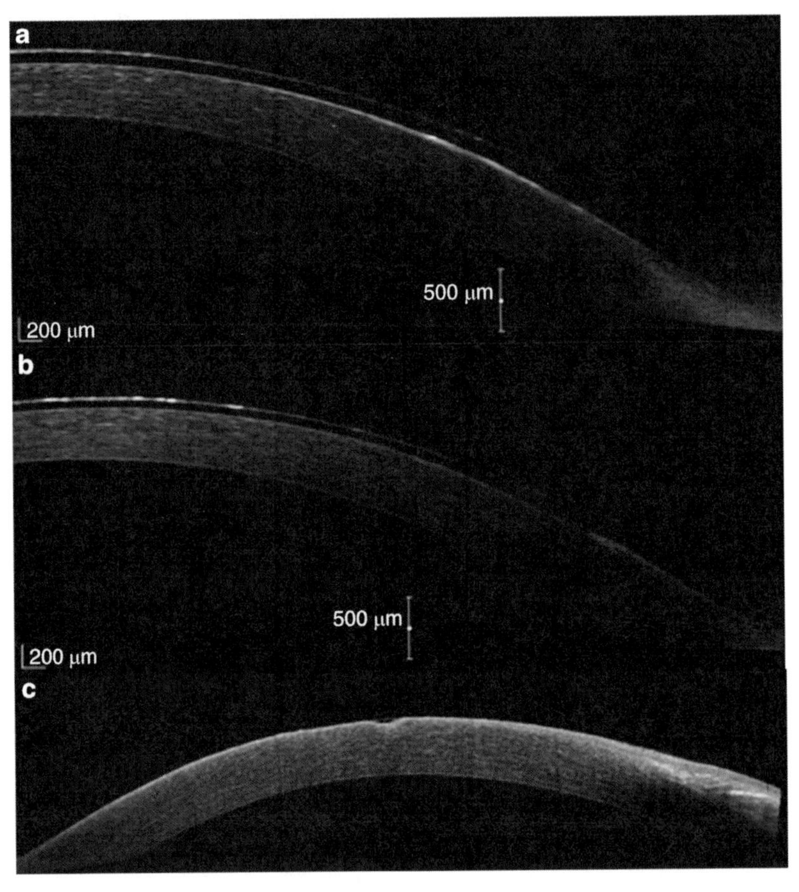

Fig. 6.5 Epithelial wound healing following surface excimer refractive photoablation. (Spectralis®). (**a**, **b**) Normal epithelial growth under soft contact lens at day 0 (**a**) and day 1 (**b**). (**c**) Late pathological epithelial wound healing

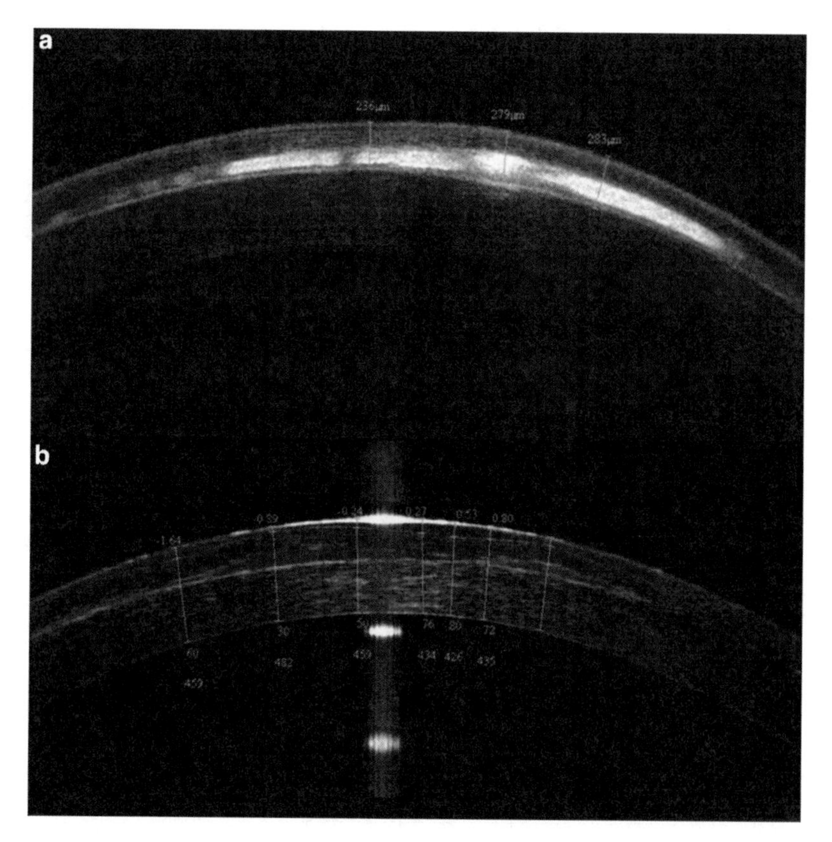

Fig. 6.6 AS-OCT imaging of post-LASIK epithelial ingrowth (Visante OCT®). (**a**) Epithelial ingrowth appears as hyperreflective deposits within the flap with an increased flap thickness. (**b**) Pathological remodeling of the flap and the posterior corneal bed after the treatment

6.4 Phakic Intraocular Lens

AS-OCT provides useful quantitative data upon static and dynamic anterior segment parameters, including anterior chamber depth, iridocorneal angle, and anterior lens movements. It has become a powerful device in phakic intraocular lens (IOL) surgery for preoperative ocular assessment as well as for patient follow-up (Fig. 6.7). Indeed, AS-OCT data are used to better determine predictive factors for surgical success in such refractive procedures.

Preoperative morphological assessment of the anterior segment is one of the key points in phakic IOL refractive surgery. Such measurements are necessary to better choose both the type and position of IOL and to fit its size on that of the target eye. Compared to other ocular imaging devices, AS-OCT performs entire eye measurements with a single acquisition in physiological conditions,

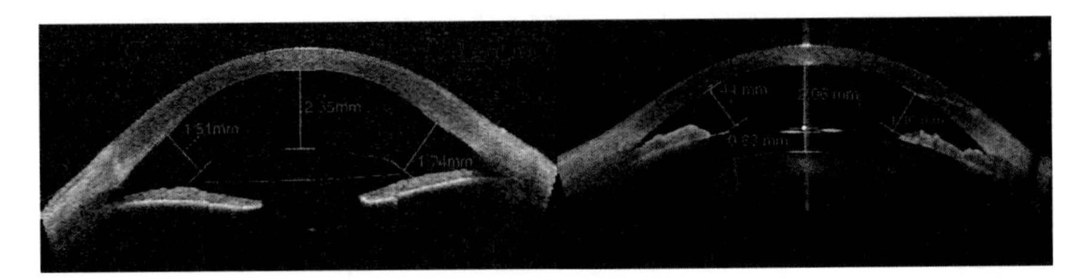

Fig. 6.7 AS-OCT (OCT-3®) imaging of phakic intraocular lenses with quantitative measures. (**a**) Iris claw anterior segment IOL. (**b**) Posterior segment IOL

i.e., in static and/or accommodating conditions and without any pharmacological mydriasis [8]. AS-OCT provides anterior chamber depth mapping in order to ensure the endothelial safety of the surgery. Recently, lens anterior curvature, namely, crystalline lens rise, as measured by AS-OCT has been suggested to be the most predictive factor to schedule patients for anterior chamber IOL implantation [9]. Moreover, angle-to-angle distance—but not sulcus-to-sulcus one that is better assessed by ultrasound biomicroscopy (UBM)—can be directly measured using AS-OCT in order to adapt anterior chamber IOL size [10].

AS-OCT is today the better way to perform dynamic imaging of the anterior segment as it allows to measure live ocular changes related to light adaptation [11] as well as to accommodation. Pupil changes and accommodation are associated with anterior segment movements that can modify the relationships between the posterior cornea, the IOL, and the lens [12]. For example, it has been determined that both iris claw IOL and crystalline move of 70 μm and 85 μm, respectively, toward the cornea during accommodation as assessed by AS-OCT [13].

Long-term follow-up of the paraxial measurements of anterior chamber depth is today one of the better ways to check the endothelial safety after phakic IOL implantation [14, 15]. Live imaging of the 360° entire iridocorneal angle is important to determine whether a phakic IOL could cause partial angle blockade [16]. Moreover, AS-OCT can diagnose abnormal contacts between the IOL and the iris causing atrophy and/or pigmentary dispersion. Last, live imaging combined with 3D anterior eye modelization has demonstrated that dynamic and abnormal touches of the anterior lens to the phakic IOL must cause cataract if the IOL is not quickly removed.

6.5 Glaucoma

Assessment of angle structures is essential in patients with OHT or glaucoma. Although gonioscopy remains the gold standard approach to evaluate clinically the morphology of the iridocorneal angle, it is a highly subjective examination technique depending on the experience of the examiner and the conditions of examination. AS-OCT provides a cross-sectional high-resolution image of the iridocorneal angle structures located anterior to the iris. Because the light is blocked by pigment, structures that are posterior to the iris are not clearly visualized. However, the root of the iris, the angle recess, and the anterior surface of the ciliary body could be discerned in some cases with low-pigmented iris.

Noncontact and easy to use, AS-OCT is an interesting diagnostic tool for the diagnosis and the screening of angle closure but also for a better understanding of the mechanisms of angle closure. The noncontact and illumination (suprluminescent light) principles of AS-OCT may provide more physiological images of angle structures and consequently may detect more cases of angle closure than does gonioscopy [17]. Inter- and intraobserver reliability has been reported to be high with AS-OCT [18], and studies comparing AS-OCT and UBM for the detection and the quantification of narrow angles have found no differences [19]. AS-OCT may also help to elucidate the mechanism in pupil block glaucoma, plateau iris configuration, or even choroidal effusion (Fig. 6.8). Moreover, a dynamic approach of the relationships between the iris, angle structures, and lens surface could be easily obtained under dark and light conditions (Fig. 6.9). This dynamic evaluation may be particularly interesting in angle closure suspects in order to study angle structures in response to light. Recently, AS-OCT has been used to estimate iris volume, and it has been observed that iris volume increases after pupil dilation in narrow-angle eyes predisposed to acute angle closure [20]. In secondary dispersion glaucoma, AS-OCT has also been used to demonstrate intermittent iris-IOL contact through a dilated pupil. The assessment of the anterior segment structures in the pediatric age group may also be facilitated by AS-OCT [21].

A large variety of biometric parameters have been developed for UBM quantification of angle structures. Some of these parameters are used for characterizing angle structures with AS-OCT such as the ACA (anterior chamber angle), the AOD (angle-opening distance, distance from cornea to iris at 500 or 750 μm from the

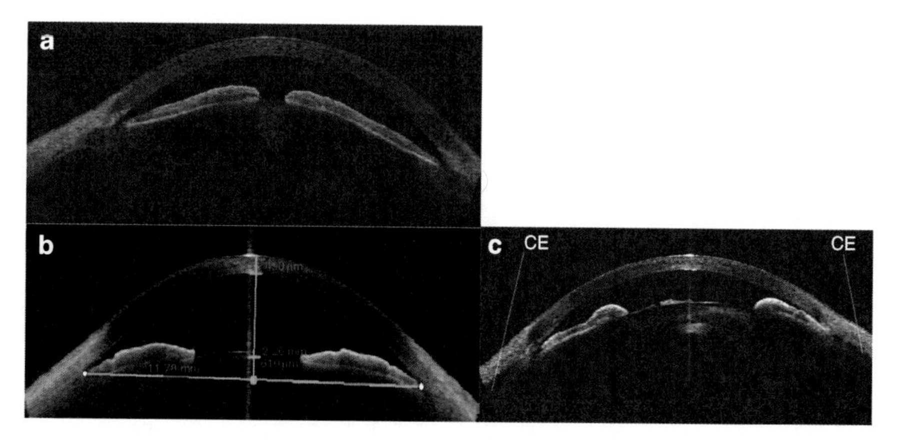

Fig. 6.8 (**a**) AS-OCT image of angle closure by intermittent pupillary block. (**b**) AS-OCT image of narrow anterior chamber secondary to an increased lens volume. (**c**) AS-OCT image of angle closure in an acute choroidal effusion syndrome (CE)

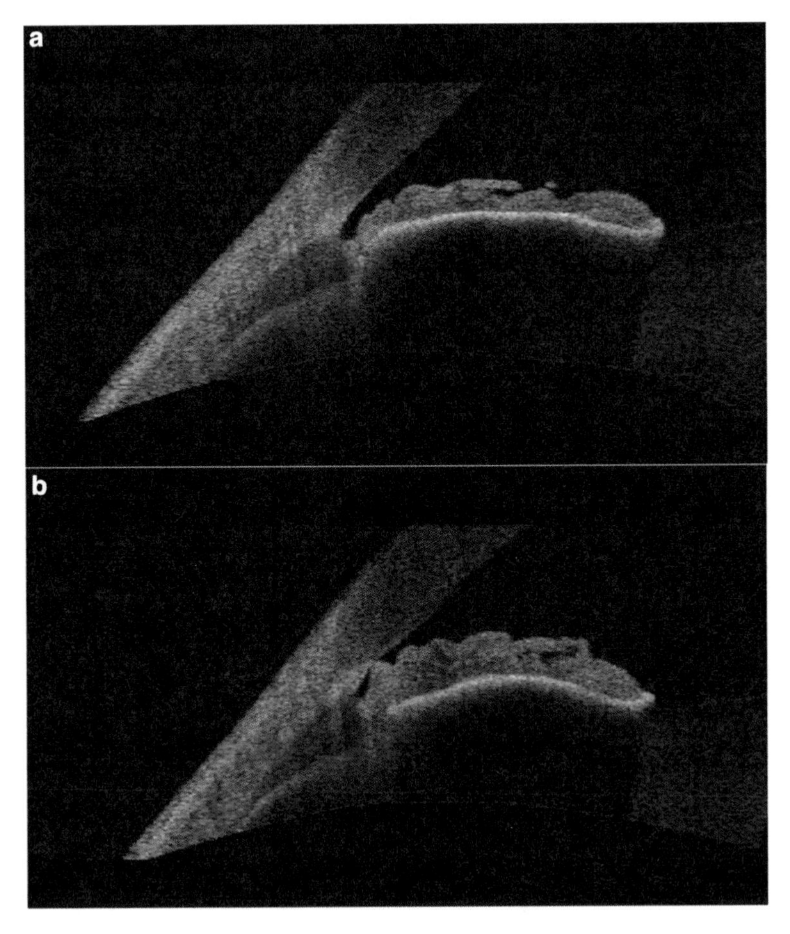

Fig. 6.9 AS-OCT images of a dynamic approach of the relationships between iris and angle structures in a case of plateau iris configuration under light (**a**) and dark (**b**) conditions

scleral spur), and the ARA (angle recess area, area of a triangle between angle recess and iris and cornea 500 or 750 μm from scleral spur). New parameters have been developed specifically for AS-OCT such as TISA (trabecular-iris space area, area of the trapezoid between iris and cornea from the scleral spur to 500 or 750 μm) or TICL (trabecular iris contact length, linear distance of contact between iris and sclera or cornea beginning at scleral spur). A rapid and objective quantification of these different parameters is obtained with software included in AS-OCT devices. The scleral spur has to be identified, and then automatically these parameters are calculated (Fig. 6.10). According to Radhakrishnan et al., the best parameters for the detection of occludable angles are the AOD 500, ARA 500, and TISA 500 [19]. These parameters may also be interesting for the screening of angle closure glaucoma in large population-based studies.

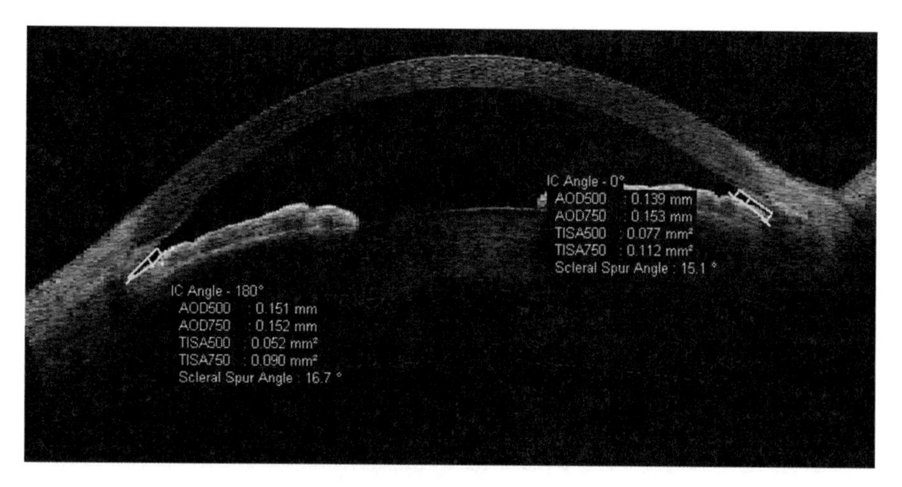

Fig. 6.10 AS-OCT image showing quantitative parameters for the evaluation of iridocorneal angle structures. *AOD* angle-opening distance 500 or 750, *TISA* trabecular-iris space area 500 or 750

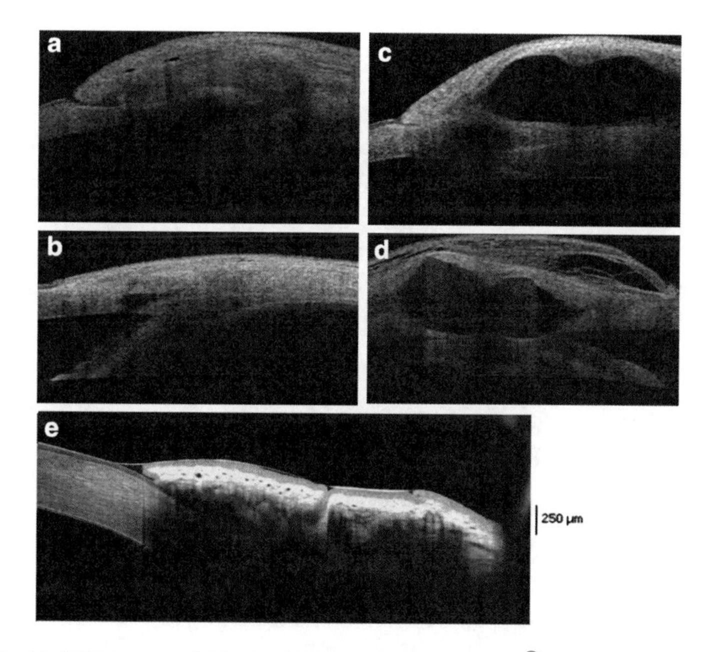

Fig. 6.11 AS-OCT images of filtering blebs. (**a–d**) Visante OCT®: (**a**) functioning bleb after trabeculectomy; (**b**) nonfunctioning flat bleb after trabeculectomy; (**c**) nonfunctioning encapsulated bleb after deep nonpenetrating sclerectomy; (**d**) collagen implants used in nonpenetrating deep sclerectomy. (**e**) SD-OCT, Spectralis®: functioning bleb after nonpenetrating sclerectomy

Last, AS-OCT may provide useful information following surgery or laser procedures in glaucoma and OHT. The development of a filtering bleb, determined by the postoperative wound healing process, is a major factor of efficiency and long-term success of surgical procedures. AS-OCT has been used to evaluate filtering blebs after trabeculectomy or deep nonpenetrating sclerectomy [22]. Trabeculectomy site as well as trabeculo-descemetic membrane (in nonpenetrating deep sclerectomy), scleral flap, conjunctival flap, iris, and the relationships between these structures could be analyzed noninvasively with this imaging technique. Functioning blebs had a hyporeflective and irregular conjunctival tissue associated with a route for aqueous humor under the scleral flap from the anterior chamber toward the subconjunctival space (Fig. 6.11). Flat and encapsulated nonfunctioning blebs had a dense and hyperreflective conjunctival tissue. In the particular case of nonfunctioning flat blebs, there is no route for aqueous humor filtration under the scleral flap. Collagen implants used in nonpenetrating deep sclerectomy could also be evaluated using AS-OCT and followed during the wound healing process. Interestingly, this noncontact technique may provide image immediately after surgical procedures and may help clinicians in the evaluation of postoperative procedures such as bleb needling, laser suture lysis after trabeculectomy [23], or goniopuncture after nonpenetrating deep sclerectomy. Other surgical techniques such as canaloplasty

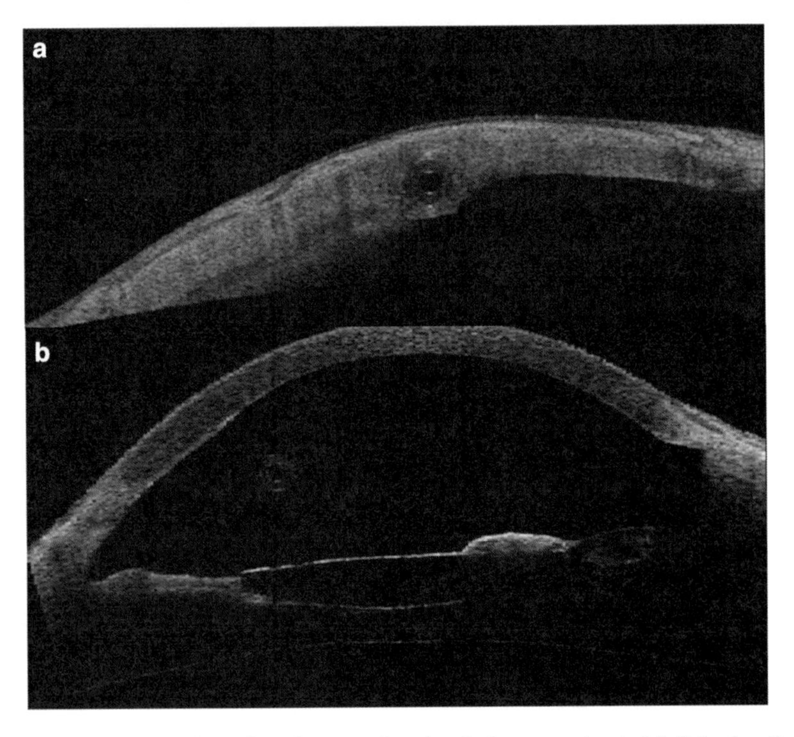

Fig. 6.12 AS-OCT imaging of a glaucoma functional glaucoma shunt. (**a**) Scleral pathway. (**b**) Shunt extremity floating into the anterior chamber

and the complications of filtering surgery could be also analyzed using AS-OCT. In glaucoma shunts, AS-OCT is able to visualize the shunt position within the anterior chamber even through corneal edema (Fig. 6.12). AS-OCT has also been used to evaluate the effect of laser peripheral iridotomy or iris trabeculoplasty on widening the iridocorneal angle in angle closure [24].

6.6 Conclusion

AS-OCT provides in vivo powerful imaging of the entire anterior segment of the eye. By offering noninvasive cross sections of the anterior segment, it has demonstrated its usefulness in corneal and intraocular refractive surgery, corneal graft, as well as in glaucoma. Developments of new systems based on different wavelengths and on better signal transduction and computerization have improved the spatial resolution as well as deep structure analyses. Even if it is a noncontact technique, it requires experienced operator to perform right acquisition and accurate image analysis. Reliability of the pachymetric measurements should be better assessed and improved in order to reach the gold standard that is still ultrasound measurement.

Until today, AS-OCT offers axial imaging only, but new experimental devices should be able to perform plane imaging in the future. The future of this technique will be a better resolution but also the development of 3D and 2D image reconstruction softwares and perhaps a combination with other techniques such as in vivo confocal microscopy in order to combine micrometric cell imaging with 3D morphological imaging of the anterior eye.

Acknowledgments Acknowledgments to Adil El Maftouhi, Marlene Francoz, and Michel Puech (Paris, France), for their participation to image collections.

References

1. G. Prakash, A. Agarwal, S. Jacob, D.A. Kumar, A. Agarwal, R. Banerjee, Comparison of Fourier-domain and time-domain optical coherence tomography for assessment of corneal thickness and intersession repeatability. Am. J. Ophthalmol. **148**(2), 282.e2–290.e2 (2009)
2. T. Simpson, D. Fonn, Optical coherence tomography of the anterior segment. Ocul. Surf. **6**(3), 117–127 (2008)
3. C.Y. Shih, D.C. Ritterband, P.M. Palmiero, J.A. Seedor, G. Papachristou, N. Harizman, J.M. Liebmann, R. Ritch, The use of postoperative slit-lamp optical coherence tomography to predict primary failure in descemet stripping automated endothelial keratoplasty. Am. J. Ophthalmol. **147**(5), 796–800, 800.e1 (2009)
4. A. Knüttel, S. Bonev, W. Knaak, New method for evaluation of in vivo scattering and refractive index properties obtained with optical coherence tomography. J. Biomed. Opt. **9**(2), 265–273 (2004)
5. Y. Li, M.V. Netto, R. Shekhar, R.R. Krueger, D. Huang, A longitudinal study of LASIK flap and stromal thickness with high-speed optical coherence tomography. Ophthalmology **114**(6), 1124–1132 (2007)

6. J.E. Stahl, D.S. Durrie, F.J. Schwendeman, A.J. Boghossian, Anterior segment OCT analysis of thin IntraLase femtosecond flaps. J. Refract. Surg. **23**(6), 555–558 (2007)
7. C. Ustundag, H. Bahcecioglu, A. Ozdamar, C. Aras, R. Yildirim, S. Ozkan, Optical coherence tomography for evaluation of anatomical changes in the cornea after laser in situ keratomileusis. J. Cataract Refract. Surg. **26**(10), 1458–1462 (2000)
8. J.A. Goldsmith, Y. Li, M.R. Chalita, V. Westphal, C.A. Patil, A.M. Rollins, J.A. Izatt, D. Huang, Anterior chamber width measurement by high-speed optical coherence tomography. Ophthalmology **112**(2), 238–244 (2005)
9. G. Baikoff, Anterior segment OCT and phakic intraocular lenses: a perspective. J. Cataract Refract. Surg. **32**(11), 1827–1835 (2006)
10. M. Bechmann, S. Ullrich, M.J. Thiel, K.R. Kenyon, K. Ludwig, Imaging of posterior chamber phakic intraocular lens by optical coherence tomography. J. Cataract Refract. Surg. **28**(2), 360–363 (2002)
11. L.P. Cruysberg, M. Doors, T.T. Berendschot, J. De Brabander, C.A. Webers, R.M. Nuijts, Irsi-fixated anterior chamber phakic intraocular lens for myopia moves posteriorly with mydriasis. J. Refract. Surg. **25**(4), 394–396 (2009)
12. A. Koivula, M. Kugelberg, Optical coherence tomography of the anterior segment in eyes with phakic refractive lenses. Ophthalmology **114**(11), 2031–2037 (2007)
13. W. Sekundo, W. Bissmann, A. Tietjen, Behaviour of the phakic iris-claw intraocular lens (Artisan/Verisyse) during accommodation: an optical coherence biometry study. Eur. J. Ophthalmol. **17**(6), 904–908 (2007)
14. C.R. Lovisolo, D.Z. Reinstein, Phakic intraocular lenses. Surv. Ophthalmol. **50**(6), 549–587 (2005)
15. R. Saxena, S.S. Boekhoorn, P.G. Mulder, B. Noordzij, G. van Rij, G.P. Luyten, Long-term follow-up of endothelial cell change after Artisan phakic intraocular lens implantation. Ophthalmology **115**(4), 608.e1–613.e1 (2008)
16. M. Doors, T.T. Berendschot, F. Hendrikse, C.A. Webers, R.M. Nuijts, Value of preoperative phakic intraocular lens simulation using optical coherence tomography. J. Cataract Refract. Surg. **35**(3), 438–443 (2009)
17. W.P. Nolan, J.L. See, P.T. Chew, D.S. Friedman, S.D. Smith, S. Radhakrishnan, C. Zheng, P.J. Foster, T. Aung, Detection of primary angle closure using anterior segment optical coherence tomography in Asian eyes. Ophthalmology **114**(1), 33–39 (2007)
18. S. Radhakrishnan, J. See, S.D. Smith, W.P. Nolan, Z. Ce, D.S. Friedman, D. Huang, Y. Li, T. Aung, P.T. Chew, Reproducibility of anterior chamber angle measurements obtained with anterior segment optical coherence tomography. Invest. Ophthalmol. Vis. Sci. **48**(8), 3683–3688 (2007)
19. S. Radhakrishnan, J. Goldsmith, D. Huang, V. Westphal, D.K. Dueker, A.M. Rollins, J.A. Izatt, S.D. Smith, Comparison of optical coherence tomography and ultrasound biomicroscopy for detection of narrow anterior chamber angles. Arch. Ophthalmol. **123**(8), 1053–1059 (2005)
20. F. Aptel, P. Denis, Optical coherence tomography quantitative analysis of iris volume changes after pharmacologic mydriasis. Ophthalmology **117**(1), 3–10 (2010)
21. A. Labbé, P. Niaudet, C. Loirat, M. Charbit, G. Guest, C. Baudouin, In vivo confocal microscopy and anterior segment optical coherence tomography analysis of the cornea in nephropathic cystinosis. Ophthalmology **116**(5), 870–876 (2009)
22. A. Labbé, P. Hamard, V. Iordanidou, S. Dupont-Monod, C. Baudouin, Utility of the Visante-OCT in the follow-up of glaucoma surgery. J. Fr. Ophthalmol. **30**(3), 225–231 (2007)
23. M. Singh, T. Aung, M.C. Aquino, P.T. Chew, Utility of bleb imaging with anterior segment optical coherence tomography in clinical decision-making after trabeculectomy. J. Glaucoma **18**(6), 492–495 (2009)
24. J.L. See, P.T. Chew, S.D. Smith, W.P. Nolan, Y.H. Chan, D. Huang, C. Zheng, P.J. Foster, T. Aung, D.S. Friedman, Changes in anterior segment morphology in response to illumination and after laser iridotomy in Asian eyes: an anterior segment OCT study. Br. J. Ophthalmol. **91**(11), 1485–1489 (2007)

Chapter 7
Optical Coherence Tomography: A Concept Review

Pedro Serranho, António Miguel Morgado and Rui Bernardes

Abstract Optical coherence tomography (OCT) is an imaging modality broadly used in biological tissue imaging. In this chapter, we review the history of OCT and its development throughout the last years. We will focus on the physical concept of OCT imaging of the eye fundus, considering several settings currently used. We also list some research directions of recent and ongoing work concerned with the future developments of the technique and its application.

7.1 Introduction

Optical coherence tomography (OCT) is a useful and common technique in ophthalmologic imaging. This technique allows to obtain a 3D representation of the eye fundus based on the different reflectivity from the different retinal layers. For clinical evaluation purposes, one gets a series of parallel 2D cross-section representations of the eye fundus. OCT image quality and resolution gave it the nickname of optical biopsy in some literature. It allows an axial resolution of 1–15 μm (e.g., [1,2]) in images of biological tissue obtained in situ and in real time. This is clearly an advantage in cases where standard excisional biopsy carries high risks or is not possible, as in the case of the human retina, therefore the popularity of OCT for human ocular fundus imaging. Previous axial ocular imaging techniques (along the depth axis) were based in ultrasound, e.g., [3]. However, OCT superior axial resolution (about one order of magnitude in comparison with ultrasound, which lies about 200 μm [4]) gives the clinician structural information of the retina

P. Serranho (✉)
Department of Science and Technology, Open University, Campus TagusPark, Av. Dr. Jacques Delors, 2740-122 Porto Salvo, Oeiras, Portugal

IBILI-Institute for Biomedical Research in Light and Image, Faculty of Medicine, University of Coimbra, Portugal
e-mail: pserranho@uab.pt

R. Bernardes and J. Cunha-Vaz (eds.), *Optical Coherence Tomography*,
Biological and Medical Physics, Biomedical Engineering,
DOI 10.1007/978-3-642-27410-7_7, © Springer-Verlag Berlin Heidelberg 2012

and its layers at a biopsy level. Moreover, OCT is a noncontact technique, which is a huge benefit in eye imaging in terms of patients' comfort during the examination, on top of not needing anesthesia.

Highly transparent tissues such as the retina can be imaged by OCT. The little backscattering of light occurring in the retinal highly transparent tissue is recovered by light interferometry, which is the physical principle behind OCT. While ultrasound is based on sound waves echo delay measurements, OCT techniques use interferometry to measure the optical path of backscattered light.

Another important aspect of OCT is that the axial or depth resolution of the system is almost totally independent from the lateral resolution. The axial resolution depends mainly on the coherence properties of the light source, namely, the length of coherence as we will detail further ahead in (7.3). For ultrahigh-resolution OCT, the axial resolution is about 1 μm for nontransparent tissue [5]. This technique was later applied in ophthalmic imaging with axial resolution of 2–3 μm [6, 7]. As for the lateral resolution, it depends on the optical quality of the optical system, including the patient's eye, being constrained by the pupil size and the system optical aberrations. For retinal imaging, transverse resolution is approximately 15 μm in every retinal OCT image obtained in an axial signal acquisition setting [6]. This lateral resolution can be improved by a full-field OCT (FF-OCT) setting as we will detail in the respective section further ahead or using adaptive optics, as discussed in detail in another chapter of the book (Chap. 10).

To obtain higher axial resolution in OCT, infrared broad bandwidth light sources are advised in order to get the required low-coherence length [as detailed in (7.3) further ahead]. However, the central wavelength for retinal imaging needs to be around 800 nm. Otherwise, due to its aqueous nature, the humor vitreous would highly absorb the light. Therefore, retinal imaging OCT is restricted in terms of bandwidth of the light source. Nonetheless, it allows unprecedented axial resolution of retinal tissue in comparison to other techniques.

Actual trends in OCT research and recent work comprehend obtaining functional information from OCT data [8–10]. These works are anchored on the fact that many diseases affect local optical properties of the retina and therefore changes of these properties should be encoded in backscattered light data gathered by OCT. Another research direction currently under consideration is to adapt incident light properties in order to image the choroid [11, 12], since at wavelengths around 800 nm, this is not possible. Imaging with 1,040–1,060 nm center wavelengths results in deeper penetration into the retina. However, water absorption limits the usable bandwidth at these wavelengths, when compared to the available bandwidth at 800 nm, leading to worse axial resolution. Adaptive optics for improving transverse resolution and polarization sensing for early detection of structural changes in retinal tissues, through birefringence assessment, are two other active research areas in ophthalmological OCT.

In this review, we intend to give a comprehensive yet accurate insight of the OCT technical formulation and concept, focusing on its application on eye fundus imaging. We will describe the early years of the technique and its development until recent years. We will also describe the main physical principles behind OCT imaging, both in the time-domain and in the frequency-domain approaches, giving a comprehensive insight on the physical concept that allows to attain structural information from backscattering of low-coherence light. In the last section, we will focus on the latest developments in the technique and in future directions aimed by several research groups.

Though OCT is a recent topic, several review papers on OCT are available. Apart from the reviews [2, 13–17] covering the OCT theory and applications as a whole, we refer to [18] for a technical review on the principles and applications of OCT, to [19] for OCT in nontransparent tissue, to [20] for the development of OCT as a clinical tool, to [21, 22] for a review on OCT as a fast nondestructive testing modality, and to [23, 24] for more recent reviews on spectral-domain OCT (SD-OCT).

7.2 The Genesis of OCT

The optical concept behind OCT, based on low-coherence interferometry, is similar to the one used about 40 years ago in [25] for thin film thickness measurement. However, most of the denominations (e.g., A-scan, B-scan, and C-scan) in OCT are borrowed from ultrasound imaging. In a hand-waiving way, while in ultrasonography depth information is obtained directly by measuring sound echo times, OCT retrieves depth through low-coherence interference between backscattered light and a reference beam. The first biomedical application of the backscattering light interferometry principle using partially coherent light was for the measurement of the length of the eye [26]. Two years later, the first approach for cross-sectional eye fundus imaging using time-domain (TD) low-coherence interferometry was presented at the ICO-15 SAT [27] and later published in [28] using the laser Doppler principle. In the latter, a 1D profile of the retinal pigment epithelium (RPE) of the human eye was obtained in vivo. This technique appeared as a higher-resolution alternative for previous developed clinical techniques based on ultrasonic echo-impulse techniques [3] that were standard for about 30 years. Among the disadvantages of the ultrasound technique [3] in comparison to low-coherence light interferometry [28] are the need for mechanical contact between the apparatus and the eye (therefore requiring anesthesia) and poorer axial resolution.

OCT as a technique for eye fundus volumetric imaging is anchored in the work by the group of Fujimoto in 1991 [29], where lateral scanning was introduced to the system and a fiber optic Michelson interferometer was used for time-domain low-coherence light interferometry. In vivo 2D tomograms were published 2 years later by Fercher et al. [30] and Swanson et al. [31]. Several studies regarding the properties of OCT were presented in the following years, regarding, for instance, its statistics of attenuation and backscattering [32] and its resolution for the anterior segment [33]. The validation of OCT for human eye fundus imaging was established in [34], in the sense that a correlation was found between OCT results and the known anatomy of the eye. In [35], 3D OCT images are reported, with an acquisition time of 20 s for the 40 frames of the volumetric data, which was prone to errors due to saccades. In [36], 3D scans (64 × 256 × 128 voxels) of the human retina were obtained in 1.2 s by the conjugation of the transverse scanning approach of confocal scanning and the depth scanning capability of OCT, making it possible to obtain OCT volumes of the human retina with minor movement errors due to the substantial reduction of the acquisition time.

Another important development was to consider the interferometry phenomena in the spectral domain, also called Fourier or frequency domain. Two ways were considered to this end. The first uses a spectrometer to measure the interferometric signal in the Fourier domain. The second employs a tunable light source that would sweep the frequency range of interest. This development allowed higher resolution of the obtained volumetric data and lower acquisition time (since the reference arm was kept fix). In a way, all the backscattering effect for each A-scan is measured simultaneously instead of moving the reference mirror to measure the backscattering light for each depth distance. The work of Fercher et al. [37] used backscattering spectral interferometry for the measurement of intraocular distances, both on a model eye and in vivo human eye. As in the first approaches in time-domain interferometry, the latter originated a 1D reflectivity profile of the human eye fundus. The work of [38] introduced a new concept for SD-OCT by considering source frequency tuning, which decreased acquisition time. Following the same principle, in [39], the in vivo acquisition time was in the order of the millisecond, and several intraocular distances were obtained simultaneously in a 1D profile. In [40], the frequency range (1,200–1,275 nm) is swept by a tunable laser in less than half the time, that is, 500 μs, allowing a scan rate of 2 kHz and an axial resolution of 15 μm. Häusler et al. [41] used the frequency-domain OCT concept for skin imaging.

Though the faster acquisition time was clear since the arrival of the Fourier domain OCT in 1995, only in 2003 [42–44] the sensitivity of time and Fourier domain setups is compared, being the performance of the latter clearly better. This was then extended to the comparison of macular thickness measurements [45] and implications on the measurement of macular edema [46]. Moreover, the first in vivo 2D tomogram of the human retina using SD-OCT was only published in 2002 [47], followed by the 3D retinal imaging in 2005 [7].

Several developments have been considered and suggested in the OCT technique and instrumentation in order to improve the results. Among these are attempts to increase depth and axial resolution and good choices of appropriate emission wavelength for choroid imaging. Several postprocessing methods for OCT images were also carried out more recently. We will refer to recent developments and future directions concerning OCT imaging in the last section of this chapter.

7.3 The Physical Principle

As already mentioned, OCT physical principle is similar to ultrasound imaging, using light instead of sound wave propagation and reflection. Ultrasound imaging is based on echo times, that is, it takes into account the time that an emitted sound

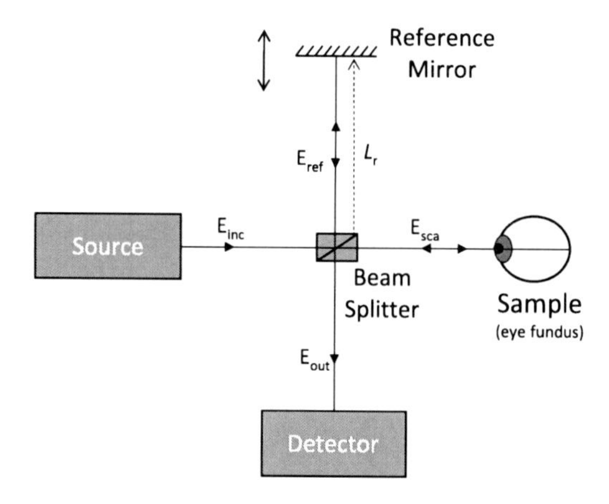

Fig. 7.1 Scheme for the principle of time-domain optical coherence tomography

signal takes to echo back from the structure to be imaged, usually called the sample. In OCT, as light travels much faster than sound, physicists had to adapt the principle in order to attain the backscattering optical path, since there is no available hardware to accurately measure the time of flight for the short values imposed by the axial resolution requirement (10^{-14} s). Therefore, low-coherence light and interferometry were considered to overcome this difficulty, finding a way to measure the differences in optical path of the detected backscattered light similarly to the sound echo time that ultrasound uses for image reconstruction.

Optical interferometry is based on the interference of two light beams. One knows that when a second wave interferes with a first one, having the same frequency (i.e., the same wavelength), their interference produces a higher intensity signal if in phase, while the signal vanishes if in opposite phase. Two light beams obtained by division of a single beam are initially in phase. If these beams are mixed after traveling some distance, the interference signal will be amplified only if the beams are still in phase. That only happens if both have traveled the same optical path (the optical path is the product of the geometrical path by the medium refractive index) or if the optical path difference is an integer multiple of their wavelength. This is only true for monochromatic (i.e., single wavelength) light beams. For high-bandwidth low-coherence light, signal amplification by interference can only happen when the optical paths differ by no more than a small value, equal to the coherence length of the light source.

The use of interferometry comes now in play in OCT, in the sense that it allows to detect the distance traveled by the backscattered light (or to be more rigorous, its optical path) in the following way (see Fig. 7.1): an incident low-coherence beam of light is split in two by a beam splitter; one of the beams travels (through a so-called sample path) to the sample and is partially backscattered, while the second beam travels (through a so-called reference path) to a reference mirror with a given adjustable distance to the splitter where it is also backscattered. The two backscattered beams are then again recombined, and the interference is measured. If the interference signal is high, it means that a lot of backscattered light from the sample comes from the same optical path distance of the reference mirror. The position of the mirror is ranged in the imaging depth range of interest in order to get a full A-scan profile. In this way, one gets a map of reflectivity of the sample along an axial depth direction. This rationale is the basis for time-domain OCT (TD-OCT).

7.4 Low-Coherence Light Interferometer

As already mentioned, the scheme for the principle of OCT imaging is presented in Fig. 7.1.

Low-coherence light is the superposition of a finite bandwidth of frequencies, so the incident field reaching the beam splitter can be characterized by

$$E_{\mathrm{inc}}(\omega, t) = A(\omega)\mathrm{e}^{-\mathrm{i}\omega t},$$

where i is the complex unit, ω is the frequency, and $A(\omega)$ is the correspondent amplitude spectrum of the source field. Since the initial phase of the incident wave is arbitrary, we have let it drop for the sake of simplicity [16]. In the beam splitter, the incident beam is divided in the reference incident beam with amplitude $A_{\mathrm{r}}(\omega)$ and the sample incident beam with amplitude $A_{\mathrm{s}}(\omega)$ with $A_{\mathrm{r}}(\omega) + A_{\mathrm{s}}(\omega) \leq 1$. The backscattered fields at the interferometer from the reference mirror and sample arm are therefore given by

$$E_{\mathrm{ref}}(\omega, t, L_{\mathrm{r}}) = A_{\mathrm{r}}(\omega)\mathrm{e}^{-\mathrm{i}(2\beta_{\mathrm{r}}(\omega)L_{\mathrm{r}} - \omega t)} \quad E_{\mathrm{sca}}(\omega, t) = \int_0^{+\infty} A_{\mathrm{s}}(\omega, z)\mathrm{e}^{-\mathrm{i}(2\beta_{\mathrm{s}}(\omega, z)z - \omega t)}\mathrm{d}z,$$

where β_{r} and β_{s} are the propagating coefficients for the reference and sample paths, respectively [48]. While the coefficient β_{r} for the reference arm (air) is considered to be independent from the depth variable z in this analysis, the sample arm propagation coefficient may vary in depth within the range of transparent ocular media and air. We now consider the output beam as

$$E_{\mathrm{out}}(\omega, t) = E_{\mathrm{ref}}(\omega, t) + E_{\mathrm{sca}}(\omega, t),$$

with intensity

$$J(\omega, L_{\mathrm{r}}) = \langle E_{\mathrm{out}} E_{\mathrm{out}}{}^{*} \rangle = |A_{\mathrm{r}}(\omega)|^2 + \left| \int_0^{+\infty} A_{\mathrm{s}}(\omega, z)\mathrm{e}^{-2\mathrm{i}\beta_{\mathrm{s}}(\omega, z)z}\mathrm{d}z \right|^2 + 2I(\omega, L_{\mathrm{r}}), \tag{7.1}$$

where $\langle\cdot\rangle$ represents a time average, that is,

$$\langle f \rangle = \lim_{T \to +\infty} \frac{1}{2T} \int_{-T}^{T} f(t)\mathrm{d}t,$$

the operator * represents the complex conjugate, L_r is the distance to the position of the reference mirror, and $I(\omega, L_r)$ is the real part of the so-called cross-interference. We point out that the two first terms of the intensity of the beam in (7.1) do not depend on the position of the reference mirror and are therefore called the self-interference. The cross-interference is therefore the important term for OCT image reconstruction and is proportional to [16]

$$I(\omega, L_r) \propto Re(\langle E_{\mathrm{ref}} E_{\mathrm{sca}}{}^* \rangle).$$

In this way, the cross-interference may be written as [48]

$$I(\omega, L_r) \propto \int_0^{+\infty} \tilde{I}(\omega, z)\mathrm{d}z, \tag{7.2}$$

where $\tilde{I}(\omega, z)$ is the cross-interference contribution from the reflectivity in the sample from distance z given by

$$\tilde{I}(\omega, z) = Re\left(A_r(\omega) A_s(\omega, z)^* \mathrm{e}^{-2\mathrm{i}\Delta\theta(\omega, z)}\right),$$

with each phase half mismatch for each backscattering distance given by

$$\Delta\theta(\omega, z) = \beta_r(\omega) L_r - \beta_s(\omega, z) z.$$

In the previous analysis, we did not yet refer to the low coherence of light sustaining the OCT principle. In fact, the low-coherence property comes from the fact that light source has several frequencies. In mathematical terms, the detected intensity is not given by (7.2) but by its integral over the range of frequencies of the light source. This integral defines the coherence length, since the periodic term of the complex exponential disappears by the integration over the frequency variable. The higher the bandwidth considered in the source, the lower the length of coherence.

For a Gaussian spectrum (i.e., a Gaussian-like distribution of frequencies), the length of coherence is given by [18]

$$l_{\mathrm{C}} = \frac{4 \ln 2}{\pi} \frac{\bar{\lambda}^2}{\Delta\lambda}, \tag{7.3}$$

where $\bar{\lambda}$ is the central wavelength and $\Delta\lambda$ is the spectral width. The axial resolution is given by half of the length of coherence [16]. It is clear that the resolution improves if the bandwidth $\Delta\lambda$ is higher and the mean wavelength $\bar{\lambda}$ is lower, suggesting the use of broad bandwidth light source. However, depending on the imaging application, these values have additional constraints. In nontransparent tissue like skin, wavelengths centered at about 1,300 nm are used to obtain better penetration, namely, about 2–3 mm [2, 19], with bandwidths ranging from 300 to 800 nm. However, for eye imaging, these wavelengths cannot be used due to water absorption in the ocular media [6]. For instance, usual OCT machines

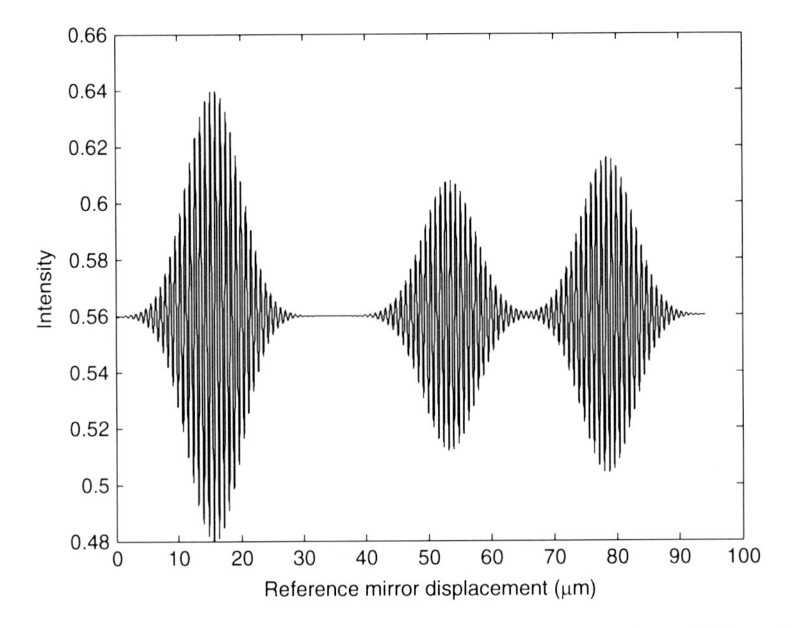

Fig. 7.2 Computer-simulated example of interference occurring at 16-, 53-, and 79-μm depth distances in the reference arm

consider values similar to $\bar{\lambda} = 820$ nm and $\Delta\lambda = 20$ nm, therefore allowing an axial resolution of about 15 μm [18]. In [16], a setting with $\bar{\lambda} = 800$ nm and $\Delta\lambda = 50$ nm is considered, which corresponds to an axial resolution of approximately 6 μm. Ultrahigh-resolution OCT considers $\bar{\lambda} = 800$ nm and $\Delta\lambda = 350$ nm, allowing an axial resolution below 1 μm [6].

We note that the coherence length determines the spatial length of interest within the sample around each considered reference arm length L_{r}. In this way, only backscattered light from within this range in the sample arm contributes to interference with the reference beam. An illustration of interference between the reference and sample beam is shown in Fig. 7.2.

7.5 Time-Domain Optical Coherence Tomography

For TD-OCT, the detector measures the sum of the intensities at each considered frequency. Therefore, from (7.1), one gets, for each position considered for the reference mirror at distance L_{r}, the intensity

$$J(L_{\mathrm{r}}) = \int_{-\infty}^{+\infty} \left(|A_{\mathrm{r}}(\omega)|^2 + \left| \int_0^{+\infty} A_{\mathrm{s}}(\omega, z) \mathrm{e}^{-2\mathrm{i}\beta_{\mathrm{s}}(\omega)z} \mathrm{d}z \right|^2 + 2I(\omega, L_{\mathrm{r}}) \right) \mathrm{d}\omega.$$

In this way, one gets a map of interferometry in depth related to the optical properties that gives information on the intensity of backscattered light from each optical path distance considered. This correlates with structural data of the sample (e.g., [34]). Instrumentation techniques are used to obtain only the cross-interference term, namely, by considering appropriate constant velocities for the movement of the reference mirror that facilitate the removal of the background self-interference (e.g., [13]).

7.6 Spectral-Domain Optical Coherence Tomography

Spectral or Fourier domain OCT also relies on interferometry, but the measurements are taken in the frequency space. In fact, in the physical setting, the detector is replaced by a spectrometer to this end. Therefore, one of the main advantages in this approach is that the reference mirror is kept fixed allowing a faster acquisition time, since no moving parts exist.

In the context of SD-OCT, one looks at (7.1) in terms of frequency for a given fixed L_r. At the receiving spectrometer, one gets the intensity spectrum given by (7.1) with respect to a frequency ω. In order to obtain the depth information in the time domain, a Fourier transform is performed, that is,

$$J(t) = \hat{J}(\omega),$$

where the hat $\wedge$ holds for Fourier transform. For faster computation, usually fast Fourier transform (FFT) is considered, since the intensity spectrum $J(\omega)$ is measured at equally spaced discrete frequency values, that is, one recovers a discrete set of intensities corresponding to a set of equally spaced frequency values. If one considers N data points in the frequency domain, the result after FFT consists of $N/2$ points in the time domain. The time resolution is given by

$$\Delta t = \frac{1}{c} \frac{\bar{\lambda}^2}{\Delta\lambda},$$

where $\bar{\lambda}$ and $\Delta\lambda$ are defined as in (7.3) and c is the speed of light [16]. A spatial step is obtained by multiplying the previous equation by the speed of light in the medium, that is,

$$\Delta z = \frac{1}{n_s} \frac{\bar{\lambda}^2}{\Delta\lambda},$$

where n_s is the constant refractive index in the sample arm. If one considers a detector array of N elements in the frequency domain, the maximum depth is given by

$$z_{\max} = \frac{N}{4n_s} \frac{\bar{\lambda}^2}{\Delta\lambda},$$

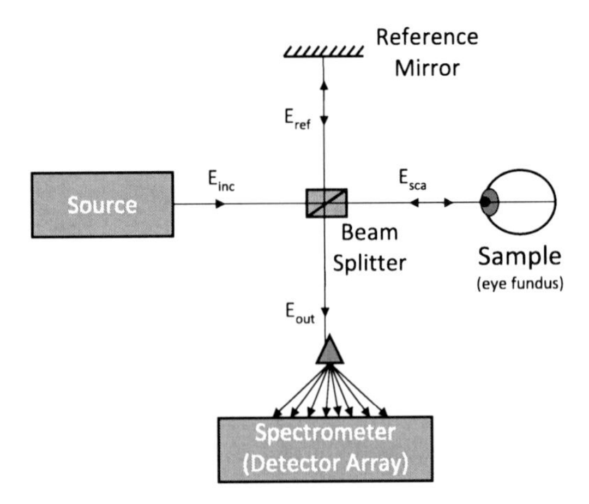

Fig. 7.3 Scheme for the principle of spectral-domain optical coherence tomography

where a division by 2 is needed in order to account for the travel through the sample arm in both directions. In an aqueous medium with $n_s = 1.33$, and considering $\bar{\lambda} = 800$ nm and $\Delta\lambda = 50$ nm as before and $N = 1,024$ elements in the detector, one gets $z_{max} \approx 2.46$ mm. For air with $n_s = 1.0008$, we get $z_{max} \approx 3.27$ mm. However, considering $\bar{\lambda} = 820$ nm and $\Delta\lambda = 20$ nm, one gets maximum depths for water and air of $z_{max} \approx 6.47$ mm and $z_{max} \approx 8.54$ mm, respectively. Therefore, one can only aim for information within this range, which for the retina is sufficient.

It is also worth mentioning that, in a spectrometer-based SD-OCT setting, the acquisition time is governed by the readout rate of the spectrometer. A major drawback of this technique is the decrease of signal level and, therefore, of sensitivity, with sample depth. This is a consequence of the convolution between the point-spread function for each wavelength, whose width increases along the detector plan, toward its edge, due to optical aberrations, and the corresponding detector pixel, whose width is finite and fixed [49]. The experimental setup is illustrated in Fig. 7.3.

7.7 Swept-Source Optical Coherence Tomography

Another possibility for Fourier domain OCT is to use a tunable light source (a tunable laser) instead of a broad bandwidth one. Though this technique is also within SD-OCT, it is usually named swept-source (SS-OCT).

For each axial direction, the frequency range of interest is rapidly swept through equally spaced frequency values. This equal spacement is necessary to allow the use of fast Fourier transform. This way, instead of using a broadband light and a spectrometer to gather information on the array of frequencies considered, in SS-OCT, only one photodetector is needed. The instrumentation process at the receptor side is simplified but that is balanced by the additional need of fast sweeping tunable lasers at the source side. One advantage of SS-OCT over SD-OCT

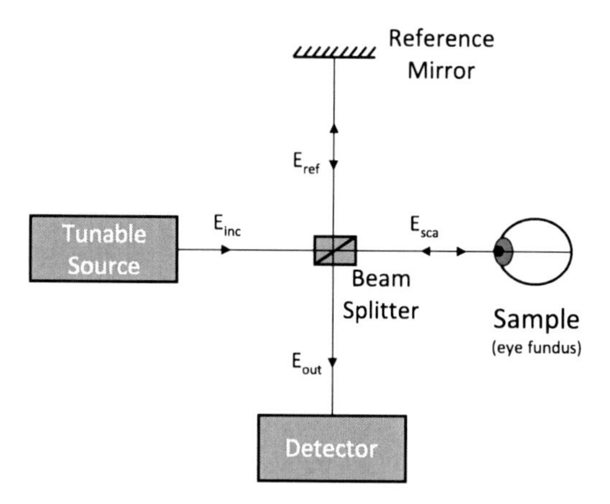

Fig. 7.4 Scheme for the principle of swept-source optical coherence tomography

is the minimal signal drop-off with depth that arises from using the same detector for all frequencies.

It is clear that the speed of acquisition of this setting is governed by the sweep speed of the laser. Several appropriate lasers to achieve this aim were developed and tested, for instance, in [39, 40, 50]. In fact, the development of tunable lasers with fast sweeping times in the range of frequencies of interest allowed to improve the acquisition time. Recent ultrahigh-speed OCT techniques were published by [51] for human retina imaging and by [52] using a Doppler OCT technique to image and quantitatively assess the retinal blood flow.

Choma and coworkers [42] compared the sensitivity of SD- and SS-OCT vs. TD-OCT. They found that TD-OCT has lower sensitivity, but the sensitivities of standard spectrometer-based SD- and SS-OCT are similar.

The development of ultrahigh-speed SS-OCT also allowed in vivo imaging of larger areas of the human retina. Actual acquisition times allow OCT imaging covering 70° field of view of 1,900 × 1,900 A-scans under 2 s [53], using this technique. The experimental setup is illustrated in Fig. 7.4.

7.8 Polarization-Sensitive Optical Coherence Tomography

Polarization-sensitive OCT (PS-OCT) gathers polarization information on top of the intensity of backscattered light. In this way, instead of recovering a scalar value of the intensity of the backscattered electromagnetic field, the measurement of polarization accounts for birefringence effects within the sample. Birefringence occurs when a medium has different refractive indexes for the two orthogonal plans of polarization. PS-OCT retrieves depth correlation information based on the birefringence within the sample.

This book has a full chapter dedicated to this modality of OCT, so we refer to Chap. 9 for more details.

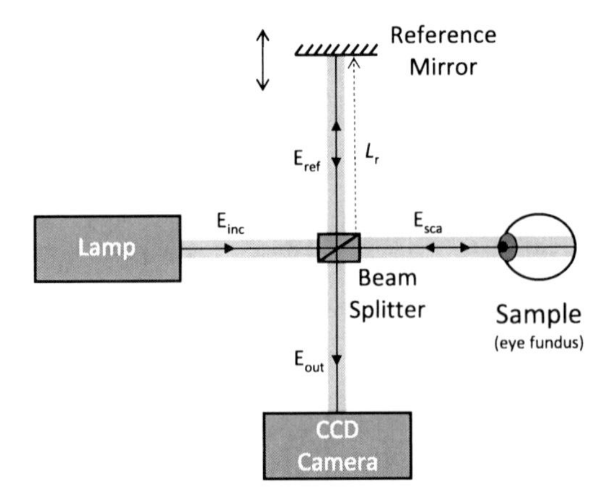

Fig. 7.5 Scheme for the principle of full-field optical coherence tomography

7.9 Full-Field Optical Coherence Tomography

FF-OCT takes advantages of both time- and spectral-domain approaches. It is based on the time-domain approach, namely, the reference mirror is shifted during acquisition time, but the acquisition time is much lower, competing with spectral-domain acquisition time. This is due to the fact that FF-OCT acquires enface images at once. Instead of acquiring information from an axial direction as both TD- and SD-OCT do, FF-OCT acquires backscattering data from the same (optical path) depth simultaneously. Therefore, though one needs the reference mirror to be moved in the spirit of TD-OCT, one does not need to move the incident light source in lateral directions over the sample.

The enface image data is acquired through the use of a charge-coupled device (CCD) camera instead of a single detector. The CCD camera captures the two-dimensional enface data from a single exposure from a thermal lamp light source instead of an axial oriented beam source. Thermal lamps are a good choice due to their relatively low price and low spatial coherence [16]. Moreover, this approach allows to improve lateral resolution. Reported axial and lateral resolutions are of 0.9 and 1.8 μm, respectively, [54] with an image time of 4 s and sensitivity of 90 dB. On the same year [55], both values of axial and lateral resolution were reduced for <1 μm, with an image acquisition time of 1 s and sensitivity of 80 dB. Later, the same principle was adapted for spectroscopic OCT [56]. It is clear that the lateral resolution of these apparatus improved significantly in comparison with the axial OCT setting, which is limited to 15 μm [6] due to the numerical aperture of the pupil. Moreover, this high-resolution performance in such short amount of time opens the path for in vivo subcellular level imaging in the future. The experimental setup is illustrated in Fig. 7.5.

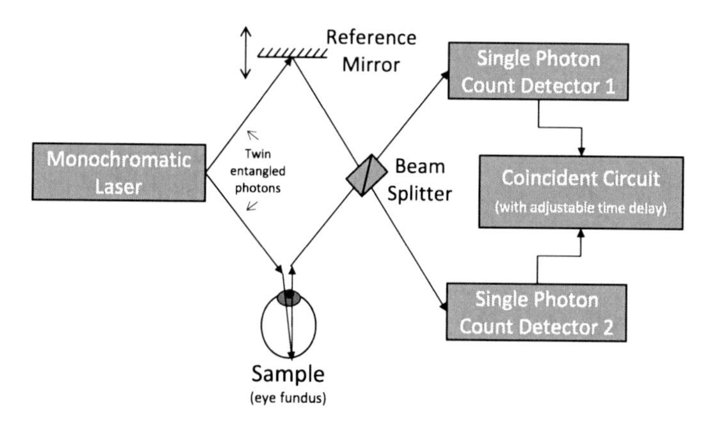

Fig. 7.6 Scheme for the principle of quantum optical coherence tomography

7.10 Quantum Optical Coherence Tomography

The axial resolution of both TD- and SD-OCT is based on the coherence length (7.3) considered. However, a new approach has been considered that promises an enhancement of twice the axial resolution as a by-product. The main difference is to take advantage of the quantum nature of light instead of its harmonic classical behavior.

Quantum OCT (Q-OCT) takes advantage of quantum light sources which light state is known. In particular, the concept takes advantage of entangled photon emission and in the principle based on the work of Hong et al. [57] that the detection probability amplitudes of an entangled state interfere destructively if two paths are equal in a properly adapted interferometer. While in classical interferometry, the interference from the same optical path distance is constructive (see Fig. 7.2), in Q-OCT, the opposite occurs. Only more recently this principle has been used in OCT [58, 59]. While dispersion of light in classical OCT originates degradation of axial resolution that needs to be overcome through numerical [60] or experimental [61] methods, quantum interference of an entangled state allows a twice axial resolution enhancement, since it is immune to dispersion. However, this needs to rely on highly efficient entangled state sources and detectors. Several works were developed subsequently to improve Q-OCT. For instance, Carrasco et al. [62] suggested improving axial resolution by chirped quasi-phase matching. The work of Nasr et al. [63] considered dispersion-canceled and dispersion-sensitive approaches in Q-OCT, in order to improve axial resolution and penetration depth. For more details on the physical principle of Q-OCT, we refer to the reviews [16, 64, 65].

Developments to consider polarization in Q-OCT were also achieved [66]. In this way, advantages of PS-OCT were added to standard Q-OCT quantum-polarized. Being immune to dispersion, polarized sensitive quantum OCT can be used to image tissue deeper than usual OCT. Details on the experimental setup can be seen in [67]. The experimental setup is illustrated in Fig. 7.6.

7.11 Latest Developments and Future Directions

OCT is a very promising imaging technique in ophthalmology due to its noninvasiveness and high-resolution characteristics. Being about 2 decades old in what concerns ophthalmology applications, research in OCT has nowadays various fronts and interests.

For instance, Bilenca et al. [68] measured the information gathered by OCT in comparison with the total information to characterize the sample in terms of Shannon information. In this way, these authors established the Shannon information limits of OCT and compared it with the current level of information that is gathered by actual OCT apparatus. The paper shows results depending on the length of coherence of light and speckle noise formation.

Another new direction of research is to use OCT data to acquire functional information of the retina status. Until recently, OCT was used to recover the internal structure of the retina, namely, information on the thickness of the whole retina or some of its layers as, for instance, the retinal nerve fiber layer. However, a new paradigm in the way researchers consider OCT data was introduced, based on the fact that it gathers information on the optical properties of the retina. Therefore, if some disease condition affects these optical properties, then OCT data should reflect these changes from the healthy state. In [8], alterations in OCT instrumentation were suggested to gather functional information on the photoreceptors from OCT data. Bernardes et al. [69] introduced the proof of concept that alterations on the blood–retinal barrier affect standard OCT data, namely, the histogram of intensities. Supposedly, leakage due to the malfunction of the blood–retinal barrier changes the refractive index of the retinal medium and therefore affects OCT data. This hypothesis is compatible with the fact that histogram changes are more stressed in upper retinal layers, where blood vessels lay. This was the purpose of a recent study by the same group [10]. In between, Grzywacz et al. [9] characterized the statistical properties of OCT data from the human retina and suggest that methods considering changes in this characterization might be used to distinguish between healthy and disease states. This characterization was also used to train an algorithm based on support vector machines (SVM) to automatically distinguish between OCT data from healthy, early stage diabetic retinopathy or macular edema diabetic retinopathy patients [70]. Accuracy by a leave-one-out cross-validation method was about 68%, which is promising in terms of allowing future automatic classification from OCT data of these three conditions. More details on these methods can be seen in Chap. 8.

Faster scanning techniques (mainly using SS-OCT) have also been a matter of interest recently [51, 52]. Ultrahigh-speed techniques allow to scan wider areas of the retina in vivo. For instance, in [53, 71], results with OCT scans covering 70° field of view are presented. The OCT scanning technique is based on swept-source OCT. Moreover, fundus images of such a wide field of view are provided by a postprocessing technique based on the mean intensity of the OCT signal between the RPE and the choroid [53].

Common wavelengths used in OCT to avoid absorption in the anterior segment, vitreous humor, and retinal media do not allow choroid imaging due to absorption. Therefore, an appropriate choice of the incident light source in order to allow choroid imaging has also been a matter of interest in recent developments. A source wavelength around 1,040 nm is more appropriate to this end [11]. Several techniques to use OCT to image both the retina and the choroid were developed thereafter [12, 72]. Ikuno and coworkers [73] used OCT high-penetration imaging techniques to test the reproducibility of choroidal thickness measurements and to correlate it with retinal thickness measurements.

References

1. J.G. Fujimoto, Optical coherence tomography for ultrahigh resolution *in vivo* imaging. Nat. Biotechnol. **21**(11), 1361–1367 (2003)
2. J.G. Fujimoto, Optical coherence tomography: principles and applications. Rev. Laser Eng. **31**, 635–642 (2003)
3. G.H. Mundt Jr., W.F. Hughes Jr., Ultrasonics in ocular diagnosis. Am. J. Ophthalmol. **41**(3), 488–498 (1956)
4. J.C. Bamber, M. Tristam, Diagnostic ultrasound, in *The Physics of Medical Imaging*, ed. by S. Webb (Adam Hilger, Bristol, 1988), pp. 319–388
5. W. Drexler, U. Morgner, F.X. Kärtner, C. Pitris, S.A. Boppart, X.D. Li, E.P. Ippen, J.G. Fujimoto, In vivo ultrahigh-resolution optical coherence tomography. Opt. Lett. **24**(17), 1221–1223 (1999)
6. W. Drexler, U. Morgner, R.K. Ghanta, F.X. Kärtner, J.S. Schuman, J.G. Fujimoto, Ultrahigh-resolution ophthalmic optical coherence tomography. Nat. Med. **7**(4), 502–507 (2001)
7. U. Schmidt-Erfurth, R.A. Leitgeb, S. Michels, B. Považay, S. Sacu, B. Hermann, C. Ahlers, H. Sattmann, C. Scholda, A.F. Fercher, W. Drexler, Three-dimensional ultrahigh-resolution optical coherence tomography of macular diseases. Invest. Ophthalmol. Vis. Sci. **46**(9), 3393–3402 (2005)
8. V.J. Srinivasan, Y. Chen, J.S. Duker, J.G. Fujimoto, In vivo functional imaging of intrinsic scattering changes in the human retina with high-speed ultrahigh resolution OCT. Opt. Express **17**(5), 3861–3877 (2009)
9. N. Grzywacz, J. de Juan, C. Ferrone, D. Giannini, D. Huang, G. Koch, V. Russo, O. Tan, C. Bruni, Statistics of optical coherence tomography data from human retina. IEEE Trans. Med. Imaging **29**(6), 1224–1237 (2010)
10. R. Bernardes, T. Santos, P. Serranho, C. Lobo, J. Cunha-Vaz, On invasive evaluation of retinal leakage using optical coherence tomography. Ophthalmologica **226**(2), 29–36 (2011)
11. A. Unterhuber, B. Považay, B. Hermann, H. Sattmann, A. Chavez-Pirson, W. Drexler, In vivo retinal optical coherence tomography at 1040 nm—enhanced penetration into the choroid. Opt. Express **13**(9), 3252–3258 (2005)
12. K. Kurokawa, K. Sasaki, S. Makita, M. Yamanari, B. Cense, Y. Yasuno, Simultaneous high-resolution retinal imaging and high-penetration, choroidal imaging by one-micrometer adaptive optics optical coherence tomography. Opt. Express **18**(8), 8515–8527 (2010)
13. J. Schmitt, Optical coherence tomography (OCT): a review. IEEE J. Sel. Top. Quant. Electron. **5**(4), 1205–1215 (1999)
14. W. Drexler, Ultrahigh-resolution optical coherence tomography. J. Biomed. Opt. **9**(1), 47–74 (2004)
15. A.G. Podoleanu, Optical coherence tomography. Br. J. Radiol. **78**(935), 976–988 (2005)

16. P.H. Tomlins, R.K. Wang, Theory, developments and applications of optical coherence tomography. J. Phys. D Appl. Phys. **38**(15), 2519–2535 (2005)
17. M.L. Gabriele, G. Wollstein, H. Ishikawa, L. Kagemann, J. Xu, L.S. Folio, J.S. Schuman. Optical coherence tomography: history, current status, and laboratory work. Invest. Ophthalmol. Vis. Sci. **52**(5), 2425–2436 (2011)
18. A.F. Fercher, W. Drexler, C.K. Hitzenberger, T. Lasser. Optical coherence tomography—principles and applications. Rep. Prog. Phys. **66**, 239–303 (2003)
19. M. Brezinski, J. Fujimoto, Optical coherence tomography: high-resolution imaging in non-transparent tissue. IEEE J. Sel. Top. Quant, Electron. **5**(4), 1185–1192 (1999)
20. A. Zysk, F. Nguyen, A. Oldenburg, D. Marks, S. Boppart, Optical coherence tomography: a review of clinical development from bench to bedside. J. Biomed. Opt. **12**(5), 051403 (2007)
21. M. Wurm, K. Wiesauer, K. Nagel, M. Pircher, E. Götzinger, C.K. Hitzenberger, D. Stifter, in *Spectral Domain Optical Coherence Tomography: A Novel And Fast Tool For NDT. IVth NDT in Progress*, Prague, Czech Republic, 5–7 November 2007
22. D. Stifter, K. Wiesauer, M. Wurm, E. Leiss, M. Plircher, E. Götzinger, B. Baumann, C.K. Hitzenberger, in *Advanced Optical Coherence Tomography Techniques: Novel and Fast Imaging Tools for Non-destructive Testing. 17th World Conference on Nondestructive Testing*, Shanghai, China, 25–28 October 2008
23. Z. Yaqoob, J. Wu, C. Yang, Spectral domain optical coherence tomography: a better OCT imaging strategy. BioTechniques **39**(6), 6–13 (2005)
24. S. Wolf, U. Wolf-Schnurrbusch, Spectral-domain optical coherence tomography use in macular diseases: a review. Ophthalmologica **224**(6), 333–340 (2010)
25. P.A. Flournoy, R.W. McClure, G. Wyntjes, White-light interferometric thickness gauge. Appl. Opt. **11**(9), 1907–1915 (1972)
26. A.F. Fercher, K. Mengedoht, W. Werner, Eyelength measurement by interferometry with partially coherent light. Opt. Lett. **13**(3), 186–188 (1988)
27. F. Fercher, Ophthalmic interferometry, in *Proceedings of the International Conference on Optics in Life Sciences*. ed. by G. von Bally, S. Khanna, Garmisch-Partenkirchen, Germany, 12–16 August 1990, pp. 221–228, (ISBN 0–444–89860–3)
28. C.K. Hitzenberger, Optical measurement of the axial eye length by laser Doppler interferometry. Invest. Ophthalmol. Vis. Sci. **32**(3), 616–624 (1991)
29. D. Huang, E.A. Swanson, C.P. Lin, J.S. Schuman, W.G. Stinson, W. Chang, M.R. Hee, T. Flotte, K. Gregory, C.A. Puliafito, Optical coherence tomography. Science **254**(5035), 1178–1181 (1991)
30. A.F. Fercher, C.K. Hitzenberger, W. Drexler, G. Kamp, H. Sattmann, In vivo optical coherence tomography. Am. J. Ophthalmol. **116**(1), 113–114 (1993)
31. E.A. Swanson, J.A. Izatt, M.R. Hee, D. Huang, C.P. Lin, J.S. Schuman, C.A. Puliafito, J.G. Fujimoto, In vivo retinal imaging by optical coherence tomography. Opt. Lett. **18**(21), 1864–1866 (1993)
32. J.M. Schmitt, A. Knüttel, M. Yadlowsky, M.A. Eckhaus, Optical-coherence tomography of a dense tissue: statistics of attenuation and backscattering. Phys. Med. Biol. **39**(10), 1705–1720 (1994)
33. J.A. Izatt, M.R. Hee, E.A. Swanson, C.P. Lin, D. Huang, J.S. Schuman, C.A. Puliafito, J.G. Fujimoto, Micrometer-scale resolution imaging of the anterior eye in vivo with optical coherence tomography. Arch. Ophthalmol. **112**(12), 1584–1589 (1994)
34. M.R. Hee, J.A. Izatt, E.A. Swanson, D. Huang, J.S. Schuman, C.P. Lin, C.A. Puliafito, J.G. Fujimoto, Optical coherence tomography of the human retina. Arch. Ophthalmol. **113**(3), 325–332 (1995)
35. A. Podoleanu, J. Rogers, D. Jackson, S. Dunne, Three dimensional OCT images from retina and skin. Opt. Express **7**(9), 292–298 (2000)
36. C. Hitzenberger, P. Trost, P.W. Lo, Q. Zhou, Three-dimensional imaging of the human retina by high-speed optical coherence tomography. Opt. Express **11**(21), 2753–2761 (2003)
37. A.F. Fercher, C.K. Hitzenberger, G. Kamp, S.Y. El-Zaiat, Measurement of intraocular distances by backscattering spectral interferometry. Opt. Commun. **117**, 43–48 (1995)

38. S.R. Chinn, E.A. Swanson, J.G. Fujimoto, Optical coherence tomography using a frequency-tunable optical source. Opt. Lett. **22**(5), 340–342 (1997)
39. F. Lexer, C.K. Hitzenberger, A.F. Fercher, M. Kulhavy, Wavelength-tuning interferometry of intraocular distances. Appl. Opt. **36**(25), 6548–6553 (1997)
40. B. Golubovic, B.E. Bouma, G.J. Tearney, J.G. Fujimoto, Optical frequency-domain reflectometry using rapid wavelength tuning of a Cr4+: forsterite laser. Opt. Lett. **22**(22), 1704–1706 (1997)
41. G. Häusler, M.W. Lindner, "Coherence radar" and "spectral radar"—new tools for dermatological diagnosis. J. Biomed. Opt. **3**(1), 21–31 (1998)
42. M. Choma, M. Sarunic, C. Yang, J. Izatt, Sensitivity advantage of swept source and Fourier domain optical coherence tomography. Opt. Express **11**(18), 2183–2189 (2003)
43. J.F. de Boer, B. Cense, B.H. Park, M.C. Pierce, G.J. Tearney, B.E. Bouma, Improved signal-to-noise ratio in spectral-domain compared with time-domain optical coherence tomography. Opt. Lett. **28**(21), 2067–2069 (2003)
44. R. Leitgeb, C. Hitzenberger, A. Fercher, Performance of Fourier domain vs. time domain optical coherence tomography. Opt. Express **11**(8), 889–894 (2003)
45. C.K. Leung, C.Y. Cheung, R.N. Weinreb, G. Lee, D. Lin, C.P. Pan, D.S. Lam, Comparison of macular thickness measurements between time domain and spectral domain optical coherence tomography. Invest. Ophthalmol. Vis. Sci. **49**(11), 4893–4897 (2008)
46. F. Forooghian, C. Cukras, C.B. Meyerle, E.Y. Chew, W.T. Wong, Evaluation of time domain and spectral domain optical coherence tomography in the measurement of diabetic macular edema. Invest. Ophthalmol. Vis. Sci. **49**(10), 4290–4296 (2008)
47. M. Wojtkowski, R. Leitgeb, A. Kowalczyk, T. Bajraszewski, A.F. Fercher, In vivo human retinal imaging by Fourier domain optical coherence tomography. J. Biomed. Opt. **7**(3), 457–463 (2002)
48. B.E. Bouma, G.J. Tearnley (eds.), *Handbook of Optical Coherence Tomography* (Marcel Dekker, New York, 2002)
49. Z. Hu, Y. Pan, A.M. Rollins, Analytical model of spectrometer-based two-beam spectral interferometry. Appl. Opt. **46**(35), 8499–8505 (2007)
50. M.A. Choma, K. Hsu, J.A. Izatt, Swept source optical coherence tomography using an all-fiber 1300-nm ring laser source. J. Biomed. Opt. **10**(4), 44009 (2005)
51. B. Potsaid, B. Baumann, D. Huang, S. Barry, A.E. Cable, J.S. Schuman, J.S. Duker, J.G. Fujimoto, Ultrahigh speed 1050 nm swept source/Fourier domain OCT retinal and anterior segment imaging at 100,000 to 400,000 axial scans per second. Opt. Express **18**(19), 20029–20048 (2010)
52. B. Baumann, B. Potsaid, M.F. Kraus, J.J. Liu, D. Huang, J. Hornegger, A.E. Cable, J.S. Duker, J.G. Fujimoto, Total retinal blood flow measurement with ultrahigh speed swept source/Fourier domain OCT. Biomed. Opt. Express **2**(6), 1539–1552 (2011)
53. A.S. Neubauer, L. Reznicek, T. Klein, W. Wieser, C.M. Eigenwillig, B. Biedermann, A. Kampik, R. Huber, Ultra-High-Speed Ultrawide Field Swept Source OCT Reconstructed Fundus Image Quality (ARVO, Fort Lauderdale, USA, 1–5 May, 2011) (Program/Poster # 1327/A264)
54. A. Dubois, K. Grieve, G. Moneron, R. Lecaque, L. Vabre, C. Boccara, Ultrahigh-resolution full-field optical coherence tomography. Appl. Opt. **43**(14), 2874–2883 (2004)
55. A. Dubois, G. Moneron, K. Grieve, A.C. Boccara, Three-dimensional cellular-level imaging using full-field optical coherence tomography. Phys. Med. Biol. **49**(7), 1227–1234 (2004)
56. A. Dubois, J. Moreau, C. Boccara, Spectroscopic ultrahigh-resolution full-field optical coherence microscopy. Opt. Express **16**(21), 17082–17091 (2008)
57. C.K. Hong, Z.Y. Ou, L. Mandel, Measurement of subpicosecond time intervals between two photons by interference. Phys. Rev. Lett. **59**(18), 2044–2046 (1987)
58. A.F. Abouraddy, M.B. Nasr, B.E.A. Saleh, A.V. Sergienko, M.C. Teich, Quantum-optical coherence tomography with dispersion cancellation. Phys. Rev. A **65**(5), 053817 (2002)
59. M.B. Nasr, B.E. Saleh, A.V. Sergienko, M.C. Teich, Demonstration of dispersion-canceled quantum-optical coherence tomography. Phys. Rev. Lett. **91**(8), 083601 (2003)

60. A. Fercher, C. Hitzenberger, M. Sticker, R. Zawadzki, B. Karamata, T. Lasser, Numerical dispersion compensation for partial coherence interferometry and optical coherence tomography. Opt. Express **9**(12), 610–615 (2001)
61. E.D.J. Smith, A.V. Zvyagin, D.D. Sampson, Real-time dispersion compensation in scanning interferometry. Opt. Lett. **27**(22), 1998–2000 (2002)
62. S. Carrasco, J.P. Torres, L. Torner, A. Sergienko, B.E. Saleh, M.C. Teich, Enhancing the axial resolution of quantum optical coherence tomography by chirped quasi-phase matching. Opt. Lett. **29**(20), 2429–2431 (2004)
63. M. Nasr, B. Saleh, A. Sergienko, M. Teich, Dispersion-cancelled and dispersion-sensitive quantum optical coherence tomography. Opt. Express **12**(7), 1353–1362 (2004)
64. M.B. Nasr, D.P. Goode, N. Nguyen, G. Rong, L. Yang, B.M. Reinhard, B.E. Saleh, M.C. Teich, Quantum optical coherence tomography of a biological sample. Opt. Commun. **282**, 1154–1159 (2009)
65. M.C. Teich, B.E.A. Saleh, F.N.C. Wong, J.H. Shapiro, Quantum optical coherence tomography: a review. Quant. Inf. Process (2012, in press), http://people.bu.edu/teich/abstracts/quantum-opt-archive.html
66. M.C. Booth, G. Di Giuseppe, B.E.A. Saleh, A.V. Sergienko, M.C. Teich, Polarization-sensitive quantum-optical coherence tomography. Phys. Rev. A **69**(4), 043815 (2004)
67. M.C. Booth, B.E. Saleh, M.C. Teich, Polarization-sensitive quantum optical coherence tomography: experiment. Opt. Commun. **284**, 2542–2549 (2011)
68. A. Bilenca, T. Lasser, B. Bouma, R.A. Leitgeb, G.J. Tearney, Information limits of optical coherence imaging through scattering media. Photon. J. IEEE **1**(2), 119–127 (2009)
69. R. Bernardes, T. Santos, J. Cunha-Vaz, in *Evaluation of Blood–Retinal Barrier Function from Fourier Domain High-Definition Optical Coherence Tomography*, ed. by O. Dössel, W.C. Schlegel. World Congress on Medical Physics and Biomedical Engineering, vol. 25/11, Munich, Germany, 7–12 September 2009, pp. 316–319 (Springer, Heidelberg, 2009)
70. R. Bernardes, Optical coherence tomography: health information embedded on OCT signal statistics, in *Proceedings of the 33rd Annual International Conference of the IEEE EMBS*, Boston, USA, 30 August–3 September 2011, pp. 6131–6133
71. T. Klein, L. Reznicek, W. Wieser, C.M. Eigenwillig, B. Biedermann, A. Kampik, R. Huber, A.S. Neubauer, *Extraction of Arbitrary OCT Scan Paths from 3D Ultra-High-Speed Ultra Wide-Field Swept Source OCT* (ARVO, Fort Lauderdale, USA, 1–5 May 2011) (Program/Poster # 1328/A265)
72. V.J. Srinivasan, D.C. Adler, Y. Chen, E. Gorczynska, R. Huber, J.S. Duker, J.S. Schuman, J.G. Fujimoto, Ultrahigh-speed optical coherence tomography for three-dimensional and en face imaging of the retina and optic nerve head. Invest. Ophthalmol. Vis. Sci. **49**(11), 5103–5110 (2008)
73. Y. Ikuno, I. Maruko, Y. Yasuno, M. Miura, T. Sekiryu, K. Nishida, T. Iida, Reproducibility of retinal and choroidal thickness measurements in enhanced depth imaging and high-penetration optical coherence tomography. Invest. Ophthalmol. Vis. Sci. **52**(8), 5536–5540 (2011)

Chapter 8
Evaluation of the Blood–Retinal Barrier with Optical Coherence Tomography

Rui Bernardes and José Cunha-Vaz

Abstract The imaging technique of optical coherence tomography (OCT) has opened a unique world in the area of ophthalmology, specifically in imaging the human retina. The findings based on the use of OCT have brought new knowledge of the retina and the changes that occur in the retina that previously have not been available in vivo.

In addition to the current uses of OCT, new fields of application have recently emerged, and it is believed that additional ones will gain maturity in the near future.

In this chapter, we demonstrate the potential use of OCT in the assessment of changes on the blood–retinal barrier and in the neuronal tissue related to the aging process.

8.1 Introduction

Vision is one of the most important senses for the human being and is fundamental to the way we interact with each other and to the surrounding environment.

Age-related macular degeneration (AMD), diabetic retinopathy (DR), and diabetic macular edema (DME) are the major ocular diseases leading to vision loss.

Recent figures for the incidence and prevalence of vision loss point to over 38 million Americans age 40 and older to experience blindness, low vision, or age-related eye diseases [1].

R. Bernardes (✉)
IBILI-Institute for Biomedical Research in Light and Image, Faculty of Medicine, University of Coimbra, Azinhaga de Santa Comba, Celas 3000-548 Coimbra, Portugal
e-mail: rcb@aibili.pt; rmbernardes@fmed.uc.pt

AIBILI-Association for Innovation and Biomedical Research on Light and Image, Azinhaga Santa Comba, 3000-548 Coimbra, Portugal

R. Bernardes and J. Cunha-Vaz (eds.), *Optical Coherence Tomography*, Biological and Medical Physics, Biomedical Engineering,
DOI 10.1007/978-3-642-27410-7_8, © Springer-Verlag Berlin Heidelberg 2012

Two hundred thousand Americans develop advance AMD every year, while 7.3 million are at the risk for vision loss, and over 4 million Americans age 40 and older have diabetic retinopathy, and 1 in 12 diabetics for the same age range has advanced diabetic retinopathy threatening their vision [1].

Finally, macular edema is likely to increase as the prevalence of diabetes mellitus is expected to raise by more than 50% from 2000 to 2030 [2] and the fact the prevalence of DME in the USA to be about 10% among the diabetic population.

These ocular diseases are closely linked to the functional status of the blood–retinal barrier (BRB), and in this chapter, we demonstrate the potential use of the optical coherence tomography (OCT) to assess changes in the BRB function noninvasively.

The BRB and the blood–aqueous barrier (BAB), "... are fundamental to keep the eye as a privileged site in the body by regulating the contents of its inner fluids and preserving the internal ocular tissues from variations which occur constantly in the whole circulation" [3].

These barriers exist to prevent toxic substances from crossing from the bloodstream into the neuronal tissue and putting at risk the photoreceptors in the retina and neurons in the brain.

The concept of BRB and their inner and outer components were described for the first time by Cunha-Vaz et al. in 1966 [4]. Additionally, the major role of tight junctions in the BRB was demonstrated for the first time by Shakib and Cunha-Vaz [5], a finding later confirmed to apply also to the blood–brain barrier (BBB) [6].

8.2 Blood–Retinal Barrier

The BRB has two components, and they separate retinal nerve cells from the blood. The inner blood–retinal barrier (iBRB) is established by the tight junctions between the endothelial cells of the retinal capillaries, while the outer blood–retinal barrier (oBRB) is established by the tight junctions of the retinal pigment epithelium (RPE) cells [3].

The BRB regulates ion, protein, and water flux into the retina and thus allows the neural retina to establish and maintain its proper neuronal function. Moreover, it regulates the infiltration of blood-borne toxins and several hormones that could negatively affect neuronal function and survival [7].

At about the same time the blood-retinal barrier was discovered, fluorescein angiography (FA) was introduced [8]. It soon became the most important functional imaging modality in the field of ophthalmology, and it has held that position for decades. Even now, fluorescein angiography has been surpassed only in numbers by color fundus photography, a noninvasive technique.

8.3 Clinical Evaluation of the Blood–Retinal Barrier

Fluorescein angiography was introduced in 1961 by Novotny and Alvis [8], and it has contributed considerably to the current knowledge of diabetic retinopathy. In addition, fluorescein angiography has allowed the regions of perfusion and ischemia in the retina to be identified.

Fluorescein angiography is based on emitted fluorescence from intravenously administered sodium fluorescein molecules (NaFl, $C_{20}H_{10}O_5Na_2$), a fluorescent dye.

Two fluorophores coexist in the eye fundus, being fluorescein angiography and fundus autofluorescence based on the exogenous and endogenous fluorophores, respectively.

After administration, 80% of sodium fluorescein binds to albumin. The 10–20% that does not bind to protein remains free and diffuses through the choriocapillary network, Bruch's membrane, optic nerve, and sclera [9].

In general, fluorescein angiography cannot distinguish leakage from the iBRB and oBRB components [10]. Still, the technique allows for the qualitative assessment of BRB function, and it has been the only method suited for this task.

Because many factors can influence the outcome of fluorescein angiography, a method to quantitatively measure the status of the BRB is a major need.

8.4 Quantitative Evaluation of the Blood–Retinal Barrier

According to Waltman and Kaufman [11], the initial work to quantify fluorescence dates back to the research of David Maurice in 1963, although it was restricted to measuring fluorescence in the cornea and anterior chamber.

Work by Zeimer et al. [12, 13] led to the introduction of vitreous fluorometry through the development of the Coherent Fluorotron Master (CFM—OcuMetrics Inc., Mountain View, CA, USA), which is still available commercially.

In brief, vitreous fluorometry allows the measurement of fluorescence, the fluorescein concentration, across the entire eye from the cornea to the choroid along the optical path while patients focus on a fixation target. A single A-scan reading is obtained to convey information on the BRB of the whole eye.

Currently, this technique is used mainly for imaging in research of small animals, serving as an important technique for the study of disease models.

Although it represented a leap forward from fluorescein angiography, vitreous fluorometry did not allow the location of BRB changes to be identified within the eye fundus. Moreover, because only a central A-scan fluorescence profile was produced, only major changes could be identified.

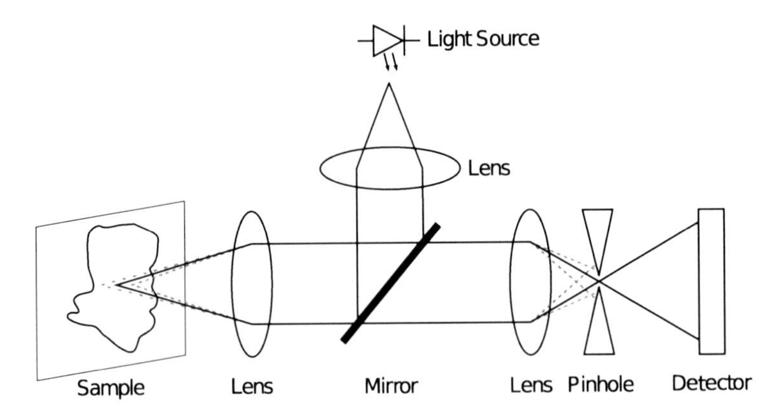

Fig. 8.1 A set of lenses (*left*) allows focusing light in the sample, and the same set allows collecting reflected and emitted fluorescence light. An additional set of lenses (*right*) allows focusing collected light into the detector, in front of which a pinhole allows blocking out-of-focus light

As a result, the development of a technique that followed the same principles and aimed to identify the location within the eye fundus of BRB changes became the next sequence in the CFM system.

8.4.1 Confocal Scanning Laser Ophthalmoscopy

The confocal principle used in the confocal scanning laser ophthalmoscope is based on the method patented by Marvin Minsky (USA Patent 3,013,467, 1961) that allows light from regions anterior and posterior to a selected (confocal) plane to be rejected.

The typical scheme for the apparatus is shown in Fig. 8.1. When used for imaging the ocular fundus, an extra set of lenses (eye lenses) are placed in front of the sample.

A laser beam allows for a point source of light to be focused on one spot of the eye fundus. The reflected light (and the fluorescence light for the current application) is imaged by objective lenses that focus it on a pinhole (in a plane optically conjugated to the focal plane) in front of a detector. Only light passing through this pinhole reaches the detector and is quantified, while defocused light, from both anterior and posterior planes, is rejected [14].

The scanning principle is based on the work of Webb and Hughes [15] who integrated both principles in the first confocal scanning laser ophthalmoscope system [16].

While in the traditional fluorescein angiography all pixels are imaged simultaneously, in confocal scanning laser ophthalmoscopy (CSLO), a single pixel/voxel is imaged at a time. In this way, the system's lateral resolution is increased, and the laser power at the retina is reduced.

8.5 Retinal Leakage Analyzer

The retinal leakage analyzer (RLA) is an imaging technique that can compute fluorescein leakage from the blood circulation into the human retina and vitreous in vivo; thus, it can provide information on the status of BRB function.

In this technique, sodium fluorescein is intravenously administered to patients, then their eyes are scanned through a confocal scanning laser ophthalmoscope to gather the three-dimensional distribution of fluorescein across the retina and vitreous.

The initial RLA system resorted to a prototype of the confocal scanning laser ophthalmoscope from Carl Zeiss (Carl Zeiss, Jena, Germany), which was modified in-house for a tailored scanning protocol.

This work led to several clinical findings in relation to diabetic retinopathy and the status of BRB function [17–19].

Following the same underlying principles, and making use of a later introduced instrumentation, the Heidelberg Retina Angiograph (HRA, Heidelberg Engineering, Heidelberg, Germany), a newer version of the RLA technique, was presented [20].

This new confocal scanning laser ophthalmoscope (HRA) was modified to increase confocality by reducing the pinhole size and to automatically set the initial confocal plane on the start of acquisition to cover the vitreous, the retina, and the choroid.

Advantages of this over the prior system included the increased confocal resolution and the requirement of a single ocular scan.

In this system, each fluorescence profile undergoes a deconvolution process through the system point-spread function (PSF) to compute the real fluorescence distribution and thus that of sodium fluorescein concentration.

A set of 32 confocal planes along a 7-mm depth were gathered, each with an image resolution of 256 × 256 pixels, which covered the central 20° field of view of the human macula.

In this system, one-third of the confocal planes covered the retinal and choroidal regions, while the remaining two-thirds were within the vitreous.

The system PSF presents a Lorentzian profile:

$$y(z) = \frac{2Aw}{\pi\left[w^2 + 4(z - z_c)^2\right]}, \tag{8.1}$$

with w the full width at half maximum (FWHM), A the area under the curve, and z_c the abscissa where the maximum $(2A/w\pi)$ occurs.

The set of confocal plane images are brought into alignment through an image coregistration process to correct for saccades. After correcting for the whole volume, it was now possible to compute the fluorescence intensity profile for each location within the eye fundus.

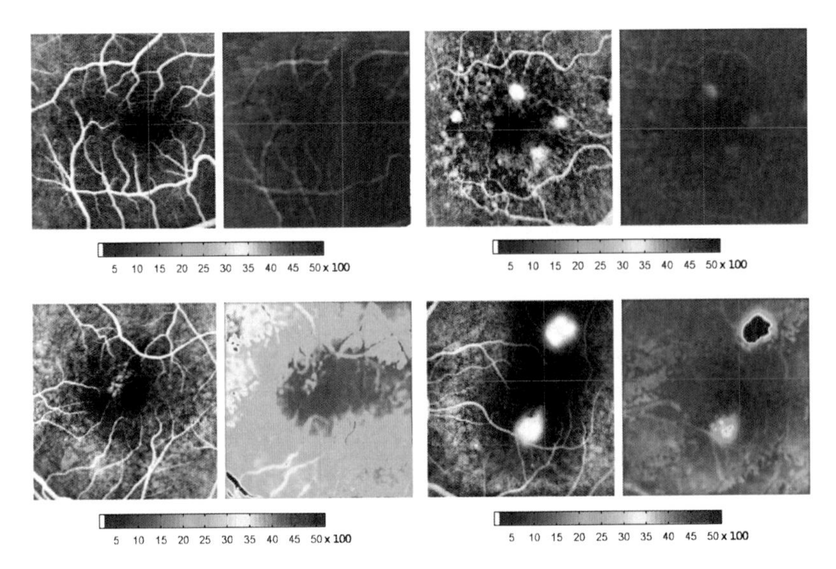

Fig. 8.2 Leakage maps (in leakage units—LU) shown on a false color code and respective fundus reference. *Top row*: two cases with intact blood–retinal barrier, a healthy volunteer's eye (on the *left*), and an eye with a small drusen (on the *right*). *Bottom row* (*left*): eye with nonproliferative diabetic retinopathy. *Bottom row* (*right*): eye three days after laser photocoagulation. Adapted from Bernardes, R; et al. [20]

To compute fluorescein leakage into the vitreous, based on a single ocular scan, it was mandatory to solve for the inverse problem (deconvolution) for each of the fluorescence profiles.

From the deconvolution process, the distribution of fluorescence, i.e., fluorescein concentration, within the vitreous becomes known and given by $C_v(x, y, z)$.

The amount of leakage within the vitreous in front of pixel with coordinates (x, y) in the eye fundus is

$$\mathrm{LU}(x, y) = \int_{z_0}^{z_1} C_v(x, y, z)\mathrm{d}z, \tag{8.2}$$

with $C_v(x, y, z)$ the fluorescein concentration and LU stands for leakage units. Leakage maps are shown in Fig. 8.2. Details can be found in [20].

Following this procedure, a healthy reference population allowed for the threshold of leakage to be computed as the mean plus two standard deviations ($m + 2\mathrm{SD}$).

Any amount of leakage (*LU*) above this threshold was therefore considered as abnormal, indicating increased BRB permeability to fluorescein, i.e., a disrupted BRB.

Both fluorescein angiography and RLA, with special emphasis on the latter and its association with complementary techniques, allowed for a better insight into changes occurring in the diabetic retina. Moreover, the RLA allowed for monitoring, for establishing patterns of progression [19] and recently to the proposed pattern of progression to DME that necessitates laser photocoagulation [21].

Nevertheless, the basic requirement since the introduction of fluorescein angiography has remained the same, i.e., the need for the intravenous administration of sodium fluorescein as a dye.

This mandatory step requires fluorescein angiography and RLA techniques to be performed in the presence of a medical staff, necessitating dedicated instrumentation and skilled personnel. Moreover, adverse reactions and death may still occur [22], contrary to the current trend of using minimally or noninvasive medical imaging techniques.

8.6 Noninvasive Evaluation of Blood–Retinal Barrier with OCT

The rationale for the use of the RLA, and in part for the use of fluorescein angiography, is that substances (sodium fluorescein molecules for these cases) cross the BRB to enter the retina and vitreous. These techniques take advantage of this mechanism and the fluorescence properties of sodium fluorescein to determine the dye distribution after its administration. Nevertheless, other substances in the blood circulation will also cross the BRB, as a result of its loss of integrity, to modify the contents of the retina from its healthy status.

Consequently, the retinal contents are modified and so are its optical properties. This information should be reflected in OCT data from the ocular fundus, as this technique has proved to be very sensitive to optical changes. Hence, once inside the retina, substances present in the blood circulation and able to cross the BRB act as an OCT dye.

8.6.1 Data of Healthy Volunteers vs. Patient Data

Following the rationale noted, our initial approach consisted of looking for differences in OCT data (values) distribution when comparing eyes of healthy volunteers and eyes of patients with diabetic retinopathy, AMD, or cystoid macular edema. All of the OCT data obtained between the inner limiting membrane (ILM) and the RPE were evaluated for differences in the respective distribution profiles. These profiles were scaled and aligned, so only differences in the shape of the distribution were considered.

This initial step aimed to demonstrate potential differences, with special emphasis on those between eyes from healthy volunteers and eyes from the groups with diabetic retinopathy; the eyes of the groups with AMD or cystoid macular edema acted as controls (Fig. 8.3).

The differences confirmed that potential information was embedded in OCT data regarding the status of the retina [23].

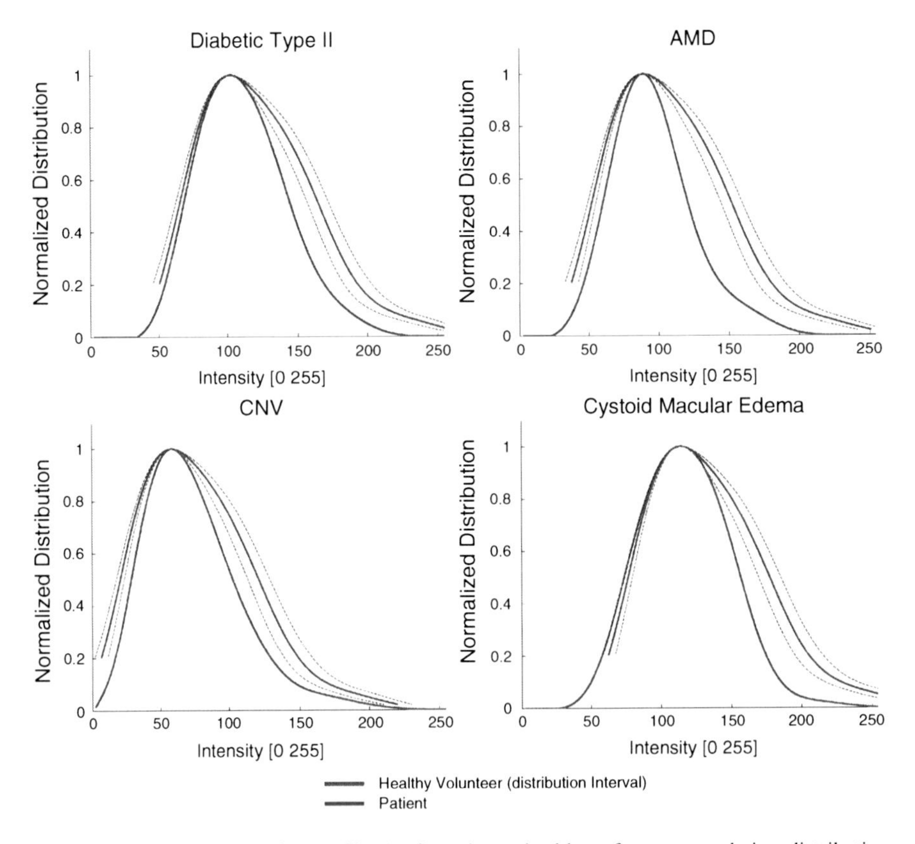

Fig. 8.3 Intensity distribution profiles (*red*) against a healthy reference population distribution profiles ($m \pm \mathrm{SD}$) (blue) for diabetes type 2 (*top-left*), age-related macular degeneration (*top-right*), choroidal neo- vascularization (*bottom-left*) and cystoid macular edema (*bottom-right*). Adapted from Bernardes, R; et al. [23]

Even though this proof of concept established the difference between eyes from a group of healthy volunteers and those from a group of patients, it did not exclude the possibility of these differences being due to changes in the acquisition, e.g., pupil size.

8.6.2 Differences Within the Same Eye and Scan

To assure that the differences found were solely due to changes in retinal tissue and not to the acquisition process, it was mandatory to demonstrate it within the same eye and scan. Consequently, patients underwent assessment with the RLA to identify regions of intact and regions of disrupted BRB within the same eye and scan [24].

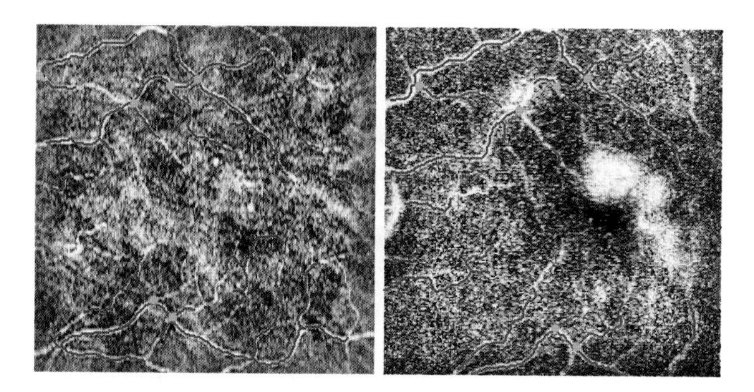

Fig. 8.4 Ocular fundus references from the same eye for the optical coherence tomogram (*left*) and retinal leakage analyzer image (*right*). Automatically computed vessel centerlines (*blue*) and bifurcations/crossover (in *green dots*). Adapted from Bernardes, R; et al. [23]

Because the RLA allows an ocular fundus reference to be computed, in addition to the leakage map, a map of intact or disrupted BRB status could be coregistered to any ocular fundus reference from the same eye, e.g., the one from OCT.

Resorting to an image coregistration process, fundus references from the OCT and RLA techniques were brought into alignment, thereby allowing OCT data to be locally correlated with the intact or disrupted status of the BRB.

For each fundus reference, the retinal vascular network was computed, and characteristic fiducial markers, vessels bifurcations and crossovers, were determined (Fig. 8.4).

By identifying the same fiducial marker on both ocular fundus references, and thus their respective image coordinates, it became possible to compute a transformation matrix, allowing both images to be brought into coregistration.

These steps allowed for each OCT A-scan to be classified into either the intact or the disrupted BRB type.

Eyes of patients with diabetes presenting regions of intact and regions of disrupted BRB within the same eye scan (through the RLA) were further analyzed.

OCT data distributions were therefore considered according to the BRB status classification, thus allowing for a comparison of (1) intact-to-intact, (2) disrupted-to-disrupted, and (3) intact-to-disrupted regions.

Two distinct types of comparisons could be performed, comparisons between (A) regions receiving the same classification (intact-to-intact and disrupted-to-disrupted) and comparisons between (B) regions receiving distinct classifications, i.e., intact-to-disrupted BRB status (Fig. 8.5).

In this way, sets of 5,000 A-scans receiving the same classification were considered. Each set allowed the computing of the optical coherence data distribution that was representative of the respective classification region (again considering only OCT data in between the ILM and the RPE).

From the comparison between these sets, it was found that sets from regions receiving the same classification were more similar to each other than those from regions receiving distinct classifications.

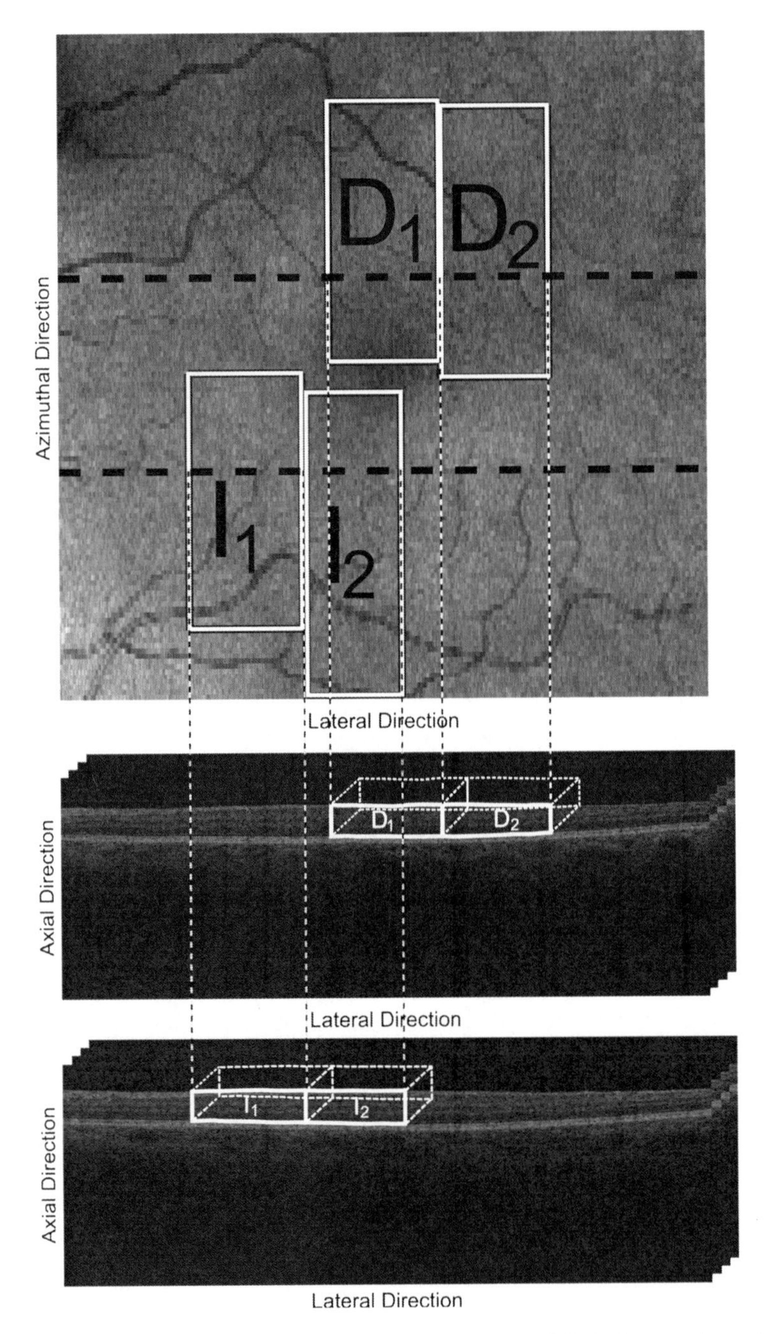

Fig. 8.5 Optical coherence tomography fundus reference with delimited areas (intact blood–retinal barrier: I; disrupted blood–retinal barrier: D). No significant differences can be seen in either the fundus reference or B-scans passing through intact and disrupted areas. Adapted from Bernardes, R; et al. [24]

Table 8.1 Summary of differences by type

Patient ID	Sum of squared differences ($\times 10^{-7}$)	
	$\mathrm{Diff}_{similar}$	$\mathrm{Diff}_{distinct}$
DR1	63.92 ± 53.98(21)	1,525.52 ± 87.57(7)
DR2	22.85 ± 31.57(22)	156.93 ± 37.02(14)
AMD1	19.69 ± 24.02(25)	508.15 ± 90.49(30)
AMD2	4.56 ± 3.32(9)	207.52 ± 33.50(12)

Reprinted with permission [24]

$\mathrm{Diff}_{similar}$ indicates between two areas of the same region type; $\mathrm{Diff}_{distinct}$, between two areas of distinct region types. The lower the SSD, the higher the similarity between profiles. Figures are mean ± SD (number of comparisons)

Table 8.1 summarizes these differences, using the sum of the squared differences (SSD) between respective distribution profiles.

These findings are in agreement with the rationale, suggesting that the presence of information on the BRB integrity is embedded within OCT data of the human retina.

Relative to the former approach, the comparison between healthy and diseased eyes, this method has the advantage of using the same eye and OCT scan, therefore removing potential bias due to differences in the acquisition process.

8.6.2.1 Differences by Retinal Layer

Another step beyond the approach described could now be performed. Because the retinal vascular network exists in the top layers of the human retina, any change in the retinal content due to the breakdown of the iBRB should be found closer to this vascular network.

To assess this potential link, changes in the retina content, and thus to differences in OCT and the vascular network, the retina was evenly split (depthwise) into seven layers.

For each of these layers, the cited procedure was repeated to compute differences in the distribution of OCT data between regions of similar and distinct region types.

Although the structural arrangement of the human retina was not taken into consideration when splitting it into different layers, the procedure allowed the retinal vascular network to be present within the top three layers, i.e., the ones closer to the ILM.

Table 8.2 summarizes these differences by retinal layer, using the sum of the squared differences (SSD) between respective distribution profiles.

From the sum of squared differences by retinal layer, it became clear that the SSD values were higher when comparing distinct region types ($\mathrm{Diff}_{distinct}$) relative to similar region types ($\mathrm{Diff}_{similar}$).

In addition, the higher SSD values were found predominantly in the top layers, therefore pointing to the potential link between these differences and the iBRB status, hence confirming the rationale.

Table 8.2 Summary of differences by layer and region type

Patient ID	Sum of squared differences ($\times 10^{-7}$)		
	Layer	$Diff_{similar}$	$Diff_{distinct}$
DR1	1	27.13 ± 25.02(21)	4,007.98 ± 264.23(7)
	2	64.70 ± 54.06(21)	4,878.04 ± 315.35(7)
	3	41.87 ± 30.71(21)	540.39 ± 103.53(7)
	4	54.72 ± 52.54(21)	2,188.84 ± 125.47(7)
	5	64.86 ± 56.27(21)	585.96 ± 144.36(7)
	6	98.33 ± 99.34(21)	883.67 ± 147.96(7)
	7	87.59 ± 83.70(21)	332.20 ± 64.54(7)
DR2	1	43.07 ± 37.35(22)	1,199.56 ± 253.91(14)
	2	34.04 ± 27.28(22)	2,040.31 ± 215.44(14)
	3	17.55 ± 11.18(22)	398.53 ± 81.62(14)
	4	35.92 ± 34.07(22)	269.89 ± 71.61(14)
	5	41.30 ± 34.12(22)	142.18 ± 43.65(14)
	6	45.72 ± 36.81(22)	478.97 ± 67.94(14)
	7	9.33 ± 5.79(22)	162.36 ± 29.17(14)
AMD1	1	26.10 ± 20.79(25)	701.18 ± 199.02(30)
	2	29.24 ± 21.41(25)	2,079.88 ± 268.88(30)
	3	21.13 ± 23.78(25)	24.23 ± 19.63(30)
	4	44.15 ± 38.46(25)	151.34 ± 75.59(30)
	5	26.00 ± 26.65(25)	81.56 ± 29.24(30)
	6	34.04 ± 39.66(25)	121.55 ± 47.64(30)
	7	16.52 ± 10.30(25)	34.30 ± 17.18(30)
AMD2	1	38.25 ± 37.81(9)	483.69 ± 168.13(12)
	2	33.15 ± 34.18(9)	465.69 ± 104.94(12)
	3	8.03 ± 5.93(9)	851.30 ± 61.05(12)
	4	13.80 ± 7.40(9)	154.18 ± 37.02(12)
	5	66.41 ± 62.00(9)	59.51 ± 37.88(12)
	6	38.62 ± 47.22(9)	79.73 ± 61.09(12)
	7	17.91 ± 9.35(9)	164.79 ± 39.71(12)

Reprinted with permission [24]

$Diff_{similar}$ indicates between two areas of the same region type; $Diff_{distinct}$, between two areas of distinct region types. The lower the SSD, the higher the similarity between profiles. Figures are mean ± SD (number of comparisons)

8.6.3 *Automatic Classification*

The cited findings suggest that it is possible to discriminate between healthy eyes and eyes with diseased retinas based solely on OCT. To do so automatically, fundamental characteristics should be computed from OCT data.

Due to the high dynamic range of OCT data, they are typically presented on the logarithmic scale. Hence, optical coherence data undergo a logarithmic compression before being saved and displayed, followed by a linear mapping into the dynamic range of the system in use, typically for the range [0...255] for one-byte systems, such as the one used in this work, the Cirrus HD-OCT.

In this way, OCT data can be analyzed in the logarithmic space or in the linear space, after the proper mapping, or in both spaces.

According to the work of Grzywacz et al. [25], these authors stated that the best distribution model for the OCT data from the human retina, in the linear space, is a stretched exponential distribution:

$$p_s(I) = k(\lambda, \beta)\mathrm{e}^{-\left(\frac{I}{\lambda}\right)^{\beta}}, \tag{8.3}$$

where I is the intensity (OCT data) and

$$k(\lambda, \beta) = \frac{1}{\lambda \Gamma\left(\frac{\beta+1}{\beta}\right)}, \tag{8.4}$$

with Γ the gamma function.

On the other hand, when dealing with OCT data in the logarithmic space, it is better described by a Gaussian function,

$$g(I) = G \exp\left(-\frac{(x-\mu)^2}{2\sigma^2}\right), \tag{8.5}$$

where G, μ, and σ are, respectively, the maximum of $g(I)$, the average, and standard deviation of the I values.

To make use of the OCT data in the linear space, it is necessary to map it accordingly [26]. In this way, each voxel value is mapped to the linear space through:

$$I_{\mathrm{LINEAR}} = K(10^{I_{\mathrm{LOG}}} - 1), \tag{8.6}$$

where I_{LINEAR} and $I_{\mathrm{LOG}} (I_{\mathrm{LOG}} \in [0, N-1], N = 256)$ are, respectively, the voxel value in the linear and logarithmic spaces, and K is a scaling factor.

Moreover, because $10^{I_{\mathrm{LOG}}}$ in (8.6) will become a quite large number, (8.7) was used instead of (8.6):

$$I_{\mathrm{LINEAR}} = 10^{\log(L)\frac{I_{\mathrm{LOG}} - \langle I_{\mathrm{LOG}} \rangle + N/2}{N-1}}, \tag{8.7}$$

with $I_{\mathrm{LOG}} \in [0, N-1]$, $N = 256$, $\langle I_{\mathrm{LOG}} \rangle$ the average of I_{LOG}, and L the nominal maximum of I_{LINEAR}.

In this way, any differences in the computed parameters from the stretched exponential distribution are solely due to the respective shape.

Consequently, a modified expression for the stretched exponential distribution in the linear space was used as defined by

$$p_s(I_{\mathrm{LINEAR}}) = k\left(\alpha, \lambda, \beta\right) \exp\left[-\left(\frac{I_{\mathrm{LINEAR}} - x_0}{\lambda}\right)^{\beta}\right], \tag{8.8}$$

where

$$k(\alpha, \lambda, \beta) = \frac{\alpha}{\lambda \Gamma\left(\frac{\beta+1}{\beta}\right)}, \tag{8.9}$$

and I_{LINEAR} as in (8.7).

From fits of the given distributions to data in both the logarithmic and linear spaces, a set of parameters could be computed, describing each ocular scan.

8.6.3.1 Supervised Classification

By using an automatic classification process to discriminate between healthy and diseased eyes solely on OCT data, specifically from the parameters computed from both the linear and logarithmic spaces, as defined above, it will demonstrate the embedded information on OCT data from the healthy status of the human retina.

For this, we used the support vector machines (SVM) classification. This family of learning algorithms was used to place objects into classes, in which the input consists of a training set (i.e., objects and the respective classes) and the output was a classifier allowing the class of any new object to be predicted.

The objects were vectors of features characterizing each of the OCT data. The respective class was the respective retina, e.g., as healthy, with DR or with cystoid macular edema.

In relation to other classification methods, e.g., Bayesian, Fisher linear discriminant, neural networks, or decision trees, the support vector machine classifier presented a good performance, was computationally efficient, and was robust in high dimensions, in addition to its strong theoretical foundations. Therefore, it became the natural option for this task.

In brief, we aim to

$$\begin{aligned} &\text{maximize } W(\alpha) = \textstyle\sum_{i=1}^{L} \alpha_i - \frac{1}{2} \sum_{i,j=1}^{L} y_i y_j \alpha_i \alpha_j K\left(x_i, x_j\right). \\ &\text{subject to } \begin{cases} \sum_{i,j=1}^{L} \quad y_i \alpha_i = 0 \\ C \geq \alpha_i \geq 0, \quad i = 1, \ldots, L \end{cases} \end{aligned} \tag{8.10}$$

For $K(x_i, x_j)$, we have used the radial basis function (RBF) kernel:

$$K\left(x, x'\right) = \exp\left(-\frac{\|x - x'\|^2}{\sigma^2}\right), \tag{8.11}$$

where x is the vector of features characterizing each OCT ocular scan. For details, see [27, 28].

Publicly available software, LIBSVM (Chang and Lin [29]), was used. All features were scaled to the range [0, 1] or [−1, 1], as appropriate, and the RBF kernel was applied.

Table 8.3 Classification performance

	Healthy	Diabetic	DME
Healthy ($n = 31$)	20 (64.5%)	5 (16.1%)	6 (19.4%)
Diabetic ($n = 31$)	4 (12.9%)	23 (74.2%)	4 (12.9%)
DME ($n = 31$)	7 (22.6%)	5 (16.1%)	19 (61.3%)

Reprinted with permission [30]
Classification performance followed the leave-one-out validation, with 62 of 93 eyes (66.7%) correctly classified. DME indicates eyes with diabetic macular edema.

Table 8.4 Healthy Eyes Classification Performance

	Age <46 years; avg (std)	Age ≥46 years; avg (std)	N; avg (std)
Age <46 years	16 (76.19%); 30.94 (6.85)	5 (23.81%); 36.07 (5.24)	21; 2.16 (6.76)
Age ≥46 years	7 (36.84%); 53.69 (4.93)	12 (63.16%); 62.47 (9.30)	19; 59.24 (8.94)

Reprinted with permission [30]
avg indicates average age (years); *std*, standard deviation (years).

8.6.3.2 Classification in Healthy, Diabetic, and Diabetic Macular Edema Groups

Using this initial approach, eyes from a group of healthy volunteers ($n = 31$), patients with diabetes ($n = 31$), and a group diagnosed with DME ($n = 31$) were classified according to the SVM classification process [26]. The results are shown in Table 8.3 [30].

The level of correct classification that was achieved (over 66% overall) suggested again that information on the healthy condition of the human retina is embedded within the OCT data, as gathered from the ocular fundus of the human eye. In this classification, a total of 43 distinct parameters were used to characterize each OCT ocular scan [26].

8.6.3.3 Aging

An additional set of features were recently considered, raising the number of features to 51 from the previously cited 43.

To assess the dependency of the classification with age, two groups of healthy volunteers and two groups of patients with diabetes were classified according to their respective age group.

Tables 8.4 and 8.5 [30] summarize the results achieved, discriminating between the age groups of both the healthy volunteers and those with diabetes, respectively, following the leave-one-out approach.

Of fundamental importance was the demonstration that for both groups, healthy volunteers and patients, there was a dependence on age, with more than 63% of the cases correctly classified to the respective age group.

Table 8.5 Diabetic Patients' Eyes Classification Performance

	Age <63.5 years; age: avg (std)	Age ≥65.3 years; age: avg (std)	*N*; age: avg (std)
Age <63.5 years	53 (71.62%); 55.24 (5.94)	21 (28.38%); 58.54 (5.09)	74; 56.17 (5.87)
Age ≥63.5 years	21 (29.17%); 68.56 (2.72)	51 (70.83%); 68.91 (3.75)	72; 68.80 (3.47)

Reprinted with permission [30]
avg indicates average age (years); *std*, standard deviation (years).

Table 8.6 Eyes Classification Performance

	Healthy; age: avg (std)	Diabetic; age: avg (std)	*N*; age: avg (std)
Healthy	22 (64.71%); 53.95 (11.18)	12 (35.29%); 56.63 (13.52)	34; 54.90 (11.93)
Diabetic	10 (29.41%); 57.83 (8.80)	24 (70.59%); 54.74 (8.67)	34; 55.65 (8.69)

Reprinted with permission [30]
avg indicates average age (years); *std*, standard deviation (years).

8.6.3.4 Discrimination of Healthy Eyes from Those with Diabetes in Age-Matched Populations

Finally, eyes from an age-matched population of patients with diabetes were compared to those from healthy volunteers to assess the potential discrimination between these groups, thus removing any potential age effect.

Table 8.6 summarizes the discrimination between eyes of the age-matched healthy volunteers and those of the patients with diabetes [30].

These results confirmed the ones previously found, suggesting the presence of two distinct levels of information on the statistics of OCT data from the human retina. On the one hand, there appeared to exist information on the retinal tissue age, both on healthy and diseased (diabetes) patients' eyes, and, on the other hand, there appeared to exist information specific to the disease process, enabling discrimination between groups of age-matched healthy volunteers and patients with diabetes.

8.7 Conclusions

At present, the findings based on OCT data suggest that information on the status of the BRB and on the aging process of the retina itself is embedded in OCT data.

A series of successive approaches have been taken in which each one confirmed the previous one and added further information.

The initial approach, comparing the data distribution between eyes of healthy volunteers and those of patients, showed the first measurable difference in the statistics of OCT data.

The next step assessed the same differences within the same eye and scan. Moreover, it showed that differences by retinal layer appeared to be linked to the retinal vascular network distribution within the retinal tissue (depthwise).

An automatic classification system then was implemented and tested to demonstrate the possibility of discriminating between healthy eyes and those with diabetes, even though no signs of diabetic retinopathy could be found in the ocular fundus.

Finally, it was demonstrated that it was possible to discriminate between age groups, suggesting the presence of changes associated to the aging of neuronal tissue, both in eyes of healthy volunteers and in those of patients.

The potential impact in using the eye as a window to the brain is highly important and requires further (ongoing) detailed studies.

Acknowledgments The authors would like to thank Dr. Melissa Horne and Carl Zeiss Meditec (Dublin, CA, USA) for their support on obtaining access to OCT data and AIBILI - The Centre for Clinical Trials (CEC) technicians for their support in managing data, working with patients, and performing scans. This study is supported in part by the Fundação para a Ciência e a Tecnologia (FCT) under the research project PTDC/SAU-BEB/103151/2008 and program COMPETE (FCOMP-01-0124-FEDER-010930).

References

1. Chronic Disease and Medical Innovation in an Aging Nation, (2011) The silver bBook: vision loss, http://www.silverbook.org/VisionLossSilverbook.pdf. Accessed 20 Oct 2011
2. E. Chen, M. Looman, M. Laouri, M. Gallagher, K. Nuys, D. Lakdawalla, J. Fortuny, Burden of illness of diabetic macular edema: literature review. Curr. Med. Res. Opin. **26**, 1587–1597 (2010)
3. J. Cunha-Vaz, Blood–retinal barrier, in *Encyclopedia of the Eye*, vol. 1, ed. by D. Dartt (Academic, Oxford, 2010), pp. 209–215
4. J. Cunha-Vaz, M. Shakib, N. Ashton, Studies on the permeability of the blood–retinal barrier. I. On the existence, development, and site of blood–retinal barrier. Br. J. Ophthalmol. **50**, 411–453 (1966)
5. M. Shakib, J. Cunha-Vaz, Studies on the permeability of the blood–retinal barrier. IV. Junctional complexes of the retinal vessels and their role in the permeability of the blood–retinal barrier. Exp. Eye Res. **5**, 229–234 (1966)
6. M. Brightman, T. Reese, Junctions between intimately apposed cell membranes in the vertebrate brain. J. Cell Biol. **40**, 648–677 (1968)
7. A. Joussen, T. Gardner, B. Kirchhof, S. Ryan, *Retinal Vascular Disease* (Springer, Heidelberg, 2007)
8. H. Novotny, D. Alvis, A method of photographing fluorescence in circulating blood in the human retina. Circulation **24**, 82–86 (1961). doi: 10.1007/s001090000086
9. J. Cunha-Vaz, Retinopatia Diabética, in *Sociedade Española de Oftalmologia* (editorial MAC LINE, S.L., Madrid, 2006). ISBN: 84-89085-32-3
10. J. van Best, M. Mota, M. Larsen, in *Manual of Ocular Fluorometry: Protocols Approved Within the Framework of a Concerted Action of the European Community Biomedical Programme on Ocular Fluorometry (1989–1992)*, ed. by J. Cunha-Vaz, E. Leite, M. Ramos (Coimbra: [s.n.], Portugal, 1993), 116 pp. ISBN 972-95958-0-1
11. S. Waltman, H. Kaufman, A new objective slit lamp fluorophotometer. Invest. Ophthalmol. Vis. Sci. **9**(4), 247–249 (1970)
12. R. Zeimer, N. Blair, J. Cunha-Vaz, Vitreous fluorophotometry for the clinical research. I. Description and evaluation of a new fluorophotometer. Arch. Ophthalmol. **101**, 1753–1756 (1983)

13. R. Zeimer, N. Blair, J. Cunha-Vaz, Vitreous fluorophotometry for clinical research. II. Method of data acquisition and processing. Arch. Ophthalmol. **101**, 1757–1761 (1983)
14. B. Masters, *Confocal Microscopy and Multiphoton Excitation Microscopy—The Genesis of Live Cell Imaging* (SPIE, Washington, DC, 2006)
15. R. Webb, G. Hughes, Scanning laser ophthalmoscope. IEEE Trans. Biomed. Eng. **28**, 488–492 (1981)
16. R. Webb, Optics for laser rasters. Appl. Optics **23**, 3680–3683 (1984)
17. C. Lobo, R. Bernardes, J. Cunha-Vaz, Alterations of the blood-retinal barrier and retinal thickness in preclinical retinopathy in subjects with type 2 diabetes. Arch. Ophthalmol. **118**, 1364–1369 (2000)
18. C. Lobo, R. Bernardes, J. Abreu, J. Cunha-Vaz, One-year follow-up of blood–retinal barrier and retinal thickness alterations in patients with type 2 diabetes mellitus and mild nonproliferative retinopathy. Arch. Ophthalmol. **119**, 1469–1474 (2001)
19. C. Lobo, R. Bernardes, J. Figueira, J. Abreu, J. Cunha-Vaz, Three-year follow-up study of blood–retinal barrier and retinal thickness alterations in patients with type 2 diabetes mellitus and mild nonproliferative diabetic retinopathy. Arch. Ophthalmol. **122**, 211–217 (2004)
20. R. Bernardes, J. Dias, J. Cunha-Vaz, Mapping the human blood–retinal barrier function. IEEE Trans. Biomed. Eng. **52**, 106–116 (2005)
21. S. Nunes, M. Ribeiro, E. Geraldes, R. Bernardes, J. Cunha-Vaz, Risk markers for progression to clinically significant macular edema in mild nonproliferative retinopathy in diabetes type 2. Invest. Ophthalmol. Vis. Sci. **49**, 3511 (2008) (E-Abstract)
22. V. Alfaro, F. Gomez-Ulla, H. Quiroz-Mercado, M. Figueroa, S. Villalba, *Retinopatía Diabética—Tratado médico quirúrgico*, 1st edn. (MAC LINE, S.L., Madrid, 2006)
23. R. Bernardes, T. Santos, J. Cunha-Vaz, Evaluation of blood–retinal barrier function from Fourier domain high-definition optical coherence tomography, in *Proceedings of the IFMBE 25/XI, WC*, vol. **25**, 2009, pp. 316–319
24. R. Bernardes, T. Santos, P. Serranho, C. Lobo, J. Cunha-Vaz, Noninvasive evaluation of retinal leakage using optical coherence tomography. Ophthalmologica **226**(2), 29–36 (2011)
25. N. Grzywacz, J. Juan, C. Ferrone, D. Giannini, D. Huang, G. Koch, V. Russo, O. Tan, C. Bruni, Statistics of optical coherence tomography data from human retina. IEEE Trans. Med. Imaging **29**(6), 1224–1237 (2010)
26. R. Bernardes, Optical coherence tomography: health information embedded on OCT signal statistics, in *33rd Annual International Conference of the IEEE EMBS*, 2011, pp. 6131–6133
27. N. Cristianini, J. Shawe-Taylor, *An Introduction to Support Vector Machines and Other Kernel-Based Learning Methods* (Cambridge University Press, Cambridge, 2000)
28. B. Scholkopf, A. Smola, *Learning with Kernels* (The MIT, Cambridge, 2002)
29. C. Chang, C. Lin, LIBSVM: a library for support vector machines, ACM Trans. Intell. Syst. Technol. **2**(27), 1–27:27 (2011), http://www.csie.ntu.edu.tw/~cjlin/libsvm
30. R. Bernardes, P. Serranho, T. Santos, V. Gonçalves, J. Cunha-Vaz, *Optical coherence tomography: automatic retina classification through support vector machines. Eur. Ophthalmic Rev.* **6**(3) (2012 in press)

Chapter 9
Polarization Sensitivity

U. Schmidt-Erfurth, F. Schlanitz, M. Bolz, C. Vass, J. Lammer, C. Schütze, M. Pircher, E. Götzinger, B. Baumann, and C.K. Hitzenberger

Abstract Optical coherence tomography (OCT) (Huang et al. Science 254(5035): 1178–1181, 1991; Fercher et al. Rep Prog Phys 66:239–303, 2003; Drexler and Fujimoto Prog Retin Eye Res 27(1):45–88, 2008) is a well-established tool for high-resolution cross-sectional imaging of human ocular structures. Despite its great success in improving ocular diagnostic imaging, conventional OCT cannot directly differentiate between different tissues. However, polarization-sensitive (PS) OCT is able to generate tissue-specific contrast that can be further used to segment ocular structures and to obtain quantitative information.

9.1 Polarization-Sensitive Optical Coherence Tomography

Optical coherence tomography (OCT) [1–3] is a well-established tool for high-resolution cross-sectional imaging of human ocular structures. While the original OCT technique was based on a mechanically scanning reference mirror performing A-scans in time domain, spectral domain (SD) OCT [4–6] has caused a paradigm change in OCT technology after the discovery of its huge sensitivity advantage [7–9]. Various applications of SD-OCT to image the human retina and anterior chamber with high speed and high resolution in 2 and 3 dimensions have been successfully demonstrated [10–12].

Despite its great success in improving ocular diagnostic imaging, conventional, intensity-based OCT still has a considerable drawback: It cannot directly differentiate between different tissues. However, light has additional properties that can be measured and used to generate tissue-specific contrast. One of these properties

---

U. Schmidt-Erfurth (✉)
Department of Ophthalmology and Optometry, Medical University Vienna, Währinger Gürtel 18-20, 1090 Vienna, Austria
e-mail: ursula.schmidt-erfurth@meduniwien.ac.at

R. Bernardes and J. Cunha-Vaz (eds.), *Optical Coherence Tomography*, Biological and Medical Physics, Biomedical Engineering,
DOI 10.1007/978-3-642-27410-7_9,

is the light's polarization state. The polarization state can be changed by various light-tissue interactions. These effects are used by polarization-sensitive (PS) OCT [13,14] to generate tissue-specific contrast that can be further used to segment ocular structures, retinal layers, and to obtain quantitative information.

9.1.1 Principles of Polarization-Sensitive OCT

PS-OCT requires a more complex instrumental setup than intensity-based OCT. The sample is typically illuminated with one or more well-defined polarization states, and the detected light is split in two orthogonal polarization states that have to be measured separately, either in parallel or subsequently. This requires additional polarizing elements, polarizing beam splitters, and, in some configurations, phase modulators or frequency shifters. Also, the postprocessing of OCT signals is more demanding than that of intensity-based OCT. With exception of the simplest arrangements, not only the amplitude but also the phase of the interferometric signals has to be recorded or processed. For these reasons, PS-OCT is not yet commercially available, only a few instruments have been developed by specialized research groups in USA, Europe, and Japan.

Different methods of PS-OCT have been reported in literature. While early work measured only reflectivity and retardation [13, 14], newer work extended the measurements to various other parameters like Stokes vectors [15], Müller [16] and Jones matrix [17], birefringent axis orientation [18], and diattenuation [19,20]. It is beyond the scope of this book to provide a comprehensive overview of the various PS-OCT techniques reported in literature. Instead, we restrict the technical description to the instrumentation used for our own work.

Our group at the Center for Medical Physics and Biomedical Engineering of the Medical University of Vienna developed a method that combines the PS low-coherence interferometry setup first devised by Hee et al. [13] with a phase-sensitive recording of the interferometric signals in the two orthogonal polarization channels [18], thus allowing to measure three parameters, reflectivity, retardation, and birefringent axis orientation simultaneously. In addition, Stokes vectors can be retrieved from the measured interferometric signals, providing access to the full polarization information available in the backscattered light beams. From these Stokes vectors, also information on depolarization can be obtained [21]. Originally developed for time domain OCT, the technique was later adapted to SD-OCT, enabling the acquisition of 20,000 A-lines/s with a sensitivity of 98 dB [22].

The basic scheme of our spectral domain PS-OCT instrument is depicted in Fig. 9.1, which is taken from [22]. In that paper, also more detailed information on various technical aspects can be found. Here is a brief description of the system: Light emitted from a super luminescent diode illuminates, after being vertically polarized, a bulk optics Michelson interferometer, where it is split by a nonpolarizing beam splitter (NPBS) into a sample and a reference beam. The reference light passes a quarter wave plate (QWP) oriented at 22.5° and is reflected

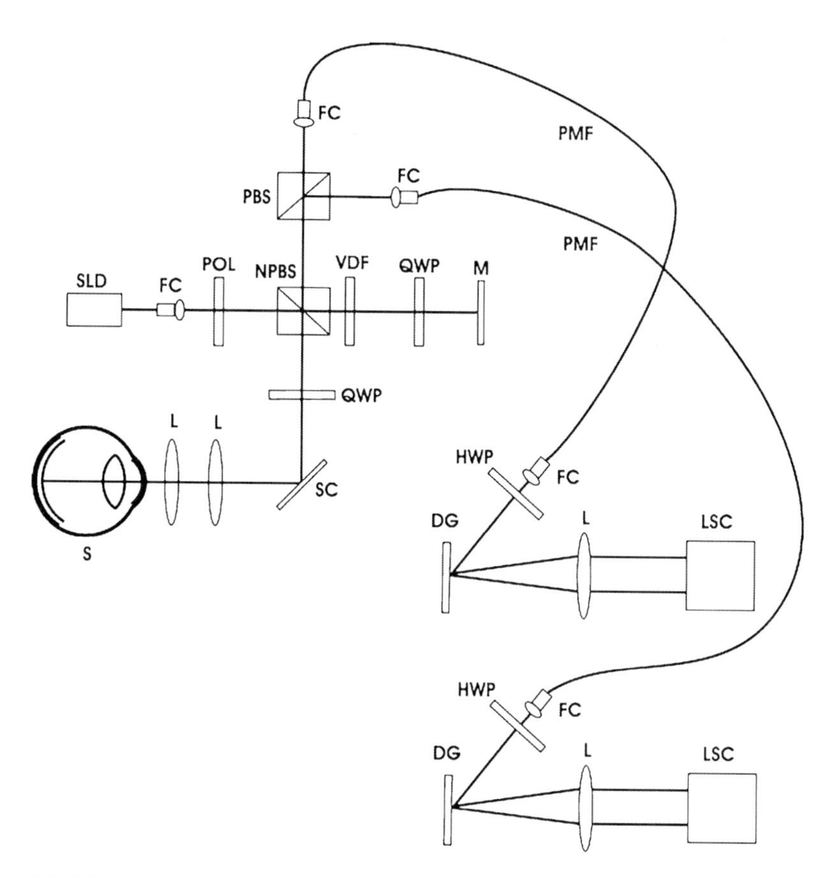

Fig. 9.1 Schematic drawing of spectral domain PS-OCT instrument. *SLD* super luminescent diode; *FC* fiber coupler; *POL* polarizer; *NPBS* nonpolarizing beam splitter; *VDF* variable density filter; *QWP* quarter wave plate; *M* mirror; *SC* galvo scanner; *L* lens; *S* sample; *PMF* polarization maintaining fiber; *HWP* half wave plate; *DG* diffraction grating; *LSC* line scan camera (From ref. [22], by permission of the Optical Society of America)

by a mirror. After double passage of the QWP, the orientation of the polarization plane is at 45° to the horizontal, providing equal reference power in both channels of the polarization-sensitive detection unit. The sample beam passes a QWP oriented at 45°, which provides circularly polarized light to the sample. An x–y galvanometer scanner scans the beam over the retina.

After recombination of the two beams at the NPBS, light is directed toward a polarization-sensitive detection unit, where it is split into orthogonal polarization states by a polarizing beam splitter. The two orthogonally polarized beam components are coupled into two polarization maintaining fibers (PMFs) and directed toward two separate spectrometers which are designed identically, consist of similar components, and have to be aligned with subpixel accuracy with respect to each other. Each spectrometer consists of a grating, a camera lens, and a line scan CCD camera. The maximum line rate of the camera used is 29 kHz, and via camera link and a high-speed frame grabber board, data are transferred continuously to a personal computer.

After standard preprocessing of the data provided by the two spectrometers (fixed pattern noise removal, zero padding, and rescaling of the spectral data from wavelength to wavenumber space), complex A-scan data containing amplitude and phase information as a function of depth are available for both, the horizontal and the vertical polarization state. In case of retinal imaging, these data are then corrected for the anterior chamber birefringence by a numerical method [23].

From the corrected horizontal and vertical channel, the following parameters can be calculated: Polarization independent reflectivity R, retardation δ, and birefringent axis orientation θ [18, 22]:

$$R \propto A_{\mathrm{H}}^2 + A_{\mathrm{V}}^2, \tag{9.1}$$

$$\delta = \arctan\left[\frac{A_{\mathrm{V}}}{A_{\mathrm{H}}}\right], \tag{9.2}$$

$$\theta = \frac{180^\circ - \Delta\Phi}{2}, \tag{9.3}$$

with A being the amplitude and Φ the phase of the respective channel and the indices H and V denoting the horizontal and the vertical polarization channel, respectively. $\Delta\Phi = \Phi_{\mathrm{V}} - \Phi_{\mathrm{H}}$ is the phase difference between the two channels. The unambiguous measurement ranges are 90° for δ and 180° for θ.

In addition to reflectivity and birefringence-related parameters, also the Stokes vector of the backscattered light beam can be calculated as a function of spatial position [21]:

$$S = \begin{pmatrix} I \\ Q \\ U \\ V \end{pmatrix} = \begin{pmatrix} A_{\mathrm{H}}^2 + A_{\mathrm{V}}^2 \\ A_{\mathrm{H}}^2 - A_{\mathrm{V}}^2 \\ 2A_{\mathrm{H}}A_{\mathrm{V}}\cos\Delta\varphi \\ 2A_{\mathrm{H}}A_{\mathrm{V}}\sin\Delta\varphi \end{pmatrix}, \tag{9.4}$$

where I, Q, U, and V denote the four Stokes vector elements. The Stokes vector carries the full information on the light's polarization state. Its elements will be used in Sect. 9.1.3 to derive information on depolarization.

Initial imaging of the retina of healthy subjects and patients with various retinal diseases was carried out with the instrument described above. Later, a modified and improved SD PS-OCT instrument was built [24]. This instrument has an improved patient interface, incorporating a scanning laser ophthalmoscope (SLO) channel for improved patient alignment, a larger scan angle (~20° × 20°), and various raster scan patterns (64 × 1,024, 128 × 512, 256 × 256 pixels). Further system parameters of this instrument are the following: center wavelength ~840 nm; wavelength bandwidth 58 nm (FWHM); axial resolution ~4 μm (in tissue); depth range 3.3 mm; illumination power 700 μW; A-scan rate 20 kHz; and max. sensitivity 97.5 dB. The majority of retinal SD PS-OCT images presented in this chapter were recorded with this instrument.

9.1.2 *Polarization Properties of Ocular Structures*

There are three interaction mechanisms between light and tissue that can change the polarization state of backscattered light and can be detected by PS-OCT: birefringence, diattenuation, and depolarization (or polarization scrambling).

Birefringence introduces a phase difference between light beams of orthogonal polarization states. The amount of this phase retardation δ is proportional to the refractive index difference Δn encountered by the two orthogonal polarization states and to the thickness of the birefringent layer the beams have passed. Two different types of birefringence can be distinguished: intrinsic birefringence, commonly found in anisotropic crystals, and form birefringence caused by structural anisotropy of otherwise optically isotropic material. Birefringence of tissue is, in most cases, of the latter type. This form birefringence is typically found in fibrous tissues where oriented fibers are surrounded by a matrix of different refractive index. It is found in several ocular tissues: in the cornea (caused by the collagen fibers of the stroma) [25, 26], sclera [27, 28], trabecular meshwork [28], ocular muscles and tendons [27], retinal nerve fiber layer (RNFL) [29, 30], Henle's fiber layer [31, 32], and in scar tissue [33].

Figure 9.2 shows an example of birefringence imaging by PS-OCT in a healthy human retina in the optic nerve head (ONH) region. A circumpapillary scan was recorded with a prototype ultrahigh-resolution SD PS-OCT instrument [34]. Figure 9.2a is a reflectivity (intensity) image. It shows increased thickness of the RNFL in the regions superior and inferior of the ONH and thin RNFL nasally and temporally. Figure 9.2b, c shows the retardation and optic axis orientation (revealing the orientation of the nerve fibers around the ONH). The increase of retardation with depth at the thickest RNFL bundles can be observed in the retardation image (color changes from blue to green). The optic axis orientation image shows two full color oscillations from left to right, in good agreement with the radial orientation of the nerve fiber bundles around the optic nerve head (corresponding to $2 \times 180°$ orientation change).

Diattenuation causes different absorption in light beams of orthogonal polarization states. However, since it is very low in most biological tissues [19, 35], it can be neglected for the applications described here.

Depolarization causes a reduction of the degree of polarization (DOP) of the backscattered light, as compared to the illuminating light beam. Depolarization can be caused by multiple scattering or by scattering on large, nonspherical particles [36]. Since PS-OCT is a coherent imaging technique, the DOP cannot be directly measured [37]. Instead, depolarization can be observed in PS-OCT images as a polarization state that varies randomly from speckle to speckle [21, 38]; therefore, this effect is often referred to as "polarization scrambling." To quantitatively describe depolarization and to display it in PS-OCT images, we introduced a quantity related to the DOP, the degree of polarization uniformity (DOPU) [21], which can be regarded as a spatially averaged DOP (cf. Sect. 1.3). Depolarization can be found not only in melanin-containing tissue like the retinal pigment epithelium (RPE) [21, 32, 38], iris pigment epithelium [39], choroidal nevi [40],

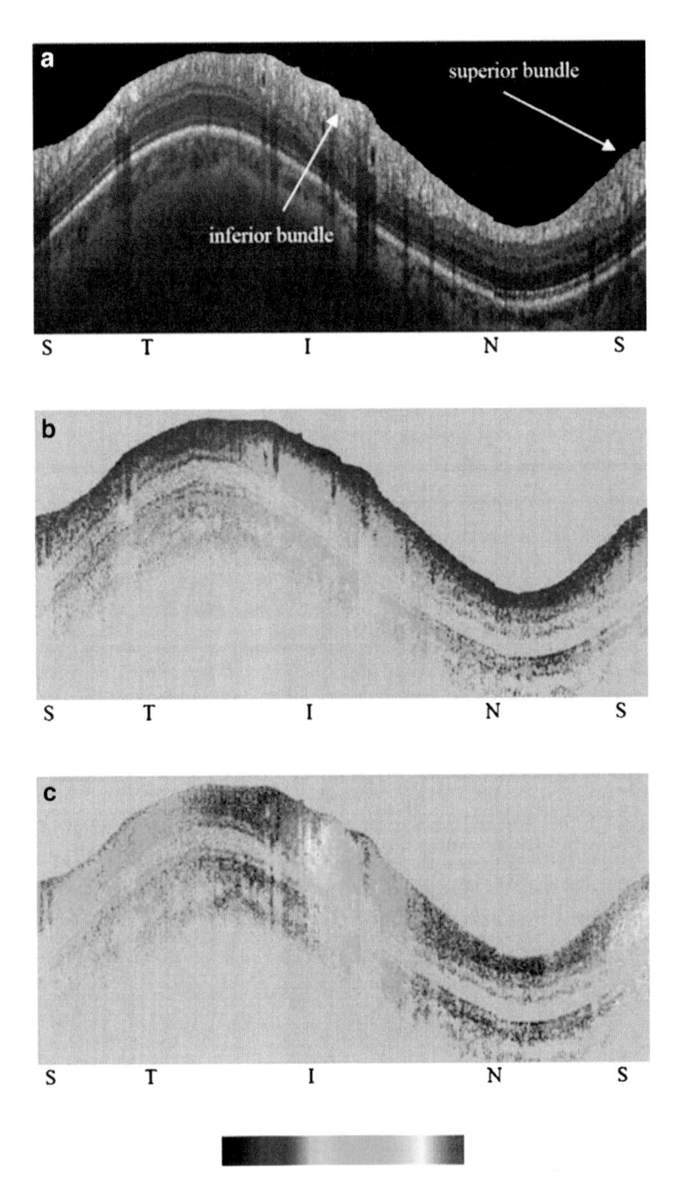

Fig. 9.2 Circumpapillary PS-OCT scan (4,000 A-scans) from healthy human retina in vivo. Scan diameter ~10° (corresponds to a circumference of ~9.4 mm, equal to horizontal image width; optical image depth 1.8 mm). (**a**) Intensity (log scale); (**b**) retardation (*color bar* 0–90°); and (**c**) optic axis orientation (*color bar* 0–180°). To avoid erroneous polarization data, areas below a certain intensity threshold are displayed in *gray*. Orientation of scan from left to right (S)uperior, (T)emporal, (I)nferior, and (N)asal (From ref. [34], by permission of the Optical Society of America)

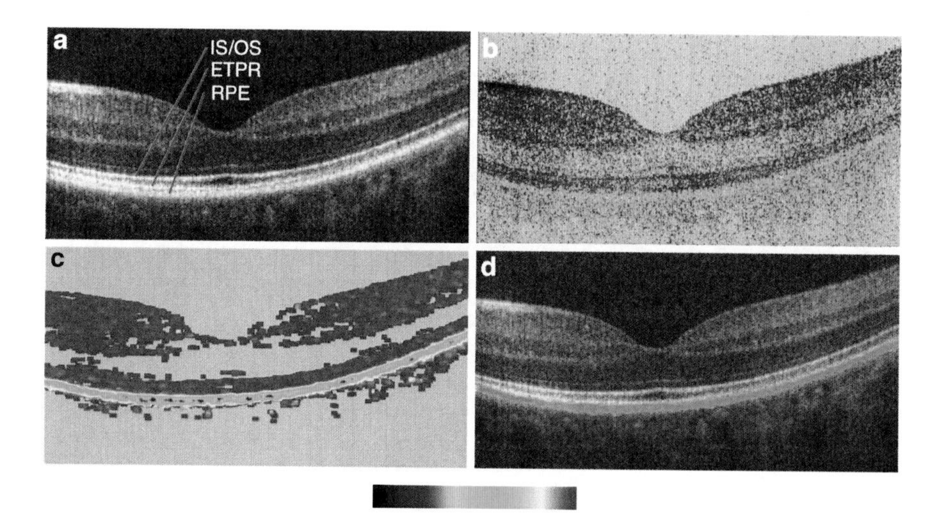

Fig. 9.3 PS-OCT B-scan images of human fovea in vivo. (**a**) Intensity (log scale); (**b**) retardation (*color bar* 0–90°); (**c**) degree of polarization uniformity DOPU (*color bar* 0–1); and (**d**) overlay image, segmented RPE shown in *red*. Image size 15° (horizontal)×0.75 mm (vertical, optical distance) (From ref. [21], by permission of the Optical Society of America)

and melanoma but also in accumulations of pigment-loaded macrophage and—to a smaller amount—in the choroid [21, 24]. The reason for depolarization in these tissues seems to be multiple scattering within particles with high melanin concentration [41,42].

Figure 9.3 shows an example of depolarization imaging in the foveal area of a healthy human retina. In the intensity image of Fig. 9.3a, it can be seen that the three posterior layers/boundaries of the retina (IS/OS, boundary between inner and outer photoreceptor segments; ETPR, end tips of photoreceptors; RPE) have rather similar reflectivity. However, the polarization-sensitive images (Fig. 9.3b, c) clearly show that the most posterior of these layers, the RPE, changes the polarization state of backscattered light, while the other retinal layers render the polarization state essentially unchanged. The retardation image (Fig. 9.3b) reveals the random nature of the polarization state backscattered by the RPE (discernible by the grainy, random appearance of the δ value within the RPE, as compared to the rather uniform blue color indicating constant δ in the other layers). The DOPU image (Fig. 9.3c) quantitatively shows the extent of the depolarization caused by the RPE, while simultaneously showing the high DOPU values retained by the other layers. It should be noted that only light directly backscattered by the RPE is depolarized. Light that traverses the RPE and is backscattered from deeper structures is still in a well-defined polarization state.

9.1.3 Segmentation Algorithms

The light-tissue interaction mechanisms that change the light's polarization state can be exploited to generate tissue-specific contrast and, in a next step, to segment tissues based on their specific effect on the polarization state. In principle, all of the mechanisms mentioned above could be used for segmentation purposes. However, since this technology is still in an early state of development, only depolarization has yet been used for segmentation purposes on a larger scale. The corresponding methods are briefly discussed in this subchapter.

Segmentation based on depolarization is typically performed in two steps. In the first step, the RPE is segmented, and in the second step, the RPE is used as a "backbone" to segment adjacent structures and lesions.

The first step starts by exploiting the depolarizing effect of the RPE. As mentioned above, depolarization can be observed in PS-OCT images because it leads to uncorrelated polarization states of adjacent speckles, i.e., their polarization state varies randomly. This effect is quantitatively described by the parameter DOPU [21] which is obtained in the following way: At first, we calculate the Stokes vector S of each pixel by (9.4). Then we average Stokes vectors over adjacent pixels in a B-scan by calculating the mean value of each Stokes vector element within a floating rectangular window (the size can vary, depending on the pixel raster; a typical value for a B-scan consisting of 1,000 A-scans is 15(x) × 6(z) pixels). Finally, we derive DOPU within this window by the following equation:

$$\mathrm{DOPU} = \sqrt{Q_{\mathrm{m}}^2 + U_{\mathrm{m}}^2 + V_{\mathrm{m}}^2}, \quad (9.5)$$

where the indices m indicate mean Stokes vector elements. As can be seen by (9.5), DOPU is closely related to DOP well known from polarization optics. However, we want to emphasize the statistical origin of DOPU, i.e., it can only be derived by local averaging of Stokes vector elements and can therefore also be regarded as a spatially averaged DOP.

In case of a polarization preserving or birefringent tissue, the value of DOPU is approximately 1, and in case of a depolarizing layer, DOPU is lower than 1. To avoid erroneous data points caused by noise (which also gives rise to random Stokes vector elements), we first apply a thresholding procedure based on the intensity data to gate out areas with low signal intensity. Depolarizing structures can now be segmented from the DOPU images by extracting those areas where DOPU is smaller than a threshold value $\mathrm{DOPU}_{\mathrm{thr}}$. Values of $\mathrm{DOPU}_{\mathrm{thr}}$ for RPE segmentation depend on evaluation window size and raster scan pattern and, to some extent, on the overall signal quality. Typical values are 0.65–0.8. After extraction of depolarizing structures, the segmented structures can be displayed in the form of an overlay image where they are, e.g., displayed in red color on top of a black and white intensity image (cf. Fig. 9.3d). It should be mentioned that the RPE segmented in that way appears thicker than it is in reality. The reason is the limited resolution caused by the nonzero extension of the evaluation window used to calculate DOPU. The RPE thickness observed in the overlay image is a convolution of the actual RPE thickness and the evaluation window size.

In a second segmentation step, the RPE found in the first step is used as a "backbone" for further data processing. We will briefly describe two of these segmentation algorithms.

9.1.3.1 Drusen Segmentation

Figure 9.4 (prepared from a 3D PS-OCT data set of a 70-year-old patient) illustrates the drusen segmentation algorithm. After finding the RPE, the depth position corresponding to the lowest DOPU value within the segmented tissue band is found for each A-scan. Thus, a thin line (thickness one pixel) is found for every B-scan, indicating the actual position of the RPE (red line in Fig. 9.4a, b). The buckles in the RPE are clearly observed. In a next step, the position where the RPE should lie in a healthy eye is found. This is done by generating and iteratively adapting a smooth function (indicated by a green line in Fig. 9.4a, b) until its position corresponds to the "should be" position of the RPE. Details of this process can be found in [24]. In addition, the internal limiting membrane (ILM) is found by an intensity-based algorithm (blue line). These segmentation lines are calculated for all of the B-scan images of the 3D data set. From these segmentation lines, various thickness maps can be calculated: a retinal thickness map excluding drusen (distance, blue to red line, Fig. 9.4d), a retinal thickness map including drusen (blue to green line, Fig. 9.4e), and a drusen thickness map (red to green line, Fig. 9.4f). From the latter map, the area and volume of the drusen can be calculated. Preliminary measurements on seven eyes indicated that the repeatability of total drusen area and volume is $\sim$7.5% [24].

9.1.3.2 Segmentation of Geographic Atrophies

In principle, atrophic zones could be detected by summing up the depolarizing pixels (the red pixels in the overlay images) along each A-scan in a 3D data set. The resulting thickness map of the depolarizing layer would reveal the existence or absence of this layer.

However, depending on the subject's pigmentation, depolarization may also be apparent in the choroid. Especially in atrophic areas, there is considerable penetration of the beam into the choroid, and depolarization from choroidal pigments can confound the images and obscure atrophic zones. To prevent integration of choroidal pixels into the thickness map of the depolarizing layer, only pixels located in a shallow band close to the photoreceptor layer and Bruch's membrane should be used. To achieve this, a smooth function approximating the normal RPE position is calculated in a similar way as described above (using additional information on the ILM position), and only depolarizing pixels in the vicinity of this line are taken into account for calculating the thickness of the depolarizing layer [24]. In this way, confounding data from the choroid are excluded.

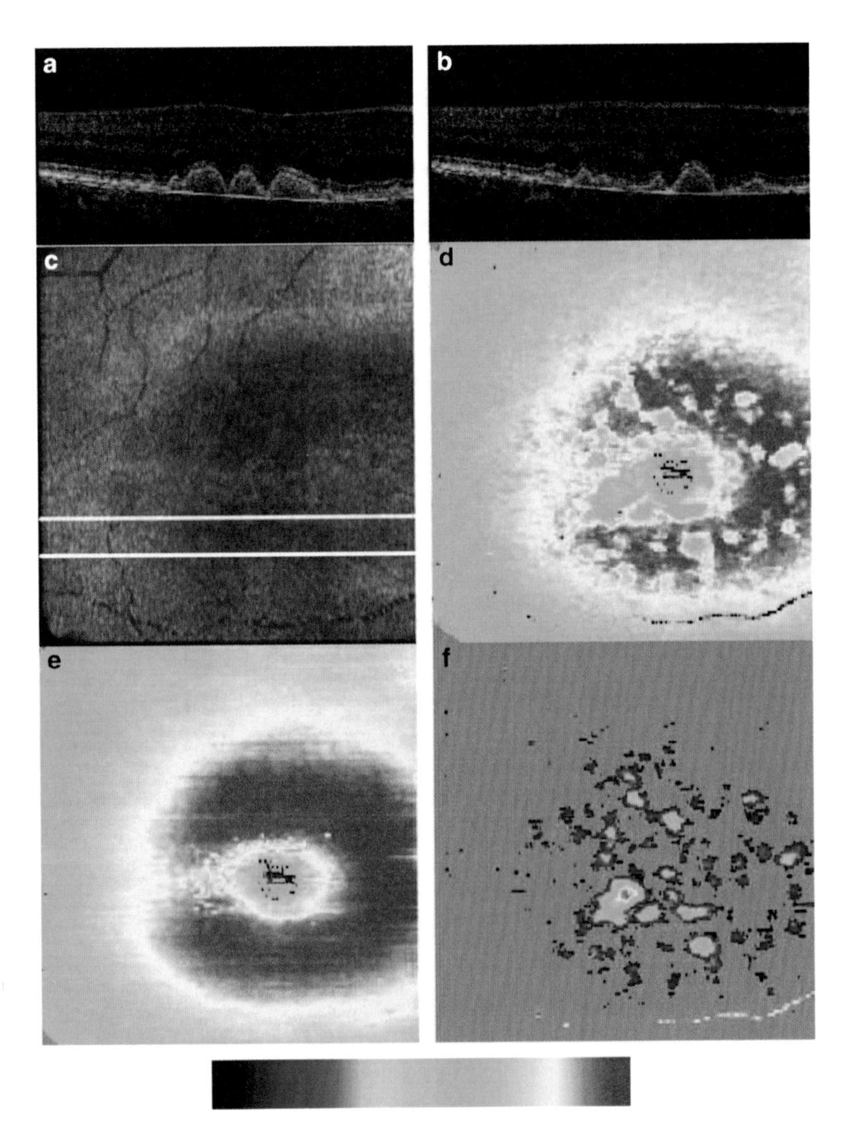

Fig. 9.4 Results from drusen segmentation. (**a**, **b**) Reflectivity B-scan images with overlaid positions of ILM (*blue*), RPE (*red*), and normal RPE position (*green*); (**c**) en face fundus image reconstructed from 3D OCT data set; (**d**) retinal thickness map excluding drusen (*color bar* 70–350 μm); (**e**) retinal thickness map including drusen (*color bar* 70–350 μm); and (**f**) drusen thickness map (*color bar* scaled to maximum elevation value of 55 pixels corresponding to ~128 μm) (From ref. [24], by permission of SPIE)

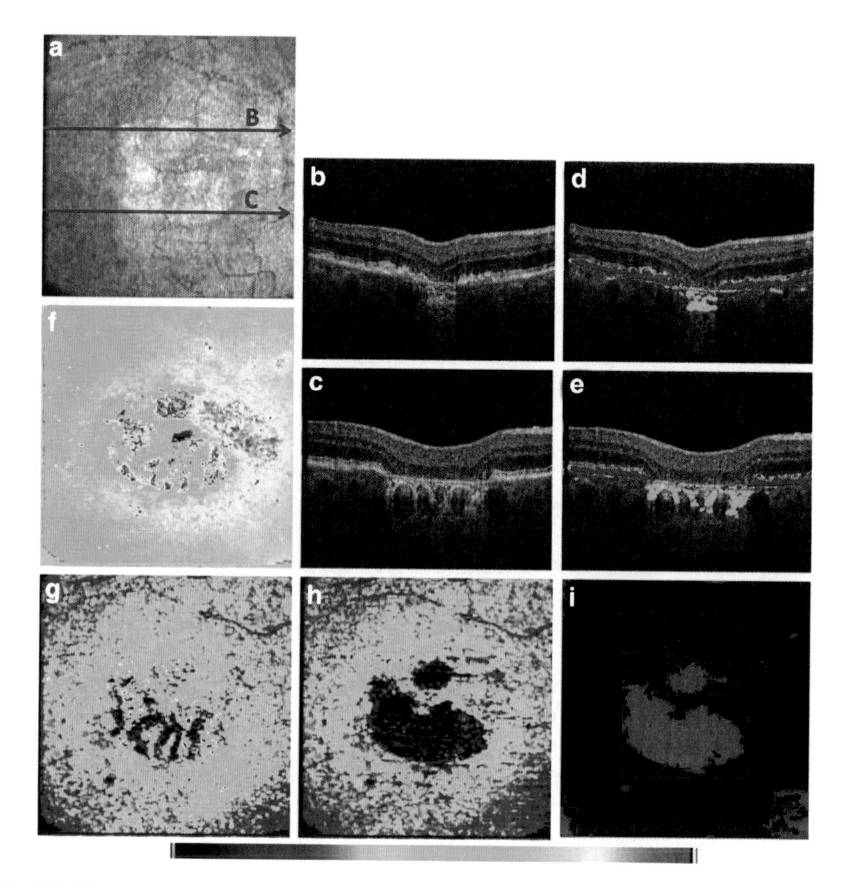

Fig. 9.5 Segmentation of geographic atrophy. (**a**) En face fundus image reconstructed from 3D OCT data set; (**b**, **c**) reflectivity B-scan images; (**d**, **e**) overlay of depolarizing structures within and outside the evaluation band shown in *red* and *green* color, respectively; (**f**) retinal thickness map (ILM to position of minimum DOPU value); (**g**) map of overall number of depolarizing pixels per A-line; (**h**) thickness map of the depolarizing layer inside the evaluation band; (**i**) binary map showing atrophic zones in *red* (*color bar* 70–325 μm for (**f**); from 0 to 39 pixels for (**g**) and (**h**)) (From ref. [24], by permission of SPIE)

Figure 9.5 shows several steps in the segmentation process of geographic atrophy (GA) (3D data set from an 80-year-old AMD patient). Figure 9.5a shows a fundus image generated from a 3D intensity data set, where the two B-scans shown in Fig. 9.5b, c are marked by red lines. Figure 9.5d, e shows the same B-scans where depolarizing RPE tissue is marked in red and depolarizing choroidal tissue is marked in green. Figure 9.5f shows a total retinal thickness map, Fig. 9.5g shows a thickness map of depolarizing tissue including choroidal signals (sum of red and green pixels of Fig. 9.5d, e), and Fig. 9.5h shows a similar thickness map with the choroidal data excluded (i.e., the sum of only the red pixels of Fig. 9.5d, e). While the atrophic zone is obscured in Fig. 9.5g, it is clearly visible in Fig. 9.5h. Figure 9.5i finally shows a binary map where atrophic zones are displayed in red. Preliminary measurements on five eyes indicate a repeatability of GA area of ~9%.

Advanced segmentation algorithms based on similar principles are presently under development for subretinal fluids, hard exudates, and other types of lesions.

9.2 Imaging of the Eye with Polarization-Sensitive OCT

The depth-resolved information provided by PS-OCT bears much potential for diagnosis and quantitative analysis of disease progression in daily clinical routine. Hence, in the second part of the chapter, the clinical value of ophthalmic PS-OCT imaging will be discussed.

PS-OCT imaging has been shown to be valuable for both, the anterior and posterior segment of the eye. Because of limited space, only posterior segment imaging will be described in detail in the remainder of this book chapter. However, a brief overview of relevant literature on anterior segment PS-OCT is provided here for the interested reader: A PS-OCT in vitro study has been carried out to analyze the birefringence distribution in the healthy cornea [43], demonstrating a characteristic, approximately radially symmetric retardation pattern. This pattern is heavily distorted in corneas with keratoconus [44], offering an interesting approach for diagnosis. In vivo imaging has been used for preliminary studies of the polarizing properties of cornea, iris, conjunctiva, sclera, trabecular meshwork, and ocular tendons [27–29]. A more sophisticated study demonstrated tissue segmentation based on a 3D feature space analysis [45]. Based on the three parameters, intensity, extinction coefficient, and birefringence, the following tissue types could be clearly differentiated: cornea, conjunctiva, sclera, trabecular meshwork, and uvea. First applications to better visualize pathologic changes and associated surgical procedures in glaucoma have recently been reported [46]. Further studies are on the way to demonstrate the clinical value of PS-OCT in the anterior segment.

However, the main field of application of PS-OCT is represented by the retina. The retina is characterized by one of the highest metabolic rates in the body and is therefore predisposed to a range of various diseases. Imaging of retinal morphology and pathomorphology is clinically important for the diagnosis of structural and functional retinal changes. The specific anatomy of the eye allows for direct examination of neuronal retinal tissue. Accordingly, retinal OCT imaging techniques have become the most clinically developed among all OCT applications [47–49].

Among the several cell layers of the retina, also the RPE is essential for vision. The RPE forms a functional unit with the overlying photoreceptor cell layer and is crucial for its metabolism and nutrient delivery. Furthermore, the RPE-Bruch's membrane complex forms the blood–retinal barrier. The RPE is composed of a cuboidal monolayer of cells comprising numerous melanosomes packed with polarization-scrambling melanin. Due to the high melanin concentration, PS-OCT is capable of depicting the exact RPE position, as well as providing qualitative information about RPE layer integrity (see Sect. 1.2).

As the RPE is involved in several retinal diseases, a detailed discussion of the potential of PS-OCT imaging for every known pathologic retinal alteration is beyond the scope of this chapter. Therefore, age-related macular degeneration (AMD), a representative disease with significant RPE involvement, is depicted to highlight the impact of the new PS-OCT imaging technique. Furthermore, the potential of PS-OCT imaging of the retina in glaucoma disease is shown. Some other diseases characterized by pathologic polarizing-scrambling material in the retina, such as diabetic retinopathy, are also mentioned.

9.2.1 Age-Related Macular Degeneration

AMD is the leading cause of irreversible blindness in people over 50 years of age in the developed world [50,51]. Moreover, the overall prevalence of advanced AMD is projected to increase by more than 50% by the year 2020, mostly due to the increase in the population of people more than 65 years of age [52].

The early stages of AMD are characterized by the presence of soft drusen and/or alterations in the pigmentation of the retina. Late stages of AMD are characterized by large atrophic retinal areas accompanied by a severe loss of vision and can be further divided into nonexudative (dry) and exudative (wet) AMD. In contrast to dry AMD, also called geographic atrophy (GA) and characterized by a slow progression of the atrophic areas, wet AMD refers to the formation of choroidal neovascularization (CNV). Due to increased vascular permeability and fragility, CNV itself may lead to subretinal hemorrhage, fluid exudation, lipid deposition, and detachment of the RPE from the choroid, as well as fibrotic scars, and is responsible for a very fast and profound loss of central vision [53,54].

Despite recent scientific progress, the pathophysiologic mechanisms leading to AMD are still not fully understood. Several studies, however, indicate that the RPE has a central role in the onset and progression of AMD. One sensitive marker for attenuation of the RPE is the formation of drusen.

9.2.1.1 The Early Forms: Drusen and Pigmentary Alterations

The presence of drusen is a hallmark of AMD and is usually the first clinical finding [55]. Drusen comprise focal depositions of acellular, polymorphous debris between the RPE and Bruch's membrane [56]. Although the biogenesis of drusen remains largely unknown, it is generally accepted that the RPE has a pivotal role in this process [57]. It is hypothesized that a preinjured RPE provides the basis for immune-mediated and inflammatory events in the subretinal space, which lead to the deposition of drusen-associated constituents [58,59]. As drusen become larger, the overlying RPE degenerates, accompanied by a loss of photoreceptors [60–62].

Therefore, the total drusen area and size is considered a significant clinical risk factor for the progression toward advanced AMD [63]. Furthermore, it is hypothesized that pigmentary alterations, which are often accompanied by drusen, are associated with a worse prognosis [60, 64]. These alterations, however, are difficult to grade using digital fundus photography [65].

Based on the fact that drusen cause a subtle focal bending of the RPE into the retina, the RPE-specific data of PS-OCT can be utilized for an automated segmentation of drusen (see Sect. 1.3). A recent study demonstrated that PS-OCT can correctly delineate more than 96% of all drusen [66]. Therefore, drusen size and area can be measured reliably. Interestingly, the amount of drusen observed in SD-OCT data is significantly larger than that in fundus photography, mostly based on a more accurate detection of very small lesions [67].

Moreover, PS-OCT can also provide qualitative information about the condition of the RPE, pigmentary alterations, and drusen constituents. Histologic [68] and OCT-based studies [69] have attempted to classify different drusen types on the basis of their appearance and the quality of the overlying RPE, including pigmentary alterations. Using PS-OCT, this classification scheme can be extended to include polarization-scrambling characteristics.

Figure 9.6 illustrates a selection of different drusen types seen in PS-OCT. Two main groups can be identified: drusen with attenuated overlying RPE (Fig. 9.6c) and drusen with depolarizing contents (Fig. 9.6d). The druse in the center of

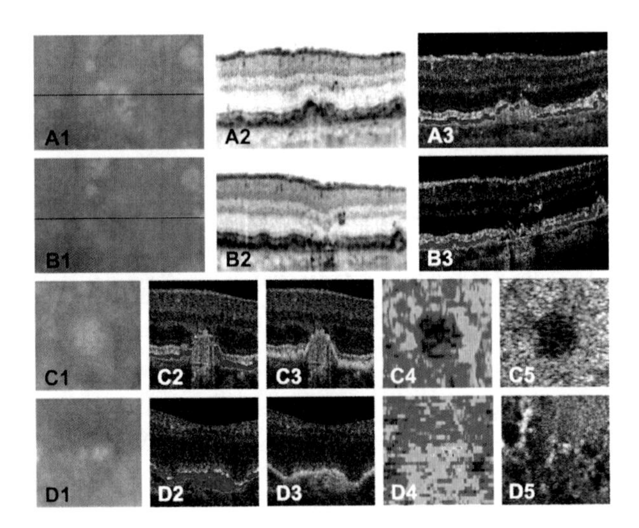

Fig. 9.6 Different drusen types. (**A1**) Fundus photography in a 76-year-old woman. The *black* lines indicate the position of the B-scans. (**A2**) SD-OCT and (**A3**) PS-OCT images of two confluent drusen in the center, including pigmentary alterations in the inner retinal layers. (**B1**) shows the same eye 7 months later. (**C1**) Fundus of a 67-year-old woman, with corresponding PS-OCT image (**C2**). (**C4**) depicts the thickness map of the depolarizing layer of the PS-OCT, and (**C5**) is the corresponding near-infrared autofluorescence (NIR-FAF) image. (**D1**) Fundus of a 60-year-old woman. The calcified druse in the center is filled with depolarizing material, which can be identified in the thickness map (**D4**). (**D5**) is the corresponding NIR-FAF image

Fig. 9.6c (fundus of a 67-year-old woman) is characterized by a pale appearance in fundus photography (Fig. 9.6C1) and shows an attenuated RPE in PS-OCT images (Fig. 9.6C2). The integrity of the overlying RPE can be evaluated in the thickness map of the depolarizing layer from the PS-OCT data (Fig. 9.6C4) as well as the near-infrared autofluorescence image (NIR-FAF) (Fig. 9.6C5), an imaging technique that is specific for melanin [70]. Figure 9.6d illustrates a calcified druse in the macula of a 60-year-old woman. Interestingly, this calcification produces a depolarizing signal in PS-OCT images (Fig. 9.6D2). This depolarizing content can be identified in the thickness map of the depolarizing layer (Fig. 9.6D4), but not in the melanin-specific NIR-FAF image (Fig. 9.6D5).

It will be of clinical significance to investigate the progression of the different drusen types toward AMD. Figure 9.6a, b shows an example of the natural course of two drusen with attenuated RPE and pigmentary alterations. Seven months later, both disappeared, leaving behind a possible initiation site of GA.

9.2.1.2 Geographic Atrophy

Geographic atrophy, also called nonexudative or dry AMD, is defined by any sharply delineated roughly round or oval area of hypopigmentation or depigmentation or an apparent absence of the RPE in which choroidal vessels are more visible than in surrounding areas that must be at least 175 μm in diameter [71]. Natural history studies have identified various characteristics of this disease and show different progression rates and functional outcomes [72–74]. While GA has a relatively uniform appearance on 2D images such as fundus photography, a broad spectrum of retinal alterations has been identified using SD-OCT imaging, a finding that might reflect the different characteristics of GA [75–77].

Due to its pigment specificity, PS-OCT can identify the area of pigment loss in the affected regions [21]. The depolarization-based segmentation algorithm (see Sect. 1.3) provides reliable results in patients with GA [78]. Even in the presence of perimarginal alterations, like an irregularly shaped and partially elevated RPE, PS-OCT is able to reliably calculate the atrophic areas.

Another imaging technique for quantitative assessment of RPE changes, autofluorescence imaging (FAF), has become widely used for grading the atrophic zones [79, 80]. Recent studies showed a good correlation between changes of the junctional zone seen in SD-OCT and FAF images, indicating that the combined use of SD-OCT and FAF can reliably detect hypertrophic and damaged RPE cells [81]. The key benefit of PS-OCT is its ability to provide depth-resolved information about polarization properties to identify the integrity of the RPE. Therefore, PS-OCT allows for detailed analysis of the margin of the atrophic areas. Figure 9.7 illustrates the benefits of such an analysis. Figure 9.7a shows the fundus of a 70-year-old woman recorded with short-wavelength autofluorescence (A1), near-infrared autofluorescence (A2), and PS-OCT (A3–5). Small differences in the shape and size of the central atrophic area can be observed between each imaging technique. Figure 9.7A4, A5, however, depicts the depth-resolved information, here

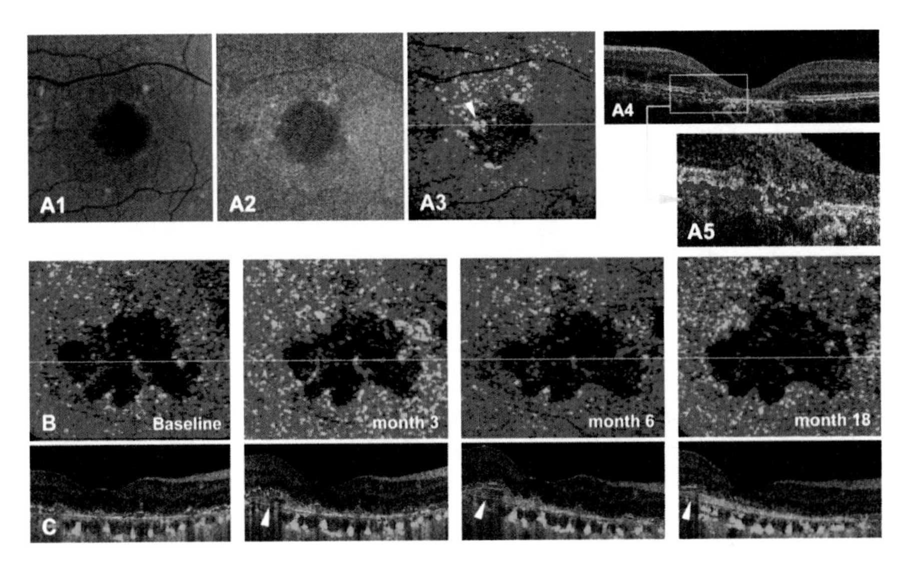

Fig. 9.7 (**a**) Fundus of a 70-year-old woman with central GA. (**A1**) Short-wavelength autofluorescence image. (**A2**) near-infrared autofluorescence image. (**A3**) Thickness map of the depolarizing layer out of PS-OCT. *Yellow* lines indicate B-scans (**A4**). Note that a small peninsula is visible in the thickness map (*arrow*), which corresponds to an irregularly pigmented margin zone (**A5**). (**b**) Shows the natural history of GA in a 70-year-old woman, thickness maps out of PS-OCT. *White* lines represent macular B-scans (**c**). Areal changes are slowly progressive, and disease activity can be visualized in detail in the margin zones (*white arrows*)

shown for the cryptic peninsula seen in the thickness map of depolarizing material of the PS-OCT (white arrow in A3). Recent studies confirmed that the different structures of the marginal zones play important roles in the progression of the atrophic areas [77]. Figure 9.7b, c shows an example of the natural history of GA in a 70-year-old woman. Areal changes are slowly progressive, but disease activity can be seen in the B-scans (white arrows in Fig. 9.7c).

Further studies are in progress to investigate the possible clinical benefits of such analyses. This is of high importance, as PS-OCT might be useful for determining possible intervention targets in GA, for which appropriate therapy is currently still lacking.

9.2.1.3 Exudative AMD

Exudative Changes: Intraretinal Hemorrhages and Subretinal Fluid

Subretinal macular hemorrhage commonly occurs as a secondary manifestation of CNV [82]. The toxicity of subretinal hemorrhage on the neurosensory retina has been demonstrated in animal models and in clinical studies [83, 84].

Subretinal hemorrhage that affects the central macular area represents a severe threat to visual acuity while also hindering possible treatment interventions, such as the use of focal laser photocoagulation or the administration of photodynamic therapy. Studies have shown a poor natural prognosis of submacular hemorrhage [85–87]. Several therapeutic approaches have been developed to clear the submacular blood to diminish permanent damage to the photoreceptors and the RPE [88].

With OCT, subretinal hemorrhage usually appears optically empty. PS-OCT may be applied to specifically evaluate the disintegrity of the RPE in patients with subretinal macular hemorrhage potentially caused by mechanical effects such as fibrotic shearing of photoreceptors, metabolic disruption imposed by the clot as a diffusion barrier, hypoxia, and direct neurotoxicity induced by blood components such as iron [89].

Subretinal Fluid

Exudative or neovascular AMD (NVAMD) is characterized by neovascular membranes invading the subpigment epithelial and subretinal spaces characterizing CNV. CNV itself may lead to the accumulation of subretinal fluid (SRF) and bleeding [90].

With disease progression, NVAMD may lead to fibrosis, which is an irreversible pathologic manifestation.

The presence of SRF appears to be an important indicator for the exudative activity of a neovascular lesion and causal for a frequently reversible loss of visual acuity [91]. RPE lesions are frequently observed in patients with NVAMD (Fig. 9.8).

Several treatment strategies have recently been established for the management of NVAMD. The most commonly used therapeutic approach is the intravitreal application of antivascular endothelial growth factor (anti-VEGF) substances to inhibit VEGF, one of the key growth factors identified in CNV formation [92, 93].

SRF is a clinically relevant parameter as the SRF volume significantly correlates with visual acuity following treatment with anti-VEGF [91]. Efforts have recently concentrated on the automatic segmentation and detection of SRF, due to the fact that manual segmentation depends on the expertise of the individual grader. Further, automatic segmentation procedures are potentially valid for obtaining data in larger patient populations.

Entire 3D retinal datasets of patients with NVAMD presenting with SRF can be extracted and processed by a specifically designed software algorithm based

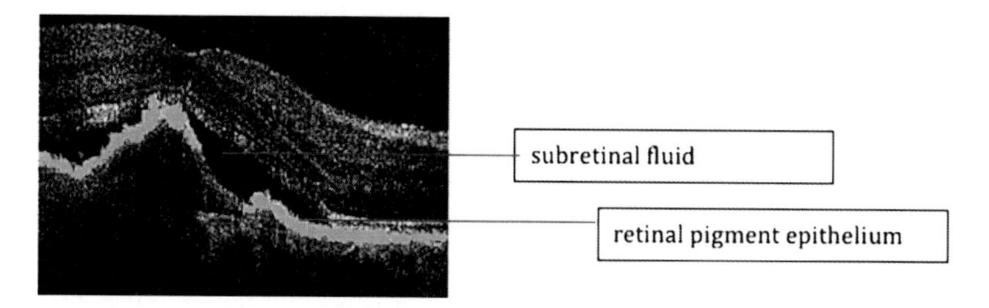

Fig. 9.8 PS-OCT image of a patient with neovascular AMD and related subretinal fluid accumulation

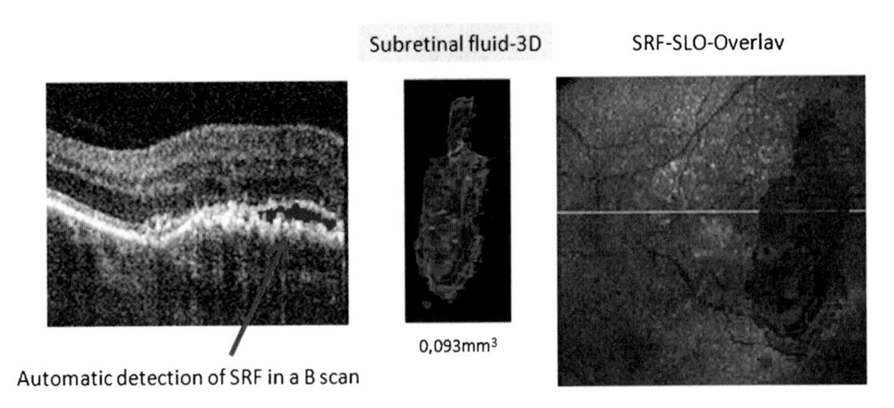

Fig. 9.9 Automatic detection and quantification of subretinal fluid (SRF) in a patient with neovascular AMD. The software algorithm allows for the automatic detection of SRF, the respective volume can be quantified and visualized in three dimensions

on a classification system using intensity data and polarization information [94]. Thereby, automatic SRF detection and quantification is possible in three dimensions. Following this procedure, alterations in the extent and quantity of SRF can be analyzed over time (Fig. 9.9).

By applying the method discussed above, PS-OCT may be clinically useful for treatment follow-up analysis and evaluation of the therapeutic success, as SRF reduction may potentially be quantified, due in part to the ability to detect RPE alterations.

Pigment Epithelial Detachment

Pigment epithelial detachments (PEDs) are frequently associated with AMD [56, 95]. A broad spectrum of pathologic changes occurs with certain types of PED, but retinal elevation is characteristic. One common cause of PED is CNV, which frequently elevates the retina [96, 97]. The extent of PED formation strongly depends on the location of CNV formation [98]. Figure 9.10 shows characteristic features of a PED represented in PS- and SD-OCT images.

Fibrosis

The formation of fibrin-like deposits in the retina marks the end stage of exudative AMD [70]. In standard SD-OCT images, a nonspecific thickening of the former RPE-photoreceptor band can be observed, accompanied by other severe intraretinal alterations including cysts and fluids. The PS-OCT images reveal the destroyed RPE

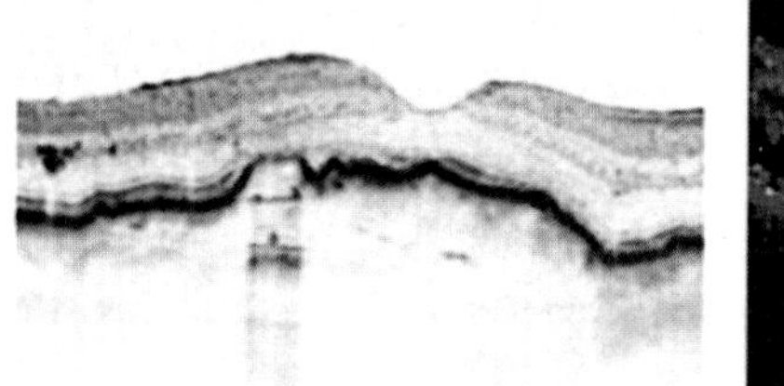
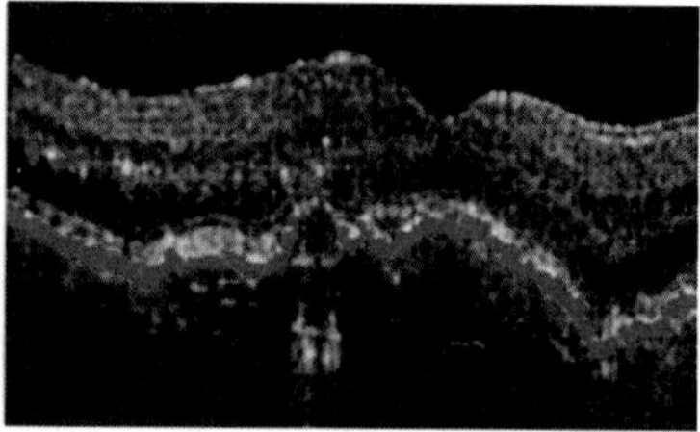

Fig. 9.10 SD-OCT intensity image (*left*) of a patient with a pigment epithelial detachment (PED). The RPE appears elevated, and the PED reveals membrane-like structures beneath the atrophic RPE. One can observe pigmentary alterations, which also depolarize in addition to the RPE, observable in the PS-OCT image (*right*)

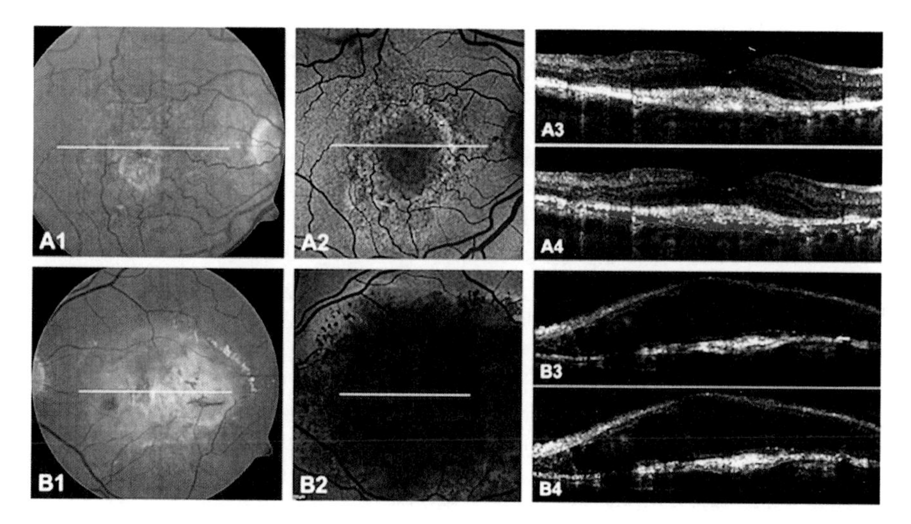

Fig. 9.11 (A1) Fundus of a 70-year-old woman with fibrin-like deposits in subretinal space caused by CNV. *White bars* represent the position of the B-scans. (**A2**) Autofluorescence image with weak macular RPE signal. Using PS-OCT, the RPE can be localized below the fibrinoid tissue (**A4**). (**B1**) Fundus of a 73-year-old man with a fibrinoid scar. The autofluorescence image (**B2**) indicates the complete loss of RPE in the affected region. Residual islands of depolarizing material can be identified in the PS-OCT images (**B4**)

band (Fig. 9.11). In most patients, however, small residual islands of depolarizing tissue within the scar tissue complex can be observed [77].

9.2.2 *Glaucoma*

Glaucoma is a chronic slowly progressive optic neuropathy and one of the leading causes of irreversible blindness in all parts of the world [99, 100]. The underlying process is an apoptotic cell death of the retinal ganglion cells and their axons

[101–103]. Since treatment cannot reverse and may not even completely stop the disease process [104], an early diagnosis is essential to preserve vision in an aging population.

Glaucoma is characterized by a specific appearance of the optic nerve head, by a loss of the RNFL and by visual field defects [105–107]. Considerable variation in physiologic optic disk appearance frequently complicates the diagnosis of glaucoma. Additionally, visual field tests may be normal until 25–50% of the retinal nerve fibers have been lost [108, 109]. To bridge the gap between incipient glaucomatous optic neuropathy and the ability to definitely confirm the diagnosis, several diagnostic technologies have been used, such as the scanning laser ophthalmoscope HRT®, the scanning laser polarimeter (SLP) GDx®, and also the OCTs. Unfortunately, all these instruments have only limited diagnostic power in an early disease stage. For glaucoma suspects with prior structural progression or subsequent visual field progression, a sensitivity of 28–35% (at 92–95% specificity) for the HRT [110, 111], of 50% (at 95% specificity) for the GDx [111], and of 50% (at 81% specificity) for the OCT [112] has been reported.

With the GDx® (Carl Zeiss Meditec), the retardation of the RNFL can be measured. The retardation values are mathematically converted to micrometer thickness using a constant birefringence value [113]. The underlying physical property of the RNFL that produces the retardation is form birefringence, which is caused by intracellular microtubules within the ganglion cell axons [114–116]. Microtubuli, which are intracellular tubular polymers with outer diameter of 25 nm and inner diameter of 15 nm, are the only relevant source of birefringence in rat retina [116].

PS-OCT allows for direct and simultaneous measurement of both the optical thickness of the RNFL and its birefringence. The birefringence of the RNFL varies around the optic nerve head by the factor 3 with a maximum at the superior and inferior poles and a minimum at the temporal and nasal aspects [30, 117, 118]. This variation might be due to different neurofilament or microtubuli content of ganglion cell axons [119] and also in part to different glial content of the RNFL [120]. For primates, there is however evidence that it is primarily the packing density of microtubuli within the axons and not the packing density of the axons that determines the variation of birefringence around the optic nerve head [121].

Using PS-OCT, both the optical thickness and the retardation of the RNFL are measured simultaneously, and birefringence can be calculated easily as the slope of increasing retardation of the backscattered light by increasing depth within the RNFL. Figure 9.12 shows an example of the circumpapillary RNFL of a healthy volunteer and a patient with mild glaucoma (MD −3.98). In this patient, the optical thickness, the retardation, and also the birefringence are apparently reduced compared to the normal subject.

Götzinger and coauthors [122] have compared a small series of 12 normal volunteers and patients with 12 early glaucoma or glaucoma suspects. Compared to healthy eyes, mean RNFL thickness of glaucoma suspect eyes was reduced by 10% (superior, 12.7%; inferior, 10.9%), whereas mean RNFL birefringence was reduced by 33% (superior, 51.7%; inferior, 34%).

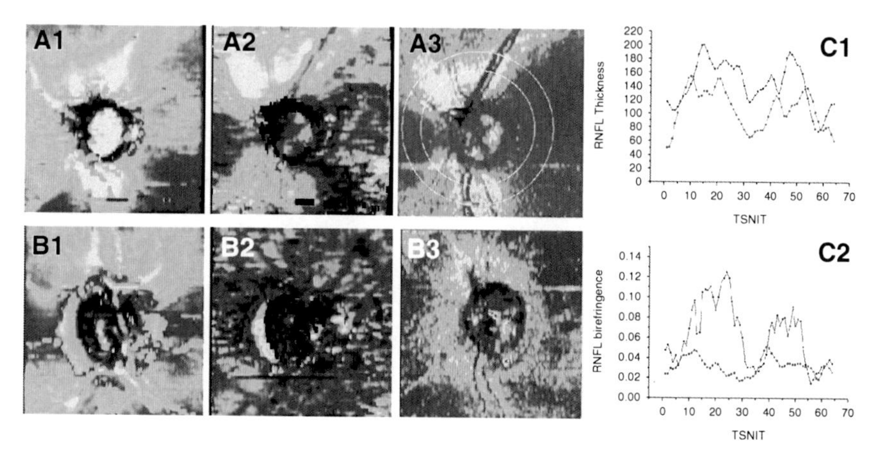

Fig. 9.12 (**a**) Circumpapillary retinal nerve fiber layer (RNFL) map of a healthy volunteer; optical thickness (**A1**) of the RNFL, birefringence (**A2**) of the RNFL, and retardation (**A3**) of the RNFL. (**b**) Circumpapillary RNFL map of a glaucoma patient with mild glaucoma (MD −3.98), optical thickness (**B1**) of the RNFL, birefringence (**B2**) of the RNFL, and retardation (**B3**) of the RNFL. It is clearly visible that the birefringence of RNFL is very markedly reduced (**B2** vs. **A2**), while the optical thickness of RNFL shows only a moderate reduction (**B1** vs. **A1**). The TSNIT diagram also shows a moderate reduction of optical RNFL thickness, (**C1**) whereas the birefringence (**C2**) is markedly reduced in the patient with mild glaucoma (From ref. [118], by permission of Wiley-VCH)

The authors concluded that reduced RNFL birefringence might be an early indicator for glaucoma and therefore might improve early diagnosis of glaucoma.

There are several possible explanations for this dissociation of optical thickness and birefringence. First, ganglion cell axons containing more microtubules might constitute a larger proportion of those ganglion cells lost during early stages of glaucoma [123]. Secondly, birefringence could be a sensitive measurement of the density of ganglion cell axons. Based on observations of early structural changes of retinal ganglion cells in glaucomatous neuropathy [124, 125], it has been speculated that ganglion cell axon density may change before a reduction of thickness of the RNFL can be measured [117]. Finally—and most likely—there might be a loss of intracellular structures like the microtubules early in the disease process and before the structural loss of the ganglion cell axons. This is supported by evidence of early loss of microtubules in the prelaminar region of the optic nerve head as a response to 12 h IOP elevation in pigs [126]. If such a loss of microtubules is propagated from the prelaminar region of the optic nerve head to the RNFL, this would result in reduced birefringence and retardation, more than in reduced optical thickness of the RNFL.

SLP measures thickness together with an aspect of structural integrity of the RNFL, but cannot separate between these two sources of retardation. PS-OCT is able to measure these two features independently. It might thus improve the early diagnosis of glaucoma. It may also be a direct measure of the health state of the RNFL at a disease stage where pathology is potentially reversible.

9.2.3 *Diabetic Maculopathy*

Diabetic macular edema (DME), an important complication secondary to diabetes mellitus, is also a leading cause of visual impairment in the Western world [127]. The Wisconsin Epidemiologic Study of Diabetic Retinopathy reported a 10-year incidence of DME between 13.9% and 25.4% with a poor long-term prognosis [128]. Randomized, controlled clinical trials with type I and type II diabetic patients demonstrated the beneficial effects of intensive glycemic control [129,130], treatment of elevated blood pressure [131, 132], and dyslipidemia [132, 133] on the prognosis of DME. Furthermore, as demonstrated by the Early Treatment Diabetic Retinopathy Study [134], retinal photocoagulation can reduce the risk of severe vision loss by at least 50% [135].

PS-OCT provides significant insight for both the morphologic changes secondary to DME and changes in the retinal architecture following therapy, such as laser coagulation or anti-VEGF agents.

9.2.3.1 Microexudates in DME

Dyslipidemia in diabetic patients is treated with lipid-lowering drugs such as statins or fibrates. These are the most effective drugs to lower low-density lipoprotein cholesterol levels, to lower high triglyceride levels, and to raise low high-density lipoprotein cholesterol levels. Several trials demonstrated that statins not only are safe and well tolerated but also significantly decrease cardiovascular morbidity and mortality in hypercholesterolemic patients in both primary and secondary prevention.

Recent reports describe distinct hyperreflective foci distributed throughout multiple layers of the retina of patients with DME [136, 137]. These foci were detected by SD-OCT at various retinal locations without corresponding changes in color fundus photography, biomicroscopy, or infrared fundus imaging. Similar, but more condensed, hyperreflective material was found in retinal areas showing clinically detectable hard exudates. These hyperreflective foci are thus consistent with precursors of hard exudates and can be interpreted as a subclinical marker visible only in OCT. Furthermore, the hyperreflective foci were often identified in the walls of retinal vessels, suggesting that they represent extravasated proteins and/or lipids shifted toward the extracellular space. This finding is of particular clinical interest because the presence of clinically detectable hard exudates characterizes the advanced stage of DME and is associated with pronounced neurosensory destruction and severe functional loss [138].

PS-OCT allows for differentiating tissues on the basis of differences in intrinsic physical properties. Especially hard exudates and their precursors can be detected by PS-OCT due to their polarization-scrambling qualities [139, 140]. This implies that hard exudates and their precursors can be automatically detected and quantified using raster scanning and 3D mapping (Fig. 9.13). In addition, a higher sensitivity

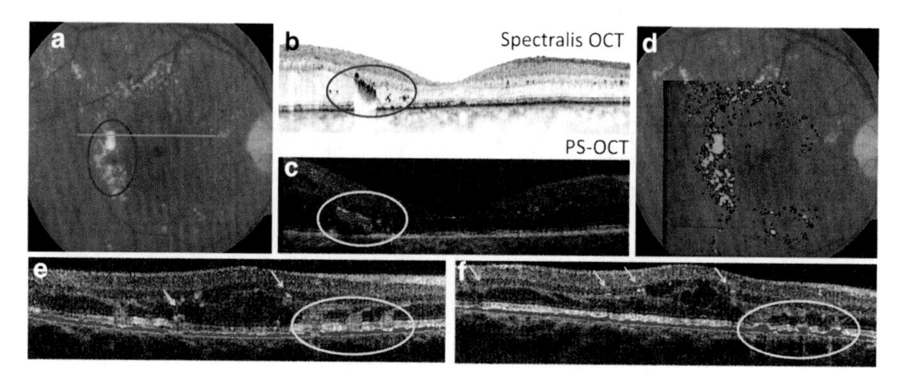

Fig. 9.13 The *red circle* in (**a**) highlights the area of interest in the color fundus photograph. The *green line* indicates the approximate position of the OCT scans. (**b**) The *red circle* shows a hard exudate in Spectralis OCT, with the typical shadow phenomenon below the hard exudate. (**c**) Shows the hard exudate segmented by PS-OCT—the fully automatic segmentation algorithm is based on the polarization-scrambling properties of the hard exudate. In image (**c**), the segmentation algorithm is assigned to detect only hard exudates without the RPE. From a volume scan (512 × 128 A-scans), a 3D thickness map of the HE can be generated: overlay of the 3D thickness map and CF image (**d**). (**e**) Shows a typical DME, the *yellow circle* indicates three coagulation spots 1 day after photocoagulation. Note the thinning/gap of the polarization-scrambling layer. One week after laser treatment, polarization-scrambling column-like formations are observed (**f**). The *arrows* mark polarization-scrambling precursors of hard exudates

of PS-OCT imaging compared to color fundus photography was reported [140]. Detailed analysis of follow-up images revealed precursors of hard exudates in PS-OCT before being detectable in other imaging modalities, such as color fundus photography. These hyperreflective foci, or microexudates, appear distinctly as isolated spots throughout all retinal layers, but obviously accumulate in the outer nuclear and outer plexiform layers with time, forming what is then visible as hard exudate in the biomicroscopic fundus examination, infrared imaging, or color fundus photography.

The phenomenon of hard exudate formation also depends on the specific dynamics of the DME. With a decrease in retinal thickness following treatment of DME, either by photocoagulation or by anti-VEGF treatment, the microexudates either resolve completely or become confluent at the apical border of the outer nuclear layer and eventually form ophthalmoscopically detectable hard exudates.

Thus, the specific treatment has an impact on the progression. While hard exudates developed in the course of several weeks after photocoagulation [136], ophthalmoscopically visible plaques can be found in a very short time period, e.g., after 3 days, according to the fast decrease in retinal thickness after anti-VEGF treatment [137]. Also, a dissemination of plaques of hard exudates into multiple, separate, hyperreflective foci is observed over time.

Based on these findings and the polarization-scrambling properties of these microexudates, further studies are in progress to quantitatively analyze the therapeutically induced effects on the amount of intraretinal lipoprotein exudation using PS-OCT [139].

9.2.3.2 Effects of Photocoagulation

The most widely accepted and, until recently, standard treatment of DME is focal or grid photocoagulation of the retina. During the last decade, several pharmacologic treatments for DME have been proposed, such as intravitreal injections of anti-VEGF agents or cortisol.

Recent studies show a paradigm shift from the former gold standard of exclusive photocoagulation to a combination or single therapy with such agents [141]. The specific mechanisms by which focal photocoagulation reduce DME, however, remain fairly unknown. The most common hypothesis on the therapeutic effect, measured as reduced retinal blood flow, includes improvements in retinal tissue oxygenation [128, 135, 142, 143], overall reduced retinal tissue, or biochemical changes at the level of the RPE [143–145]. Furthermore, until investigated by SD-OCT, it was not clear which retinal layers are altered secondary to laser treatment and if there are specific changes of laser lesions in the human retina with time [146].

PS-OCT has proved to be a valuable diagnostic tool for visualizing the effect of a grid laser treatment on retinal morphology in patients with DME in detail [147]. A significant thinning of the polarization-scrambling layers, ranging from focal thinning to the presence of a gap, could be detected 1 day after photocoagulation, depending on the previously applied laser fluence. Furthermore, the increased SD-OCT laser beam transmission into the choroid lying below the altered retinal layers indicates a local loss of RPE cells. At 1 week after laser treatment, the lesions show a column-like growth of polarization-scrambling tissue at the level of the photoreceptor layer reaching the external limiting membrane (Fig. 9.13). One month after laser treatment, the spots showed the same typical appearance as at 1 week, with an increase of polarization-scrambling tissue and hyperreflective alterations in the outer nuclear layer (ONL). A surrounding RPE atrophy, which was recently described in panretinal photocoagulation [148], however, was not observed. Also, the light transmission below the lesions returned to a degree similar to that of pretreatment findings. While the physical appearance of the ONL returned to normal within 2 months, the characteristic columns of polarization-scrambling tissue remained unchanged until 3 months following photocoagulation.

These findings support previous observations in the rodent retina following local photocoagulation. Both Paulus et al. and Li et al. reported a local RPE cell proliferation rather than fibrotic glial scarring [149, 150].

9.3 Conclusion and Future Directions

PS-OCT provides depth-resolved information about the polarization-scrambling properties of retinal tissue. Due to the depolarizing characteristics of melanin, PS-OCT can easily detect and localize the exact position of the RPE to analyze its integrity and condition in various diseases. Various studies have highlighted the potential of PS-OCT imaging for identifying, differentiating, and monitoring diseases affecting the RPE, as well as diseases that are characterized by pathologic depolarizing deposits in the retina, such as in DME.

Further studies are on the way to investigate the clinical value of PS-OCT imaging in other RPE-related diseases, including Morbus Stargardt, Morbus Best, and retinitis pigmentosa. Furthermore, imaging of choroidal nevi and melanoma might reveal characteristic features valuable for diagnosis and perhaps even prognosis. Birefringence can be utilized to measure changes in the nerve fiber layer, which might be useful in diseases such as glaucoma and multiple sclerosis.

Further technical improvements have been developed, such as an eye tracking solution for minimizing movement errors as well as advanced segmentation algorithms. Future research will also show the value of choroidal PS-OCT imaging using a light source with 1,060 nm wavelength.

Acknowledgements The authors would like to gratefully acknowledge the Austrian Science Fund (FWF grants P16776-N02 and P19624-B02) and EU-Project FUN OCT (FP7 HEALTH, Contract No. 201880) for the financial support.

References

1. D. Huang, E.A. Swanson, C.P. Lin, J.S. Schuman, W.G. Stinson, W. Chang, M.R. Hee, T. Flotte, K. Gregory, C.A. Puliafito, Optical coherence tomography. Science **254**(5035), 1178–1181 (1991)
2. A.F. Fercher, W. Drexler, C.K. Hitzenberger, T. Lasser, Optical coherence tomography—principles and applications. Rep. Prog. Phys. **66**, 239–303 (2003)
3. W. Drexler, J.G. Fujimoto, State-of-the-art retinal optical coherence tomography. Prog. Retin. Eye Res. **27**(1), 45–88 (2008)
4. A.F. Fercher, C.K. Hitzenberger, G. Kamp, S.Y. Elzaiat, Measurement of intraocular distances by backscattering spectral interferometry. Opt. Commun. **117**(1–2), 43–48 (1995)
5. G. Häusler, M.W. Lindner, "Coherence radar" and "spectral radar"—New tools for dermatological diagnosis. J. Biomed. Opt. **3**, 21–31 (1998)
6. M. Wojtkowski, R. Leitgeb, A. Kowalczyk, T. Bajraszewski, A.F. Fercher, In vivo human retinal imaging by Fourier domain optical coherence tomography. J. Biomed. Opt. **7**(3), 457–463 (2002)
7. R. Leitgeb, C.K. Hitzenberger, A.F. Fercher, Performance of fourier domain vs. time domain optical coherence tomography. Opt. Express. **11**(8), 889–894 (2003)
8. J.F. de Boer, B. Cense, B.H. Park, M.C. Pierce, G.J. Tearney, B.E. Bouma, Improved signal-to-noise ratio in spectral-domain compared with time-domain optical coherence tomography. Opt. Lett. **28**(21), 2067–2069 (2003)
9. M.A. Choma, M.V. Sarunic, C.H. Yang, J.A. Izatt, Sensitivity advantage of swept source and Fourier domain optical coherence tomography. Opt. Express. **11**(18), 2183–2189 (2003)
10. T.C. Chen, B. Cense, M.C. Pierce, N. Nassif, B.H. Park, S.H. Yun, B.R. White, B.E. Bouma, G.J. Tearney, J.F. de Boer, Spectral domain optical coherence tomography—Ultra-high speed, ultra-high resolution ophthalmic imaging. Arch. Ophthalmol. **123**(12), 1715–1720 (2005)

11. U. Schmidt-Erfurth, R.A. Leitgeb, S. Michels, B. Povazay, S. Sacu, B. Hermann, C. Ahlers, H. Sattmann, C. Scholda, A.F. Fercher, W. Drexler, Three-dimensional ultrahigh-resolution optical coherence tomography of macular diseases. Invest. Ophthalmol. Vis. Sci. **46**(9), 3393–3402 (2005)
12. M. Wojtkowski, V. Srinivasan, J.G. Fujimoto, T. Ko, J.S. Schuman, A. Kowalczyk, J.S. Duker, Three-dimensional retinal imaging with high-speed ultrahigh-resolution optical coherence tomography. Ophthalmology **112**(10), 1734–1746 (2005)
13. M.R. Hee, D. Huang, E.A. Swanson, J.G. Fujimoto, Polarization-sensitive low-coherence reflectometer for birefringence characterization and ranging. J. Opt. Soc. Am. B Opt. Phys. **9**(6), 903–908 (1992)
14. J.F. de Boer, T.E. Milner, M.J.C. van Gemert, J.S. Nelson, Two-dimensional birefringence imaging in biological tissue by polarization-sensitive optical coherence tomography. Opt. Lett. **22**(12), 934–936 (1997)
15. J.F. de Boer, T.E. Milner, J.S. Nelson, Determination of the depth-resolved Stokes parameters of light backscattered from turbid media by use of polarization-sensitive optical coherence tomography. Opt. Lett. **24**(5), 300–302 (1999)
16. G. Yao, L.V. Wang, Two-dimensional depth-resolved Mueller matrix characterization of biological tissue by optical coherence tomography. Opt. Lett. **24**(8), 537–539 (1999)
17. S.L. Jiao, L.H.V. Wang, Jones-matrix imaging of biological tissues with quadruple-channel optical coherence tomography. J. Biomed. Opt. **7**(3), 350–358 (2002)
18. C.K. Hitzenberger, E. Götzinger, M. Sticker, M. Pircher, A.F. Fercher, Measurement and imaging of birefringence and optic axis orientation by phase resolved polarization sensitive optical coherence tomography. Opt. Express. **9**(13), 780–790 (2001)
19. B.H. Park, M.C. Pierce, B. Cense, J.F. de Boer, Jones matrix analysis for a polarization-sensitive optical coherence tomography system using fiber-optic components. Opt. Lett. **29**(21), 2512–2514 (2004)
20. M. Todorovi, S. Jiao, L.V. Wang, G. Stoica, Determination of local polarization properties of biological samples in the presence of diattenuation by use of Mueller optical coherence tomography. Opt. Lett. **29**(20), 2402–2404 (2004)
21. E. Götzinger, M. Pircher, W. Geitzenauer, C. Ahlers, B. Baumann, S. Michels, U. Schmidt-Erfurth, C.K. Hitzenberger, Retinal pigment epithelium segmentation by polarization sensitive optical coherence tomography. Opt. Express. **16**(21), 16410–16422 (2008)
22. E. Götzinger, M. Pircher, C.K. Hitzenberger, High speed spectral domain polarization sensitive optical coherence tomography of the human retina. Opt. Express. **13**(25), 10217–10229 (2005)
23. M. Pircher, E. Götzinger, B. Baumann, C.K. Hitzenberger, Corneal birefringence compensation for polarization sensitive optical coherence tomography of the human retina. J. Biomed. Opt. **12**(4), 041210 (2007)
24. B. Baumann, E. Gotzinger, M. Pircher, H. Sattmann, C. Schuutze, F. Schlanitz, C. Ahlers, U. Schmidt-Erfurth, C.K. Hitzenberger, Segmentation and quantification of retinal lesions in age-related macular degeneration using polarization-sensitive optical coherence tomography. J. Biomed. Opt. **15**(6), 061704 (2010)
25. L.J. Bour, Polarized light and the eye, in *Visual Optics and Instrumentation*, ed. by W.N. Charman (CRC Press, Boca Raton, FL, 1991), pp. 310–325
26. E. Götzinger, M. Pircher, M. Sticker, A.F. Fercher, C.K. Hitzenberger, Measurement and imaging of birefringent properties of the human cornea with phase-resolved, polarization-sensitive optical coherence tomography. J. Biomed. Opt. **9**(1), 94–102 (2004)
27. B. Baumann, E. Götzinger, M. Pircher, C.K. Hitzenberger, Single camera based spectral domain polarization sensitive optical coherence tomography. Opt. Express. **15**(3), 1054–1063 (2007)
28. M. Yamanari, S. Makita, Y. Yasuno, Polarization-sensitive swept-source optical coherence tomography with continuous source polarization modulation. Opt. Express. **16**(8), 5892–5906 (2008)

29. A.W. Dreher, K. Reiter, R.N. Weinreb, Spatially resolved birefringence of the retinal nerve fiber layer assessed with a retinal laser ellipsometer. Appl. Opt. **31**(19), 3730–3735 (1992)
30. B. Cense, T.C. Chen, B.H. Park, M.C. Pierce, J.F. de Boer, Thickness and birefringence of healthy retinal nerve fiber layer tissue measured with polarization-sensitive optical coherence tomography. Invest. Ophthalmol. Vis. Sci. **45**(8), 2606–2612 (2004)
31. H.B. Brink, G.J. van Blokland, Birefringence of the human foveal area assessed in vivo with Mueller-matrix ellipsometry. J. Opt. Soc. Am. A **5**(1), 49–57 (1988)
32. M. Pircher, E. Götzinger, R. Leitgeb, H. Sattmann, O. Findl, C.K. Hitzenberger, Imaging of polarization properties of human retina in vivo with phase resolved transversal PS-OCT. Opt. Express. **12**(24), 5940–5951 (2004)
33. S. Michels, M. Pircher, W. Geitzenauer, C. Simader, E. Götzinger, O. Findl, U. Schmidt-Erfurth, C.K. Hitzenberger, Value of polarisation-sensitive optical coherence tomography in diseases affecting the retinal pigment epithelium. Br. J. Ophthalmol. **92**(2), 204–209 (2008)
34. E. Götzinger, B. Baumann, M. Pircher, C.K. Hitzenberger, Polarization maintaining fiber based ultra-high resolution spectral domain polarization sensitive optical coherence tomography. Opt. Express. **17**(25), 22704–22717 (2009)
35. N. Kemp, H. Zaatari, J. Park, H.G. Rylander Iii, T. Milner, Form-biattenuance in fibrous tissues measured with polarization-sensitive optical coherence tomography (PS-OCT). Opt. Express. **13**(12), 4611–4628 (2005)
36. J.M. Schmitt, S.H. Xiang, Cross-polarized backscatter in optical coherence tomography of biological tissue. Opt. Lett. **23**(13), 1060–1062 (1998)
37. S. Jiao, G. Yao, L.V. Wang, Depth-resolved two-dimensional Stokes vectors of backscattered light and Mueller matrices of biological tissue measured with optical coherence tomography. Appl. Opt. **39**(34), 6318–6324 (2000)
38. M. Pircher, E. Götzinger, O. Findl, S. Michels, W. Geitzenauer, C. Leydolt, U. Schmidt-Erfurth, C.K. Hitzenberger, Human macula investigated in vivo with polarization-sensitive optical coherence tomography. Invest. Ophthalmol. Vis. Sci. **47**(12), 5487–5494 (2006)
39. M. Pircher, E. Goetzinger, R. Leitgeb, C.K. Hitzenberger, Transversal phase resolved polarization sensitive optical coherence tomography. Phys. Med. Biol. **49**(7), 1257–1263 (2004)
40. E. Götzinger, M. Pircher, B. Baumann, C. Ahlers, W. Geitzenauer, U. Schmidt-Erfurth, C.K. Hitzenberger, Three-dimensional polarization sensitive OCT imaging and interactive display of the human retina. Opt. Express. **17**(5), 4151–4165 (2009)
41. B. Baumann, E. Götzinger, M. Pircher, C.K. Hitzenberger, Measurements of depolarization distribution in the healthy human macula by polarization sensitive OCT. J. Biophotonics. **2**(6–7), 426–434 (2009)
42. B. Baumann, S.O. Baumann, T. Konegger, M. Pircher, E. Götzinger, H. Sattmann, M. Litschauer, C.K. Hitzenberger, Polarization sensitive optical coherence tomography of melanin provides tissue inherent contrast based on depolarization. Proc. SPIE **7554**, 75541M (2010)
43. E. Götzinger, M. Pircher, M. Sticker, A.F. Fercher, C.K. Hitzenberger, Measurement and imaging of birefringent properties of the human cornea with phase-resolved, polarization-sensitive optical coherence tomography. J. Biomed. Opt. **9**(1), 94–102 (2004)
44. E. Götzinger, M. Pircher, I. Dejaco-Ruhswurm, S. Kaminski, C. Skorpik, C.K. Hitzenberger, Imaging of birefringent properties of keratoconus corneas by polarization-sensitive optical coherence tomography. Invest. Ophthalmol. Vis. Sci. **48**(8), 3551–3558 (2007)
45. A. Miyazawa, M. Yamanari, S. Makita, M. Miura, K. Kawana, K. Iwaya, H. Goto, Y. Yasuno, Tissue discrimination in anterior eye using three optical parameters obtained by polarization sensitive optical coherence tomography. Opt. Express. **17**(20), 17426–17440 (2009)
46. Y. Yasuno, M. Yamanari, K. Kawana, T. Oshika, M. Miura, Investigation of post-glaucoma-surgery structures by three-dimensional and polarization sensitive anterior eye segment optical coherence tomography. Opt. Express. **17**(5), 3980–3996 (2009)

47. M.R. Hee, J.A. Izatt, E.A. Swanson, D. Huang, J.S. Schuman, C.P. Lin, C.A. Puliafito, J.G. Fujimoto, Optical coherence tomography of the human retina. Arch. Ophthalmol. **113**(3), 325–332 (1995)
48. C. Ahlers, U. Schmidt-Erfurth, Three-dimensional high resolution OCT imaging of macular pathology. Opt. Express. **17**(5), 4037–4045 (2009)
49. M.L. Gabriele, G. Wollstein, H. Ishikawa, J. Xu, J. Kim, L. Kagemann, L.S. Folio, J.S. Schuman, Three dimensional optical coherence tomography imaging: advantages and advances. Prog. Retin. Eye Res. **29**(6), 556–579 (2010)
50. N. Congdon, B. O'Colmain, C.C. Klaver, R. Klein, B. Muñoz, D.S. Friedman, J. Kempen, H.R. Taylor, P. Mitchell, Eye Diseases Prevalence Research Group, Causes and prevalence of visual impairment among adults in the United States. Arch. Ophthalmol. **122**(4), 477–485 (2004)
51. S. Resnikoff, D. Pascolini, D. Etya'ale, I. Kocur, R. Pararajasegaram, G.P. Pokharel, S.P. Mariotti, Global data on visual impairment in the year 2002. Bull. World Health Organ. **82**(11), 844–851 (2004)
52. D.S. Friedman, B.J. O'Colmain, B. Muñoz, S.C. Tomany, C. McCarty, P.T. de Jong, B. Nemesure, P. Mitchell, J. Kempen, Eye Diseases Prevalence Research Group, Prevalence of age-related macular degeneration in the United States. Arch. Ophthalmol. **122**(4), 564–572 (2004)
53. M.A. Zarbin, Current concepts in the pathogenesis of age-related macular degeneration. Arch. Ophthalmol. **122**(4), 598–614 (2004)
54. H.E. Grossniklaus, W.R. Green, Choroidal neovascularization. Am. J. Ophthalmol. **137**(3), 496–503 (2004)
55. R.D. Jager, W.F. Mieler, J.W. Miller, Age-related macular degeneration. N. Engl. J. Med. **358**(24), 2606–2617 (2008)
56. W. Green, Histopathology of age-related macular degeneration. Mol. Vis. **5**, 27 (1999)
57. A. Abdelsalam, L. Del Priore, M. Zarbin, Drusen in age-related macular degeneration: pathogenesis, natural course, and laser photocoagulation-induced regression. Surv. Ophthalmol. **44**(1), 1–29 (1999)
58. L.V. Johnson, S. Ozaki, M.K. Staples, P.A. Erickson, D.H. Anderson, A potential role for immune complex pathogenesis in drusen formation. Exp. Eye Res. **70**(4), 441–449 (2000)
59. G.S. Hageman, P.J. Luthert, N.H. Victor Chong, L.V. Johnson, D.H. Anderson, R.F. Mullins, An integrated hypothesis that considers drusen as biomarkers of immune-mediated processes at the RPE-Bruch's membrane interface in aging and age-related macular degeneration. Prog. Retin. Eye Res. **20**(6), 705–732 (2001)
60. C.A. Curcio, N.E. Medeiros, C.L. Millican, Photoreceptor loss in age-related macular degeneration. Invest. Ophthalmol. Vis. Sci. **37**(7), 1236–1249 (1996)
61. S.H. Sarks, J.J. Arnold, M.C. Killingsworth, J.P. Sarks, Early drusen formation in the normal and aging eye and their relation to age related maculopathy: a clinicopathological study. Br. J. Ophthalmol. **83**(3), 358–368 (1999)
62. H. Al-Hussaini, M. Schneiders, P. Lundh, G. Jeffery, Drusen are associated with local and distant disruptions to human retinal pigment epithelium cells. Exp. Eye Res. **88**(3), 610–612 (2009)
63. The Age-Related Eye Disease Study Research Group, Risk factors associated with age-related macular degeneration. A case-control study in the age-related eye disease study: age-related eye disease study report number 3. Ophthalmology **107**(12), 2224–2232 (2000)
64. L. Zhao, Z. Wang, Y. Liu, Y. Song, Y. Li, A.M. Laties, R. Wen, Translocation of the retinal pigment epithelium and formation of sub-retinal pigment epithelium deposit induced by subretinal deposit. Mol. Vis. **13**, 873–880 (2007)
65. H.P. Scholl, S.S. Dandekar, T. Peto, C. Bunce, W. Xing, S. Jenkins, A.C. Bird, What is lost by digitizing stereoscopic fundus color slides for macular grading in age-related maculopathy and degeneration? Ophthalmology **111**(1), 125–132 (2004)

66. F.G. Schlanitz, B. Baumann, C. Ahlers, T. Spalek, S.M. Schriefl, I. Golbaz, M. Pircher, E. Gotzinger, C.K. Hitzenberger, U. Schmidt-Erfurth, Automatic delineation of drusen with polarization-sensitive optical coherence tomography. ARVO Poster #1783/A532-2010
67. F.G. Schlanitz, C. Ahlers, S. Sacu, C. Schütze, M. Rodriguez, S. Schriefl, I. Golbaz, T. Spalek, G. Stock, U. Schmidt-Erfurth, Performance of drusen detection by spectral-domain optical coherence tomography. Invest. Ophthalmol. Vis. Sci. **51**(12), 6715–6721 (2010)
68. M. Rudolf, M.E. Clark, M.F. Chimento, C.M. Li, N.E. Medeiros, C.A. Curcio, Prevalence and morphology of druse types in the macula and periphery of eyes with age-related maculopathy. Invest. Ophthalmol. Vis. Sci. **49**(3), 1200–1209 (2008)
69. A.A. Khanifar, A.F. Koreishi, J.A. Izatt, C.A. Toth, Drusen ultrastructure imaging with spectral domain optical coherence tomography in age-related macular degeneration. Ophthalmology **115**(11), 1883–1890 (2008)
70. C.N. Keilhauer, F.C. Delori, Near-infrared autofluorescence imaging of the fundus: visualization of ocular melanin. Invest. Ophthalmol. Vis. Sci. **47**(8), 3556–3564 (2006)
71. A.C. Bird, N.M. Bressler, S.B. Bressler, I.H. Chisholm, G. Coscas, M.D. Davis, P.T. de Jong, C.C. Klaver, B.E. Klein, R. Klein, An international classification and grading system for age-related maculopathy and age-related macular degeneration. Surv. Ophthalmol. **39**(5), 367–374 (1995)
72. J.S. Sunness, J. Gonzalez-Baron, C.A. Applegate, N.M. Bressler, Y. Tian, B. Hawkins, Y. Barron, A. Bergman, Enlargement of atrophy and visual acuity loss in the geographic atrophy form of age-related macular degeneration. Ophthalmology **106**(9), 1768–1779 (1999)
73. J.S. Sunness, The natural history of geographic atrophy, the advanced atrophic form of age-related macular degeneration. Mol. Vis. **5**, 25 (1999)
74. J.S. Sunness, E. Margalit, D. Srikumaran, C.A. Applegate, Y. Tian, D. Perry, B.S. Hawkins, N.M. Bressler, The long-term natural history of geographic atrophy from age-related macular degeneration: enlargement of atrophy and implications for interventional clinical trials. Ophthalmology **114**(2), 271–277 (2007)
75. M. Fleckenstein, P. Charbel Issa, H.M. Helb, S. Schmitz-Valckenberg, R.P. Finger, H.P. Scholl, K.U. Loeffler, F.G. Holz, High-resolution spectral domain-OCT imaging in geographic atrophy associated with age-related macular degeneration. Invest. Ophthalmol. Vis. Sci. **49**(9), 4137–4144 (2008)
76. B.J. Lujan, P.J. Rosenfeld, G. Gregori, F. Wang, R.W. Knighton, W.J. Feuer, C.A. Puliafito, Spectral domain optical coherence tomographic imaging of geographic atrophy. Ophthalmic Surg. Lasers Imaging **40**(2), 96–101 (2009)
77. M. Fleckenstein, S. Schmitz-Valckenberg, C. Adrion, I. Krämer, N. Eter, H.M. Helb, C.K. Brinkmann, P. Charbel Issa, U. Mansmann, F.G. Holz, Tracking progression with spectral-domain optical coherence tomography in geographic atrophy caused by age-related macular degeneration. Invest. Ophthalmol. Vis. Sci. **51**(8), 3846–3852 (2010)
78. C. Ahlers, E. Götzinger, M. Pircher, I. Golbaz, F. Prager, C. Schütze, B. Baumann, C.K. Hitzenberger, U. Schmidt-Erfurth, Imaging of the retinal pigment epithelium in age-related macular degeneration using polarization sensitive optical coherence tomography. Invest. Ophthalmol. Vis. Sci. **51**(4), 2149–2157 (2010)
79. F.G. Holz, C. Bellman, S. Staudt, F. Schütt, H.E. Völcker, Fundus autofluorescence and development of geographic atrophy in age-related macular degeneration. Invest. Ophthalmol. Vis. Sci. **42**(5), 1051–1056 (2001)
80. F.G. Holz, A. Bindewald-Wittich, M. Fleckenstein, J. Dreyhaupt, H.P. Scholl, S. Schmitz-Valckenberg, FAM-Study Group, Progression of geographic atrophy and impact of fundus autofluorescence patterns in age-related macular degeneration. Am. J. Ophthalmol. **143**(3), 463–472 (2007)
81. U.E.K. Wolf-Schnurrbusch, V. Enzmann, C.K. Brinkmann, S. Wolf, Morphologic changes in patients with geographic atrophy assessed with a novel spectral OCT-SLO combination. Invest. Ophthalmol. Vis. Sci. **49**(7), 3095–3099 (2008)

82. R.P. Singh, C. Patel, J.E. Sears, Management of subretinal macular haemorrhage by direct administration of tissue plasminogen activator. Br. J. Ophthalmol. **90**(4), 429–431 (2006)
83. C.A. Toth, L.S. Morse, L.M. Hjelmeland, M.B. Landers 3rd, Fibrin directs early retinal damage after experimental subretinal hemorrhage. Arch. Ophthalmol. **109**(5), 723–729 (1991)
84. J.D. Benner, A. Hay, M.B. Landers 3rd, L.M. Hjelmeland, L.S. Morse, Fibrinolytic-assisted removal of experimental subretinal hemorrhage within seven days reduces outer retinal degeneration. Ophthalmology **101**(4), 672–681 (1994)
85. R.L. Avery, S. Fekrat, B.S. Hawkins, N.M. Bressler, Natural history of subfoveal subretinal hemorrhage in age-related macular degeneration. Retina **16**(3), 183–189 (1996)
86. S.R. Bennett, J.C. Folk, C.F. Blodi, M. Klugman, Factors prognostic of visual outcome in patients with subretinal hemorrhage. Am. J. Ophthalmol. **109**(1), 33–37 (1990)
87. M.H. Berrocal, M.L. Lewis, H.W. Flynn Jr, Variations in the clinical course of submacular hemorrhage. Am. J. Ophthalmol. **122**(4), 486–493 (1996)
88. M.T.S. Tennant, J.L. Borrillo, C.D. Regillo, Management of submacular hemorrhage. Ophthalmol. Clin. North Am. **15**(4), 445vi–452vi (2002)
89. M.A. Hochman, C.M. Seery, M.A. Zarbin, Pathophysiology and management of subretinal hemorrhage. Surv. Ophthalmol. **42**(3), 195–213 (1997)
90. U.M. Schmidt-Erfurth, C. Pruente, Management of neovascular age-related macular degeneration. Prog. Retin. Eye Res. **26**(4), 437–451 (2007)
91. C. Ahlers, I. Golbaz, G. Stock, A. Fous, S. Kolar, C. Pruente, U. Schmidt-Erfurth, Time course of morphologic effects on different retinal compartments after ranibizumab therapy in age-related macular degeneration. Ophthalmology **115**(8), e39–e46 (2008)
92. N. Shams, T. Ianchulev, Role of vascular endothelial growth factor in ocular angiogenesis. Ophthalmol. Clin. North Am. **19**(3), 335–344 (2006)
93. A. Kvanta, P.V. Algvere, L. Berglin, S. Seregard, Subfoveal fibrovascular membranes in age-related macular degeneration express vascular endothelial growth factor. Invest. Ophthalmol. Vis. Sci. **37**(9), 1929–1934 (1996)
94. C. Schuetze, C. Ahlers, B. Baumann, M. Pircher, E. Götzinger, R. Donner, J. Ofner, C. Hitzenberger, U. Schmidt-Erfurth, Automatic segmentation of subretinal fluid in choroidal neovascularization using polarization-sensitive optical coherence tomography. ARVO Poster #4935/A552-2010
95. A.C. Bird, Doyne Lecture. Pathogenesis of retinal pigment epithelial detachment in the elderly; the relevance of Bruch's membrane change. Eye (Lond). **5**(Pt 1), 1–12 (1991)
96. S.H. Sarks, New vessel formation beneath the retinal pigment epithelium in senile eyes. Br. J. Ophthalmol. **57**(12), 951–965 (1973)
97. L.J. Singerman, J.H. Stockfish, Natural history of subfoveal pigment epithelial detachments associated with subfoveal or unidentifiable choroidal neovascularization complicating age-related macular degeneration. Graefes Arch. Clin. Exp. Ophthalmol. **227**(6), 501–507 (1989)
98. C. Kunze, A.E. Elsner, E. Beausencourt, L. Moraes, M.E. Hartnett, C.L. Trempe, Spatial extent of pigment epithelial detachments in age-related macular degeneration. Ophthalmology **106**(9), 1830–1840 (1999)
99. C. Cedrone, C. Nucci, G. Scuderi, F. Ricci, A. Cerulli, F. Culasso, Prevalence of blindness and low vision in an Italian population: a comparison with other European studies. Eye (Lond). **20**(6), 661–667 (2006)
100. H.A. Quigley, A.T. Broman, The number of people with glaucoma worldwide in 2010 and 2020. Br. J. Ophthalmol. **90**(3), 262–267 (2006)
101. H.A. Quigley, R.W. Nickells, L.A. Kerrigan, M.E. Pease, D.J. Thibault, D.J. Zack, Retinal ganglion cell death in experimental glaucoma and after axotomy occurs by apoptosis. Invest. Ophthalmol. Vis. Sci. **36**(5), 774–786 (1995)
102. L.A. Kerrigan, D.J. Zack, H.A. Quigley, S.D. Smith, M.E. Pease, TUNEL-positive ganglion cells in human primary open-angle glaucoma. Arch. Ophthalmol. **115**(8), 1031–1035 (1997)

103. H.A. Quigley, G.R. Dunkelberger, W.R. Green, Chronic human glaucoma causing selectively greater loss of large optic nerve fibers. Ophthalmology **95**(3), 357–363 (1988)
104. A. Heijl, M.C. Leske, B. Bengtsson, L. Hyman, M. Hussein, Early Manifest Glaucoma Trial Group, Reduction of intraocular pressure and glaucoma progression: results from the early manifest glaucoma trial. Arch. Ophthalmol. **120**(10), 1268–1279 (2002)
105. A. Tuulonen, P.J. Airaksinen, Initial glaucomatous optic disk and retinal nerve fiber layer abnormalities and their progression. Am. J. Ophthalmol. **111**(4), 485–490 (1991)
106. H.A. Quigley, Early detection of glaucomatous damage. II. Changes in the appearance of the optic disk. Surv. Ophthalmol. **30**(2), 111, 117–126 (1985)
107. W.F. Hoyt, L. Frisén, N.M. Newman, Fundoscopy of nerve fiber layer defects in glaucoma. Invest. Ophthalmol. **12**(11), 814–829 (1973)
108. R.S. Harwerth, L. Carter-Dawson, F. Shen, E.L. Smith 3rd, M.L. Crawford, Ganglion cell losses underlying visual field defects from experimental glaucoma. Invest. Ophthalmol. Vis. Sci. **40**(10), 2242–2250 (1999)
109. L.A. Kerrigan-Baumrind, H.A. Quigley, M.E. Pease, D.F. Kerrigan, R.S. Mitchell, Number of ganglion cells in glaucoma eyes compared with threshold visual field tests in the same persons. Invest. Ophthalmol. Vis. Sci. **41**(3), 741–748 (2000)
110. R.N. Weinreb, L.M. Zangwill, S. Jain, L.M. Becerra, K. Dirkes, J.R. Piltz-Seymour, G.A. Cioffi, G.L. Trick, A.L. Coleman, J.D. Brandt, J.M. Liebmann, M.O. Gordon, M.A. Kass, OHTS CSLO Ancillary Study Group, Predicting the onset of glaucoma: the confocal scanning laser ophthalmoscopy ancillary study to the ocular hypertension treatment study. Ophthalmology **117**(9), 1674–1683 (2010)
111. F.A. Medeiros, G. Vizzeri, L.M. Zangwill, L.M. Alencar, P.A. Sample, R.N. Weinreb, Comparison of retinal nerve fiber layer and optic disc imaging for diagnosing glaucoma in patients suspected of having the disease. Ophthalmology **115**(8), 1340–1346 (2008)
112. M. Lalezary, F.A. Medeiros, R.N. Weinreb, C. Bowd, P.A. Sample, I.M. Tavares, A. Tafreshi, L.M. Zangwill, Baseline optical coherence tomography predicts the development of glaucomatous change in glaucoma suspects. Am. J. Ophthalmol. **142**(4), 576–582 (2006)
113. Q. Zhou, J. Reed, R.W. Betts, P.K. Trost, P.W. Lo, C. Wallace, R.H. Bienias, G. Li, R. Winnick, W.A. Papworth, M. Sinai, Detection of glaucomatous retinal nerve fiber layer damage by scanning laser polarimetry with custom corneal compensation. Proc. SPIE **4951**, 32 (2003)
114. R.P. Hemenger, Birefringence of a medium of tenuous parallel cylinders. Appl. Opt. **28**(18), 4030–4034 (1989)
115. Q. Zhou, R.W. Knighton, Light scattering and form birefringence of parallel cylindrical arrays that represent cellular organelles of the retinal nerve fiber layer. Appl. Opt **36**(10), 2273–2285 (1997)
116. X.R. Huang, R.W. Knighton, Microtubules contribute to the birefringence of the retinal nerve fiber layer. Invest. Ophthalmol. Vis. Sci. **46**(12), 4588–4593 (2005)
117. X.R. Huang, H. Bagga, D.S. Greenfield, R.W. Knighton, Variation of peripapillary retinal nerve fiber layer birefringence in normal human subjects. Invest. Ophthalmol. Vis. Sci. **45**(9), 3073–3080 (2004)
118. E. Götzinger, M. Pircher, B. Baumann, C. Hirn, C. Vass, C.K. Hitzenberger, Retinal nerve fiber layer birefringence evaluated with polarization sensitive spectral domain OCT and scanning laser polarimetry: a comparison. J. Biophotonics. **1**(2), 129–139 (2008)
119. H.G. Rylander 3rd, N.J. Kemp, J. Park, H.N. Zaatari, T.E. Milner, Birefringence of the primate retinal nerve fiber layer. Exp. Eye Res. **81**(1), 81–89 (2005)
120. T.E. Ogden, Nerve fiber layer of the primate retina: thickness and glial content. Vision Res. **23**(6), 581–587 (1983)
121. G.M. Pocock, R.G. Aranibar, N.J. Kemp, C.S. Specht, M.K. Markey, H.G. Rylander 3rd, The relationship between retinal ganglion cell axon constituents and retinal nerve fiber layer birefringence in the primate. Invest. Ophthalmol. Vis. Sci. **50**(11), 5238–5246 (2009)

122. E. Götzinger, M. Pircher, B. Baumann, H. Resch, C. Vass, C.K. Hitzenberger, Comparison of retinal nerve fiber layer birefringence and thickness of healthy and glaucoma suspect eyes measured with polarization sensitive spectral domain OCT. ARVO Poster #5823/A161-2009
123. J.C. Vickers, R.A. Schumer, S.M. Podos, R.F. Wang, B.M. Riederer, J.H. Morrison, Differential vulnerability of neurochemically identified subpopulations of retinal neurons in a monkey model of glaucoma. Brain Res. **680**(1–2), 23–35 (1995)
124. A.J. Weber, P.L. Kaufman, W.C. Hubbard, Morphology of single ganglion cells in the glaucomatous primate retina. Invest. Ophthalmol. Vis. Sci. **39**(12), 2304–2320 (1998)
125. T. Shou, J. Liu, W. Wang, Y. Zhou, K. Zhao, Differential dendritic shrinkage of alpha and beta retinal ganglion cells in cats with chronic glaucoma. Invest. Ophthalmol. Vis. Sci. **44**(7), 3005–3010 (2003)
126. C. Balaratnasingam, W.H. Morgan, L. Bass, S.J. Cringle, D.Y. Yu, Time-dependent effects of elevated intraocular pressure on optic nerve head axonal transport and cytoskeleton proteins. Invest. Ophthalmol. Vis. Sci. **49**(3), 986–999 (2008)
127. J.H. Kempen, B.J. O'Colmain, M.C. Leske, S.M. Haffner, R. Klein, S.E. Moss, H.R. Taylor, R.F. Hamman, Eye Diseases Prevalence Research Group, The prevalence of diabetic retinopathy among adults in the United States. Arch. Ophthalmol. **122**(4), 552–563 (2004)
128. R. Klein, B.E. Klein, S.E. Moss, K.J. Cruickshanks, The Wisconsin Epidemiologic Study of Diabetic Retinopathy. XV. The long-term incidence of macular edema. Ophthalmology **102**(1), 7–16 (1995)
129. Group TDCaCTR. The effect of intensive treatment of diabetes on the development and progression of long-term complications in insulin-dependent diabetes mellitus. The Diabetes Control and Complications Trial Research Group. N. Engl. J. Med. **329**(14), 977–986 (1993)
130. P. Reichard, B.Y. Nilsson, U. Rosenqvist, The effect of long-term intensified insulin treatment on the development of microvascular complications of diabetes mellitus. N. Engl. J. Med. **329**(5), 304–309 (1993)
131. UK Prospective Diabetes Study Group, Tight blood pressure control and risk of macrovascular and microvascular complications in type 2 diabetes: UKPDS 38. Br. Med. J. **317**(7160), 703–713 (1998)
132. ACCORD Study Group, ACCORD Eye Study Group, E.Y. Chew, W.T. Ambrosius, M.D. Davis, R.P. Danis , S. Gangaputra, C.M. Greven, L. Hubbard, B.A. Esser, J.F. Lovato, L.H. Perdue, D.C. Goff Jr, W.C. Cushman, H.N. Ginsberg, M.B. Elam, S. Genuth, H.C. Gerstein, U. Schubart, L.J. Fine, Effects of medical therapies on retinopathy progression in type 2 diabetes. N. Engl. J. Med. **363**(3), 233–244 (2010)
133. R. Klein, B.E. Klein, S.E. Moss, M.D. Davis, D.L. DeMets, The Wisconsin epidemiologic study of diabetic retinopathy: X. Four-year incidence and progression of diabetic retinopathy when age at diagnosis is 30 years or more. Arch. Ophthalmol. **107**(2), 244–249 (1989)
134. Early Treatment Diabetic Retinopathy Study Research Group, Photocoagulation for diabetic macular edema. ETDRS report number 1. Arch. Ophthalmol. **103**(12), 1796–1806 (1985)
135. Early Treatment Diabetic Retinopathy Study Research Group, Treatment techniques and clinical guidelines for photocoagulation of diabetic macular edema. ETDRS Report No 2. Ophthalmology **94**(7), 761–774 (1987)
136. G.G. Deák, M. Bolz, K. Kriechbaum, S. Prager, G. Mylonas, C. Scholda, U. Schmidt-Erfurth, Diabetic Retinopathy Research Group Vienna, Effect of retinal photocoagulation on intraretinal lipid exudates in diabetic macular edema documented by optical coherence tomography. Ophthalmology **117**(4), 773–779 (2010)
137. M. Bolz, U. Schmidt-Erfurth, G. Deak, G. Mylonas, K. Kriechbaum, C. Scholda, Diabetic Retinopathy Research Group Vienna, Optical coherence tomographic hyperreflective foci: a morphologic sign of lipid extravasation in diabetic macular edema. Ophthalmology **116**(5), 914–920 (2009)
138. W. Soliman, B. Sander, T.M. Jørgensen, Enhanced optical coherence patterns of diabetic macular oedema and their correlation with the pathophysiology. Acta Ophthalmol. Scand. **85**(6), 613–617 (2007)

139. M. Bolz, B. Pemp, J. Lammer, B. Baumann, B. Wetzel, M. Pircher, C.K. Hitzenbergerv, U. Schmidt-Erfurth, Diabetic Retinopathy Research Group Vienna, The influence of anti-VEGF agents on intra-retinal lipo-protein exudates assessed by spectral domain and polarisations sensitive optical coherence tomography. ARVO Poster #5060/D991-2010
140. J. Lammer, M. Bolz, B. Baumann, M. Pircher, B. Wetzel, C.K. Hitzenberger, U. Schmidt-Erfurth, Diabetic Retinopathy Research Group (DRRG) Vienna, Automated detection and quantification of hard exudates in diabetic macular edema using polarization sensitive optical coherence tomography. ARVO Poster #4660/D935-2010
141. Diabetic Retinopathy Clinical Research Network, M.J. Elman, L.P. Aiello, R.W. Beck, N.M. Bressler, S.B. Bressler, A.R. Edwards, F.L. Ferris 3rd, S.M. Friedman, A.R. Glassman, K.M. Miller, I.U. Scott, C.R. Stockdale, J.K. Sun, Randomized trial evaluating ranibizumab plus prompt or deferred laser or triamcinolone plus prompt laser for diabetic macular edema. Ophthalmology **117**(6), 1064.e35–1077.e35 (2010)
142. E.Y. Chew, M.L. Klein, F.L. Ferris 3rd, N.A. Remaley, R.P. Murphy, K. Chantry, B.J. Hoogwerf, D. Miller, Association of elevated serum lipid levels with retinal hard exudate in diabetic retinopathy: early treatment diabetic retinopathy study (ETDRS) report 22. Arch. Ophthalmol. **114**(9), 1079–1084 (1996)
143. D.J. Wilson, D. Finkelstein, H.A. Quigley, W.R. Green, Macular grid photocoagulation: an experimental study on the primate retina. Arch. Ophthalmol. **106**(1), 100–105 (1988)
144. A. Arnarsson, E. Stefánsson, Laser treatment and the mechanism of edema reduction in branch retinal vein occlusion. Invest. Ophthalmol. Vis. Sci. **41**(3), 877–879 (2000)
145. Writing Committee for the Diabetic Retinopathy Clinical Research Network, D.S. Fong, S.F. Strauber, L.P. Aiello, R.W. Beck, D.G. Callanan, R.P. Danis, M.D. Davis, S.S. Feman, F. Ferris, S.M. Friedman, C.A. Garcia, A.R. Glassman, D.P. Han, D. Le, C. Kollman, A.K. Lauer, F.M. Recchia, S.D. Solomon, Comparison of the modified early treatment diabetic retinopathy study and mild macular grid laser photocoagulation strategies for diabetic macular edema. Arch. Ophthalmol. **125**(4), 469–480 (2007)
146. M. Bolz, K. Kriechbaum, C. Simader, G. Deak, J. Lammer, C. Treu, C. Scholda, C. Prünte, U. Schmidt-Erfurth, Diabetic Retinopathy Research Group Vienna, In vivo retinal morphology after grid laser treatment in diabetic macular edema. Ophthalmology **117**(3), 538–544 (2010)
147. J. Lammer, M. Bolz, G. Deak, S.G. Prager, B. Baumann, M. Pircher, E. Götzinger, C.K. Hitzenberger, U. Schmidt-Erfurth, Diabetic Retinopathy Research Group Vienna, In vivo effects of laser treatment on retinal morphology observed by polarization sensitive OCT. ARVO abstract #2080-2009
148. K. Kriechbaum, M. Bolz, G.G. Deak, S. Prager, C. Scholda, U. Schmidt-Erfurth, High-resolution imaging of the human retina in vivo after scatter photocoagulation treatment using a semiautomated laser system. Ophthalmology **117**(3), 545–551 (2010)
149. Y.M. Paulus, A. Jain, R.F. Gariano, B.V. Stanzel, M. Marmor, M.S. Blumenkranz, D. Palanker, Healing of retinal photocoagulation lesions. Invest. Ophthalmol. Vis. Sci. **49**(12), 5540–5545 (2008)
150. T. Li, Q.L. Luo, H.Y. Wu, Histopathologic an immunohistochemical studies on retina after laser photocoagulation (abstract). Sichuan Da Xue Xue Bao Yi Xue Ban **35**(4), 512–515 (2004)

Chapter 10
Adaptive Optics in Ocular Optical Coherence Tomography

Enrique Josua Fernández and Pablo Artal

Abstract Adaptive optics (AO) is a technology for correcting aberrations in real time. When applied to the human eye, it has the potential of perfect imaging, from an optical perspective, the retina. Once aberrations from the eye have been compensated, theoretical resolution achievable in the living retina is 2–3 μm. Therefore, individual cells and most of the morphological structures on the retina could be in principle imaged. Optical coherence tomography (OCT) has benefitted from this novel technique since 2004. The singularities of OCT, mainly the confocal detection and the mandatory use of broadband spectral light sources, imposes particular methods when applying AO. In a few years, many advances in the combination of AO with OCT have emerged. The in vivo images obtained with that modality have unveiled amazing details of the intraretinal tissue. In this chapter, both the theory and the practice of merging AO with OCT, with special emphasis on ultrahigh-resolution (UHR) OCT, will be presented and discussed.

10.1 Introduction

Optical coherence tomography (OCT) is an interferometric technique [1], based on the use of low-coherence light. It has been used to image the cross-sectional structures within the living eye. In the case of the retina, it provides relevant information for the diagnosis and pathogenesis of a large variety of conditions associated to degeneration of the retinal tissue. Subtle morphological changes in the retinal structure can be detected with this technique. Due to the inherent interferometric character of OCT, a rigorous analysis of the image formation in this

E.J. Fernández (✉)
Laboratorio de Óptica, Centro de Investigación en Óptica y Nanofísica (CiOyN), Universidad de Múrcia, Campus de Espinardo, E-30071 Murcia, Spain
e-mail: enriquej@um.es; pablo@um.es

R. Bernardes and J. Cunha-Vaz (eds.), *Optical Coherence Tomography*, Biological and Medical Physics, Biomedical Engineering,
DOI 10.1007/978-3-642-27410-7_10,

modality demands considering some singular aspects which usually do not appear together in other ophthalmoscopic techniques.

It is widely extended now the use of ultrahigh-resolution OCT (UHR OCT) [2–4]. The latter refers for those situations where the employed light sources have broadband spectra, typically exhibiting widths of 80 nm FWHM (full widths at half maximum) and beyond. With broad spectral bandwidths, resolution of 2–3 μm axial in the living tissue has been demonstrated. For instance, such resolutions have been reported by employing mode-locked titanium: sapphire pulsed lasers sources in the near infrared [5]. In OCT, axial resolution is mostly driven by the width of the spectral light source. In addition, some fundamental limits are imposed by the electronics, source power, interferometer parameters, sample reflectivity properties, etc. All of them manifest mainly through noise in the detection [6]. The other key parameter for characterizing the imaging capabilities of the system is transverse or lateral resolution. It could be understood as the ability of the system for detecting fine structures which are placed in a single C-scan or normal plane regarding the incident direction of sampling light. Transverse resolution is limited by the optics of the setup and that of the eye, and it is directly related to the possibility of focusing a concentrated spot onto the patient's retina. In low-coherence interferometry, a singular situation pertaining resolution is found, since axial and transverse resolutions are nearly independent. In a simplistic model, axial resolution can be taken to be as mainly governed by the spectral bandwidth of the light source, while transverse resolution is given and limited by the optical quality of the optical system. Details on axial resolution will not be provided in the current chapter, since other section of the book addresses this particular issue. Transverse or lateral resolution obeys classic constraints through the shape and size of the exit pupil and the existing aberrations in the optical system. This simplified picture does not account for the interaction across spectral bandwidth and transverse resolution, whose importance and impact in the overall quality of retrieved images will be described and discussed in detail in this chapter.

Assuming the optical system devoted to the illumination of the retina to be well designed, the solely source of aberrations remaining, and indeed the degradation source of such signals, is the patient's eye optics. Ideally, ophthalmoscopic images should be retrieved through a large entrance pupil for projecting a minimum spot on the observer's retina. Unfortunately, such way makes ocular aberrations very important, degrading the retinal images up to unreasonable limits. Therefore, using large pupils required the simultaneous mandatory correction of ocular aberrations. Here, AO arises as the successful technique able for real-time measuring and correction of ocular aberrations. Enlarging the entrance pupil could be understood as augmenting the numerical aperture of the objective of our "biological" microscope: the eye. When larger numerical aperture is used, magnification of images increases accordingly, so we claim to own a more powerful microscope. In most of the ophthalmic OCT instruments, the illuminating beam is typically 1.5–2 mm diameter. It must be noted that for such dimensions, ocular aberrations barely affect the quality of images. For larger pupil diameters, the correction of aberrations by using adaptive optics (AO) would improve resolution.

In the following, first a simplified description of the optical characteristics of the human eye will be presented. The rest of the chapter is devoted to several aspects concerning the combination of AO and OCT for high-resolution retinal imaging, with special emphasis in UHR OCT.

10.2 The Eye as an Optical System and Adaptive Optics

The human eye is a simple, but extremely robust, optical instrument [7]. It is composed of only two positive lenses, the cornea and the crystalline lens, which produce real images of the world on the retina, initiating the visual process. When compared with artificial optical systems, often formed by many more lenses, the eye is simple but well adapted to the requirements of the visual system. The cornea is approximately a spherical section with a radius of 7.8 mm and refractive index of 1.377. The lens is a biconvex lens with radius of 10.2 mm and -6 mm for the anterior and posterior surfaces. The internal structure of the lens is layered which produces a nonhomogeneous refractive index, larger in the center than in the periphery and with a mean value of 1.42. An average eye with these dimensions having an axial length of 24.2 mm will image distance objects precisely in focus into the retina. This ideal situation is called emmetropia. The optical surfaces are not exactly spherical in shape, and they are not perfectly aligned. A common manifestation of these facts is the presence of ocular aberrations.

The retinal image of a point source is not another point but an extended distribution of light. This sets the minimum detail that can be imaged in the retina and then visual resolution. Several factors are responsible for the degradation of the retinal images: diffraction of the light in the eye's pupil, optical aberration, and intraocular scattering. Diffraction blurs the images formed through instruments with a limited aperture due to the wave nature of the light. The effect of diffraction in the eye is small and only noticeable with very small pupils. Aberrations affect all optical instruments and in particular the eye. Light rays entering the eye at different parts of the pupil are not focused at the same retinal locations causing an additional blur in the image. The impact of ocular aberrations in image quality is more significant for larger pupil diameters. The aperture of the eye ranges from f/8 to f/2. The amount of aberrations for a normal eye with about f/4 aperture is approximately equivalent to less than 0.25 diopters of defocus, a small error typically not corrected when dealing with refractive errors, although this can be quite significant in ophthalmoscopic imaging. In the normal young eye, the optical properties of the two ocular components are tuned to produce an improved overall image quality [8, 9]. In particular, two of the most important aberrations, coma and spherical aberration, are partially corrected in the eye, in a similar way as in an aplanatic optical system [10]. The optics of the eye is optimized with the crystalline lens acting as an aspheric compensator to provide a correction of corneal spherical aberration and also with the proper choice of shapes to reduce coma. During normal aging, this balance is partially disrupted leading to an increase of the eye's aberrations and a degradation of the retinal image quality [11]. An additional source of degradation of the retinal image is due to the spectral content of the white light and the dispersive nature of the eye media. Since the optical power of the eye depends on the wavelength of the light, an eye perfectly focusing green light in the retina will be myopic for blue and hyperopic for red light. This is known as chromatic aberration and causes a colored blur in the retinal images; although with a limited impact in vision, it is important in OCT applications.

Beyond the effect of aberrations, scattered light reduces the performance of any optical system in terms of imaging, including the eye [12]. In particular, intraocular scatter degrades retinal image quality and diminishes both visual acuity (VA) and the contrast sensitivity function (CSF). This is also generally related to glare which forms a veil of luminance in the eye. Although, under normal conditions, the presence of intraocular scatter in a young healthy eye is low, this may become significant with aging [13] or after refractive surgery (corneal haze). The loss of transparency of the lens (ultimately resulting in cataract formation) is the main source of scatter in the older eye.

If the aberrations of the eye are known, it is possible to correct them using a wavefront correcting device that compensates for the eye's aberrations in real time. This is a direct application of AO to the eye. In the ideal case, the system of *"corrector + eye"* becomes permanently aberration-free, producing perfect retinal images. In different laboratories, AO in the eye has been demonstrated using deformable mirrors or liquid crystal spatial light modulators as corrector devices [14–18]. These systems are still laboratory prototypes that include a wavefront sensor and a corrector, allowing investigators to perform testing as visual simulators [19, 20] or to record high-resolution retinal images through near to aberration-free ocular optics [15, 17, 21, 22].

10.3 Transverse Resolution in OCT

Since the correction of eye's aberrations would increase OCT transverse resolution, it is important point the understanding of how those aberrations degrade the quality of images. In this section, it will be shown what is actually occurring in the sample arm of the ophthalmic OCT setup and how correcting aberrations could improve the system's performance. It must be taken into account that there are many different designs for OCT instruments, each of them with its peculiarities and constraints. For accurate results, one should model the system as detailed as possible. In this section, we will refer our results to the most general situation: a fiberized ophthalmic OCT setup.

The use of optical fibers in OCT has made possible the use of this technology in the clinic. From the technical point of view, using fibers in the interferometer is a major advance. It avoids the need of precise alignment, as it is required in free-space interferometers, which would bring about continuous refinement and adjustment of the setup. When using fibers, the interference of the signals coming from the sample, the retina, and the reference arm interfere in a fiber coupler. Afterward, the modulated energy is recorded by the detector. An immediate constraint of the use of fibers arises the low efficiency coupling of light into the fiber tip, once it has been backscattered by the retina. Fine focusing of the light is requested. In addition, fibers employed in OCT are single-mode fibers, so that coherence is transmitted with low losses. Typical core diameters are in the 2–6 μm range, if near-infrared light is used for retinal imaging, with spectral bandwidths of 630–850 nm. Note that some OCT setup employs infrared light in the surroundings of 1,050 nm, obtaining deeper penetrations into the choroid tissue, so different fiber cores diameters might apply. The size of the fiber core is so small as compared with entrance pupil of the eye that in practice confocal detection must be considered in such OCT setups. Classical confocal microscopes take advantage of the small pinhole located in front of the detector from several perspectives. Possibly the most important is the increase in transverse resolution, governed by the pinhole diameter. Additionally, axial resolution is significantly increased since the pinhole avoids light from other image planes distinct of the focusing plane. Those characteristics are present in fiberized ophthalmic OCT instruments, although traditionally, little attention has been paid to their confocal character for estimating instrument resolution. Dual apparatus capable of both low-coherence interferometry and confocal imaging of the retina are however present in the literature [23, 24]. Using a single-mode fiber as pinhole in microscopy is by no means new, but it has been reported in the past [25]. The theory of optical quality and resolution through confocal detection has been studied in many different works; in particular, the effect of the size of the pinhole over resolution is a typical issue in the field [26, 27]. The goodness of the assumption of confocal detection is given by the normalized radius of the fiber, which ideally should be close to zero [28], as it is the case in most of the ophthalmic OCT setups. Under such constraints, the mathematical expression describing the intensity distribution of the image produced by the system from a point object is given by [29]

$$I = |\mathrm{PSF}_1|^2 \left\{ |\mathrm{PSF}_2|^2 \otimes D \right\}. \tag{10.1}$$

The point spread function (PSF) of the optical relay illuminating the sample, in this case the retina, is PSF1. Conversely, PSF2 corresponds to the PSF of the optical relay responsible for the detection of light backscattered from the retina. In (10.1), PSF2 is convolved with the intensity sensitivity D. For a perfect confocal detection, D would be described mathematically by a Dirac delta function. If a perfect confocal system is assumed, therefore indicating that the diameter of the pinhole is negligible as compared with the system's exit pupil, the expression of the effective PSF of the entire system simplifies to the square modulus of the product of the two involved PSFs. In addition, if the optical relay for illuminating the retina is also employed for bringing back the light to the detector, as it is the case in ophthalmic OCT, then (10.1) further simplifies since PSF 1 and PSF 2 are identical. In that case, the overall PSF is the square of the PSF in a single pass of the light. Note that previous simplifications notably reduced the complexity of (10.1), but the assumptions we have adopted could be not applicable in all OCT setups. The problem now reduces to find the expression for the PSF of the illumination relay. The general expression is

$$\mathrm{PSF}_i(\lambda,\rho) \propto \left(\frac{1}{\lambda}\right)^2 \left| \iint_{Ai} J_0(2\pi\rho r)\xi(\lambda,r)\exp[\mathrm{i}2\pi\Phi(\lambda,r)]\,r\mathrm{d}r \right|^2 \quad (10.2)$$

with the term J_0 as the Bessel function of first class and first order. The amplitude pupil transmission is given by the function ξ. This function can be assumed as constant across the pupil in most situations. The function Φ describes the optical aberrations at the exit pupil of the system. Since we have assumed a perfectly designed and assembled optical system, the source of aberrations will be the patient's eye. Ocular aberrations can be generally expressed as a mathematical expansion of independent and orthogonal Zernike polynomials Z_i [30–32]:

$$\Phi(\lambda,r) = \sum_{i=1}^{\infty} a_i Z_i. \quad (10.3)$$

The coefficient a_i is the value of the Zernike polynomial Z_i at the edge of the pupil. Note that (10.1)–(10.3) are dependent of the wavelength, which will have a profound impact in OCT. In the rest of ophthalmoscopic modalities, monochromatic light is used for illuminating the retina. The use of broad spectral light sources is however mandatory in OCT, since axial resolution is given by the temporal coherence of the source, as it has been discussed previously. The wavelength dependence of the ocular aberrations could be understood in terms of the chromatic aberration of the eye. Since point-like images are recorded in OCT sequentially, it is valid the assumption that only the longitudinal chromatic aberration is involved in this case because the transverse chromatic aberration is defined for extended objects. The longitudinal chromatic aberration in the eye is well known from 400 to 1,070 nm [33–41]. Assuming that illumination is performed in the NIR, a semiempirical equation describing the defocus, expressed as the fourth Zernike polynomial coefficient, as a function of wavelength is given by

$$a_4 = -\left(\frac{r^2}{4\sqrt{3}}\right) 0.0021(\lambda - 800), \quad (10.4)$$

where the radius of the pupil r must be in millimeters and λ in nanometers for obtaining the coefficient in micrometers. To obtain the final confocal and polychromatic PSF of the system, (10.2) must be integrated with the spectral distribution of the illumination, whose function is given by G:

$$\mathrm{ConfPSF_{Polychromatic}} = \left|\mathrm{PSF_{Polychromatic}}\right|^2 = \left|\frac{1}{\mathrm{K}}\frac{1}{\mathrm{M}}\int_{-\infty}^{+\infty} \mathrm{PSF}_i(\lambda,\rho)\,G(\lambda)\,\mathrm{d}\lambda\right|^2. \quad (10.5)$$

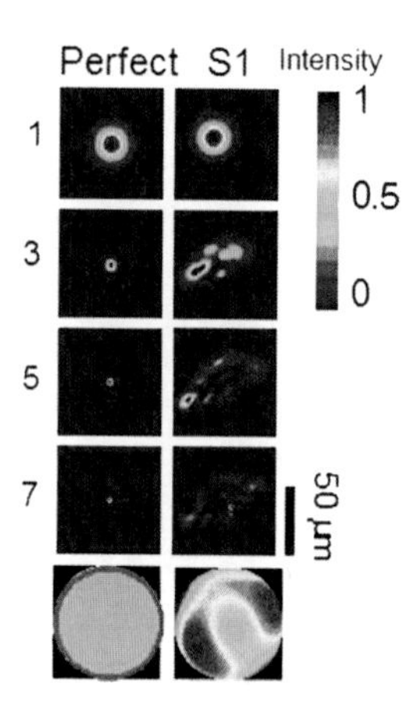

Fig. 10.1 Confocal polychromatic point spread functions through different pupil diameters in the perfect and real eye. Light source emitted a Gaussian spectrum of 80 nm FWHM. Adapted from Fernández et al. [28]

Taking G as a Gaussian function is usually appropriate for most of the spectral sources, although incorporating other spectral shapes closer to the actual situation is relatively simple in the expression (10.5). The parameters K and M are introduced for normalizing the intensity received at the detector, in terms of pupil size and spectral bandwidth of the light source, respectively. This last equation shows the impact of using polychromatic light sources for OCT on the quality of images. Even if we would neglect the impact of ocular chromatic aberration, the spectral shape of the light modulates the effective confocal PSF. Figure 10.1 shows the impact of both monochromatic and chromatic aberrations over the calculated confocal polychromatic PSF, assuming a Gaussian spectrum of 80 nm FWHM centered at 780 nm as a function of pupil diameter, compared with the perfect situation.

For the real case, the ocular aberrations of one of the authors of this chapter were used. The point images are normalized to their maxima. For small pupil sizes, the case in commercial instruments employing beam of 1 mm diameter, the images calculated from the real eye resemble the theoretical situation but for a slight tilt produced by the combined effect of all existing aberrations. For moderate pupil size, as 3 mm, the degradation of the distribution of energy in the real eye is significant. The effect is increasing as the diameter of the pupil size enlarges. The worst scenario occurs for a pupil size of 7 mm, where ocular aberrations are maxima. A curious detail of interest in this chapter arises here. Although the energy is distributed across a large retinal area in the real case as compared with the theoretical point, in the aberrated case, there is also a point approximately at the center of the image. The energy in this point is significantly smaller than in the perfect case, no doubt. Actually, the reader can perceive it just as an effect of scaling the images to their maximum value. This point, more evident for the 7-mm diameter case, arises as a consequence of confocal imaging. Incorporating aberrations distribute energy, subtracting intensity from the maximum, but the projection of the pinhole over the image plane still occurs. The previous result introduces an important idea for understanding resolution in OCT and in general in any ophthalmoscopic technique: contrast. In view of the existing central point in the degraded case for 7-mm pupil size, one would have the temptation of thinking that transverse resolution might be still somehow conserved, as an effect of confocal detection. Unfortunately, it is not the case as it will be shown in the following.

10.4 Effect of Aberration Correction in Contrast and Resolution

The existence of a remaining point-like pattern on the image in the presence of both chromatic and monochromatic ocular aberrations would in principle allow obtaining near to theoretical resolutions. That would occur if the amount of photons on the sample could be large enough, so the intensity at the detector would not be limited. But it cannot be the case, since imaging the living retina imposes exposure limits for keeping retinal tissues safe from thermal damage. Therefore, the energy introduced by the beacon light must be efficiently grouped in the sample first and then onto the detector. This is exactly the role and utility of AO in the context of ophthalmic OCT. We will present and discuss how and to what extent correcting aberrations increases resolution of retinal images. Effective resolution cannot be inferred in the retina from pure optical functions and parameters involving solely the sample arm, as PSF, MTF, etc. The properties of the retina significantly affect the achievable maximum resolution. The real scenario is even more complicated accounting for the optical properties and complexity of the different media that light passes through the retina (exhibiting different reflectivity, absorption, and scattering properties) at different wavelengths. A possible alternative for characterizing the problem is the use of contrast functions or similar approaches.

Contrast and resolution are not independent concepts. Figure 10.2 graphically shows its connection. They complement to each other for providing the effective visibility or resolution of the sample. As an example of the interaction across them, we show in Fig. 10.3 contrast transfer functions for different pupil sizes obtained assuming a fiberized UHR OCT, operating with a light source of 120 nm FWHM. In this case, we assume that the system is free of monochromatic aberrations, and solely longitudinal chromatic aberration of the eye is present. Considering two identical point objects, with their corresponding PSFs calculated from (10.5), contrast is represented as a function of the separation between the two images. Consequently, the effective resolution can be obtained as the distance at which two

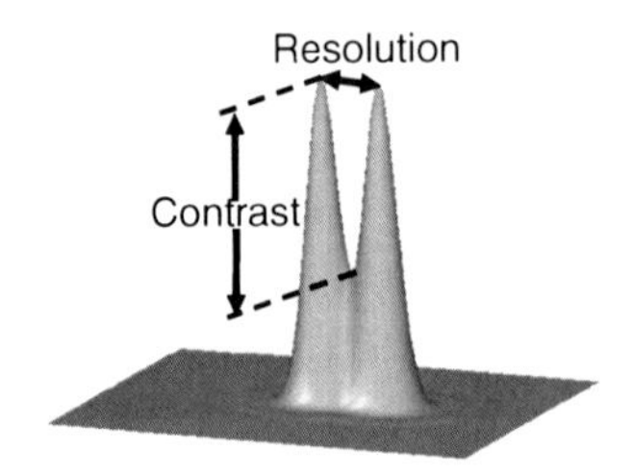

Fig. 10.2 Relationship between contrast and resolution in point images

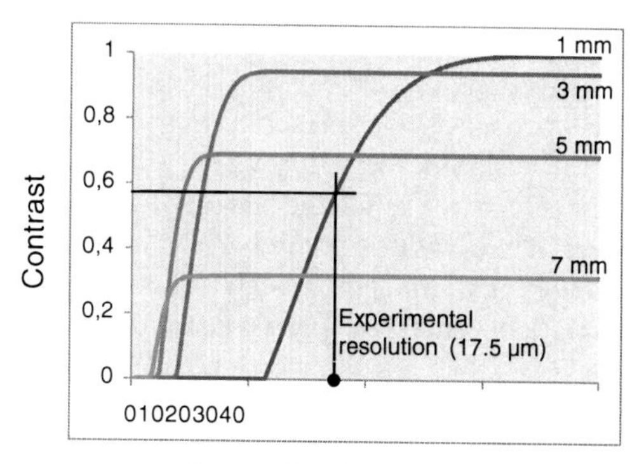

Fig. 10.3 Contrast as a function of separation of images for polychromatic illumination of 120 nm FWHM (Gaussian) and different pupil sizes. Average experimental resolution achieved with commercial systems in the human eye is depicted with *dashed line*. Adapted from Fernández et al. [28]

objects are imaged with a given value of contrast, allowing its effective detection. Note that this definition of effective resolution involves, in addition of the optical properties of the system, noise, sensitivity of the detector, etc.

To obtain the contrast functions depicted in Fig. 10.3, two identical intensity images from point objects are simulated for each pupil size to be progressively separated apart. Simultaneously, the contrast between them is calculated for each distance. The curves are monotonically increasing until they asymptotically saturate to their maximum value, beyond which there is no change at all. The maximum contrast corresponds to the peak value of the polychromatic confocal PSF, since when the images are sufficiently apart, there is no interaction between them, and no intensity is obtained in the intermediate space. In theory, resolution is given by the intersection of each curve with the separation distance axis. In practice, it is not the case since the number of photons is limited, together with the rest of existing factors. For finding the effective resolution, a particular value of contrast threshold must be selected. The intersection of the curves with this value, graphically a line parallel to the axis of separation distance, will produce the actual effective resolution. The problem is then reduced to find such value of contrast. The latter could be obtained from theoretical calculations or estimated empirically. Note that this value represents the detection threshold required for resolving two points in the instrument, so it is directly responsible of transverse resolution. For this particular example, we will adopt an experimental value for such threshold. In practice, we will use a value based on the resolution found in many commercial instruments. Transverse resolution is rather difficult to estimate in the living eye. The effective resolution test is the retina itself. Therefore, based on the knowledge of the dimensions of the different retinal features obtained from ex vivo microscopy, we can have a necessarily rough estimation of the minimum detail the system is able to correctly image and resolve. Imaging with commercial setups, using a 1-mm beam size so that spectral bandwidth is relatively unimportant for transverse resolution, intraretinal structures of size 15–20 μm are visible. Assuming a mean value of 17.5 μm, now it is possible to establish in Fig. 10.3 the contrast threshold, as it is indicated with a dashed line.

The intersection of the different curves with the contrast threshold indicates the effective transverse resolution. Note that for 7-mm pupil size, the contrast curve evolves beneath the threshold irrespectively of the separation distance. This fact indicates that no detection would occur in this case. The reason is found in the lack of correction for the longitudinal chromatic aberration of the eye. For this spectral bandwidth, 120 nm FWHM, the degradation of the confocal polychromatic PSF through 7 mm prevents for detecting any signal. Actually, correcting chromatic aberration would elevate the maximum values of contrast for any pupil size up to the level of 1 mm diameter, which was chosen as the reference value for normalization of the rest of the curves. This is an example where the correction of aberrations, in particular the chromatic aberration, would be mandatory for imaging cellular details at the retina.

The use of large pupils opens the door to enhanced resolutions, but aberration correction turns then out fundamental. The previous results introduce an important issue in UHR OCT: the need, or not, for correcting chromatic aberration in addition to traditional AO just compensating for monochromatic aberrations. The answer is not that simple as the question does. The benefit from compensating chromatic aberrations would be governed by two fundamental factors: the pupil size and the spectral bandwidth of the light source. Figure 10.4, adapted from [28], shows the interaction across these two variables, the Strehl ratio as a function of pupil size for different Gaussian spectral widths.

The curves were calculated assuming a fiberized OCT setup, working under confocal detection, and once monochromatic aberrations have been perfectly compensated. Consequently, solely longitudinal chromatic aberration from the eye is present. The curve labeled as real eyes shows the degradation produced by average ocular aberrations from four different young and healthy eyes, so this case is the only one incorporating both monochromatic and chromatic aberrations. Using 40 nm of spectral bandwidth, pupils of up to 6 mm approximately reach 90% of correction, as compared with the perfect case. The benefit of correcting monochromatic aberrations, from the average, is the difference across the considered curve and the one presenting real eyes. As the pupil size increases, the percent of improvement

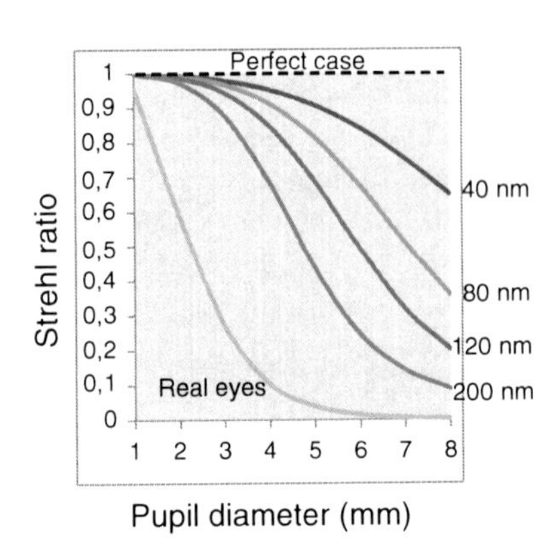

Fig. 10.4 Strehl ratio as a function of the pupil diameter for polychromatic confocal point spread function under different illuminations. *Curves* present the evolution when no chromatic aberration correction is performed confronted with the perfect case. Adapted from Fernández et al. [28]

just correcting monochromatic aberrations decreases, being progressively more important the compensation of chromatic aberrations. The latter is more evident when large spectral bandwidths are considered. The figure could be of help for those designing an AO system for UHR OCT. One should not consider very large pupils unless chromatic aberration is to be also corrected. If AO is to be performed, correcting solely monochromatic aberrations, either pupil size is constricted to reasonable limits or light source of moderate spectral bandwidth is coupled in the setup.

10.5 Correction of Ocular Monochromatic Aberrations in OCT

10.5.1 Time Domain

OCT was the last retinal ophthalmoscopic technique benefitting from the correction of aberrations. AO had already been demonstrated in scanning laser ophthalmoscopy [23] and flood illumination cameras [15, 17, 19], increasing the resolution of retinal images and even enabling the detection of individual photoreceptors at small eccentricities. The combination of AO and OCT was initially accomplished under the time-domain scheme. In this modality, retinal images were retrieved sequentially in depth through the mechanical displacement of a mirror in the reference arm of the interferometer [2]. Consequently, the acquisition velocity of the retinal images was rather limited. Attempts of merging aberration correction in OCT were proposed with a coherence-gated camera for en-face OCT imaging [42]. The first successful combination of AO and UHR OCT, still in time domain, was reported in 2004 [43]. In this work, AO was coupled to a commercial apparatus, OCT 1 instrument (Carl Zeiss Meditec AG, Dublin, California).

The available commercial instrument did not use of a broad spectral bandwidth illumination, so a Ti:sapphire laser (FEMTOLASERS Produktions GmbH, Vienna, Austria) emitting 130 nm at FWHM centered at 800 nm was employed instead. The change of the illumination was relatively simple to accomplish because of the fiberized nature of the setup. The AO system was mounted on an optical breadboard of 300 × 300 mm, attached to the head of the commercial setup. A 37-actuators electrostatic deformable mirror (OKO Technologies, Holland) was used as the correcting device [16, 18]. The device exhibited a limited capability of deformation, meaning that the amplitude or stroke of the mirrored membrane was not capable for correcting large amount of aberrations. Figure 10.5 shows a picture of the AO setup showing the main components. For measuring ocular aberrations, a Hartmann-Shack wavefront sensor was coupled to the system [14, 44, 45]. Static correction of both the system and the eye's aberrations were accomplished. In spite of the limited stroke of the corrector, still a notable increase in the signal-to-noise ratio was obtained in normal subjects, those presenting moderate and low levels of

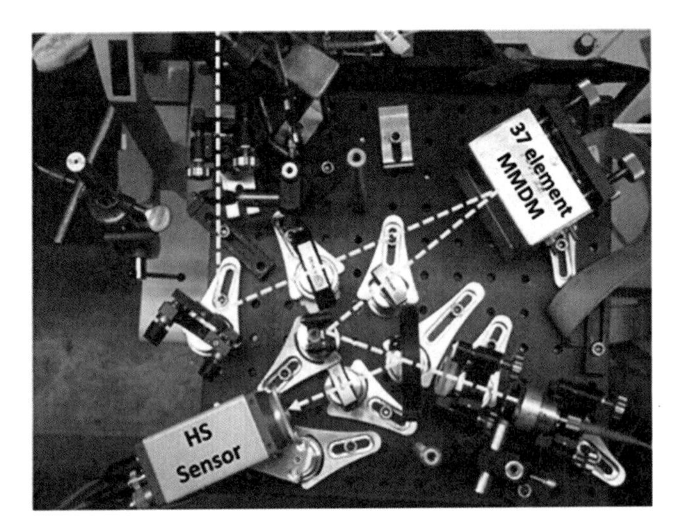

Fig. 10.5 Experimental AO setup attached to a commercial OCT instrument (OCT 1 Carl Zeiss Meditec AG, Dublin, California) used for high-resolution imaging of the living retina in 2003. The system was developed at the University of Vienna, Austria, in collaboration with the University of Murcia, Spain

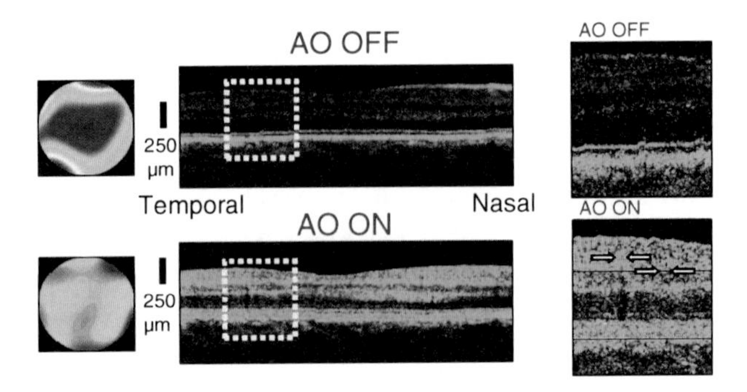

Fig. 10.6 Retinal tomograms obtained with and without adaptive optics using a 37-actuators electrostatic deformable mirror (OKO Technologies, Holland) in combination with time-domain UHR OCT. Adapted from Hermann et al. [43]

aberrations. In some cases, up to 9 dB increment was measured. The selected pupil size for correcting the ocular aberrations in this work was relatively modest: 3.63 mm diameter. The increase in signal-to-noise ratio, directly related with the augment in contrast, produced some interesting effects over the quality of the retinal images obtained with AO and UHR OCT.

Figure 10.6 illustrates the effect of AO in UHR OCT. Two tomograms, depicting cross-sectional retinal images from a normal human eye in the foveal region, are depicted. Images were retrieved across a transverse line of 2.8 mm width, with a sampling of 600 A-scans, with a lateral separation of 5 μm. In one case (top panel), the image was obtained through the natural aberrations of the subject, shown at the left side in a color-code representation. The image obtained with AO aberration correction is presented below. The correction, as discussed before, was partial because the limited stroke of the deformable mirror, as it can be noticed from the aberrations' map at the left side. The Strehl ratio of the associated PSF is improved by a factor of near 10 for the case of aberration compensation. The corrected image shows an evident increase in the contrast, allowing for a better segmentation of the different layers compounding the retinal tissue. A more detailed inspection of the image reveals interesting small features within the ganglion cells layer as well as the IPL in the corrected case. Those were very likely vessels with a diameter of 10–20 μm. These can be better identified in the right panel, where selected zones have been enlarged twofold for better visualization. Note that establishing the diameter of the vessel is an estimation based in previous knowledge of the retinal morphology from other techniques, as ex vivo microscopy. Therefore, there is a vague range of different resolutions, accounting for the minimum visible morphological detail we are able to detect, that we could claim from such images. Providing an exact number for the resolution is possibly as difficult as senseless, since different retinas might present distinct reflection or sizes in a given morphological feature. However, existing literature in most of the cases has embraced a different philosophy, sometimes paying excessive attention in providing a final and exact number for resolution. The moderate improvement in the signal-to-noise ratio (SNR) of 9 dB is clearly visible, notably enhancing the dynamic range, which was of 44 dB in the tomogram with the employed setup. The estimated axial resolution was 3 μm. Finally, the incorporation of the AO system in the apparatus introduced additional optical path, in terms of free space and material, into the interferometer. It had to be compensated in the reference arm for correctly acquiring retinal images. This fact will be a constant issue in every AO OCT system, turning more and more important as the amount of added optics in the AO system increases.

10.5.2 Frequency Domain

The implementation of Fourier domain or frequency-domain techniques in OCT was a significant advance in the field, since both the dynamic range of the retinal images and the acquisition speed experienced a notable burst [46–50]. An immediate consequence was the possibility of merging a large number of adjacent B-scans for digitally compounding three-dimensional volumetric images of the living retina. Previous time-domain OCT approaches were also affected of several artifacts arising from the relatively low acquisition rate of the A-scans, which demanded elaborated postprocessing of in vivo images. A sophisticated experimental setup was presented in 2005 as the first implementation of AO in frequency-domain OCT [51].

The system incorporated a fundus camera attached to the interferometer, for accurate control of focused plane onto the retina. The spectral bandwidth of the light source was 50 nm FWHM so it was out of the rage of UHR OCT. With a classical magnesium niobate actuators-driven deformable mirror correcting device (Xinetics, Inc. USA) of 37 independent modes, partial correction of ocular aberrations in normal eyes produced an enhancement of signal-to-noise ratio which allowed for detecting photoreceptors in the living retina. Actually, separations across individual photoreceptors at different eccentricities were estimated indirectly through Fourier analysis. An interesting concept introduced in this work, among others, was the possibility of controlling the focus plane in real time, within the retinal tissue. It was demonstrated with intensity profiles of the different retinal layers. Images were obtained through a 6-mm-diameter entrance pupil, with a sensitivity of 94 dB.

Due to the limited stroke of the deformable mirrors available at that time, we find in the previously mentioned works that aberrations were corrected in normal subjects and always partially. Ocular aberrations were not well corrected in pathological eyes or highly aberrated eyes, where the expected benefit on the resolution of retinal images should be largely more important. In this direction, an alternative correcting device was used for combining AO and UHR OCT [52]: a liquid crystal programmable phase modulator (LC-PPM [X7550, Hamamatsu Photonics K. K., Japan]) [53]. A pulsed laser source emitting a smooth Gaussian spectrum of 130 nm FWHM centered at 800 nm was coupled into the interferometer [54]. The LC-PPM consists essentially of a parallel-aligned liquid crystal layer sandwiched between a pair of transparent electrodes. One of the electrodes is in contact with an amorphous silicon wafer, a photoactive material whose impedance is a function of the external illumination. A microdisplay of 800×600 pixels is projected internally on the photoactive material, controlled as an external monitor, where the aberration profile is displayed as a gray-level image. Each gray level produces a change in impedance which translates into a local variation of the refractive index of the liquid crystal affected by the corresponding pixel. The effect is a change in the effective refractive index that modifies the optical path, resulting in a controlled wavefront shape. Once properly calibrated, the fidelity of these instruments is excellent (near to 95% and even 99%) with current generations of liquid crystal on silicon devices [55], therefore allowing open-loop operation. The calibration of these correctors is quite effective and simple, consisting on the determination of the gray level-to-phase gain [54, 55]. The amplitude of correction of the liquid crystal-based devices is not constrained by the mechanical deformation of any component, since they operate performing phase wrapping. Thus, these correctors have the potential of correcting not only moderate aberrations in healthy eyes but also aberrated cases where the use of AO would be more beneficial.

As an example, Fig. 10.7 shows an aberration of more than 14 wavelength (at 800 nm) corrected by the AO system coupled to the UHR OCT, exactly as it was internally displayed onto the LC-PPM. The associated monochromatic PSF before and after the aberration correction are also depicted in the figure. The Strehl ratio of the compensated PSF reached 0.95, which in practice is considered diffraction limited. The level of correction is far beyond the expected aberrations occurring even in the pathological cornea. The use of this technology requires of linearly polarized parallel to the original direction of the molecules of the liquid crystal. In principle, this fact introduces no limitations, since polarized light is used anyways

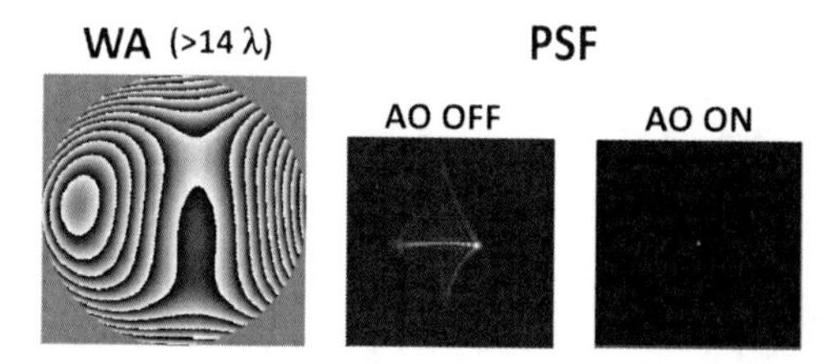

Fig. 10.7 Correction of aberrations performed by a liquid crystal programmable phase modulator (X7550, Hamamatsu KK, Japan) for frequency-domain UHR OCT. Adapted from Fernández et al. [52]

for making the interference occurs effectively at the fiber coupler. A possible issue arises for the employment of phase wrapping. A given wavelength must be selected for compensation; consequently, perfect aberration correction is solely accomplished at the center of the Gaussian spectrum. As we move to wavelengths located at the tails of the Gaussian, the correction is less effective. In practice, this fundamental limitation is not so critical, since wavelengths at the tails of the spectrum contribute significantly less than central does. Some basic calculations produce an estimation of 15–20% diminishment of correction at the tails of the spectrum for 140 nm FWHM Gaussian profile, which could be an acceptable value in many situations. Moreover, the employment of liquid crystal in moderate and regular axial resolution OCT, with spectral bandwidth of 80 nm and below, should be quite effective across the entire range. Using such technology, first volumetric images of the living retina were retrieved showing novel morphological details of certain layers. As an example, Fig. 10.8 shows interesting features detected with AO and UHR OCT with a liquid crystal corrector.

In Fig. 10.8, a sequence of adjacent en-face frames, rendered digitally after postprocessing of several B-scans, is presented. The sequence of images evolves toward posterior retina and spatially is located integrally at the external limited membrane. A detailed examination of the sequence reveals interesting structures in that particular retinal layer. Some rounded features are found in different and consecutive frames at nearly same locations, indicating that they correspond to real morphologic structures. The digital rendering allows even for estimating the length of such elongated structures. The interpretation of those filaments was not simple, since there was no reference images obtained with OCT ex vivo. The hypothesis was that these small dots corresponded to terminal bars of groups of photoreceptors, in particular rods. These terminal bars are known to be strongly reflective, as compared to the practically transparent photoreceptors. The figure incorporates an ex vivo image of this particular retinal layer obtained by standard light microscopy. The image is centered at the external limiting membrane, and the black dots, indicating higher reflectance, are the groups of terminal bars. Here arises a recurrent ambiguity often occurring in the interpretation of in vivo OCT images. There has been unfortunately paid not sufficient attention of the development UHR OCT microscopy, which should perhaps evolve in parallel to ophthalmic UHR OCT.

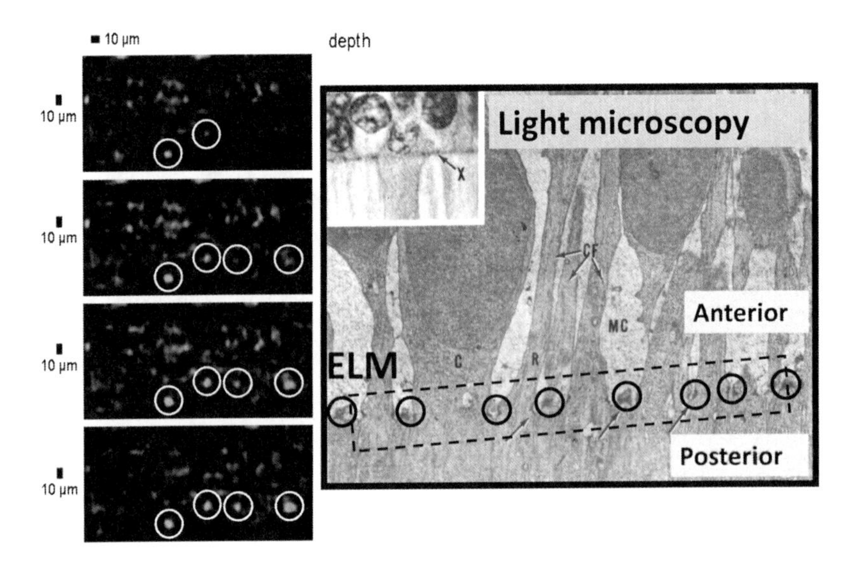

Fig. 10.8 Adjacent C-scans obtained in vivo with AO UHR OCT showing terminal bars of photoreceptors. Light microscopy image shows similar structures for comparison. Adapted from Fernández et al. [52]

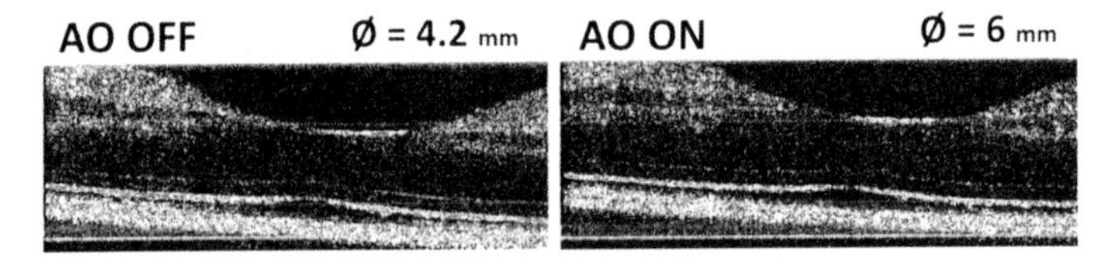

Fig. 10.9 Retinal images obtained from the same subject and location by using UHR OCT combined with AO showing the lack of benefit when performing monochromatic aberration correction through a pupil of 6 mm diameter as compared with a pupil of 4.2 mm diameter

In spite of the large number of advantages of its use, as unprecedented resolution, effective stroke for the correction and moderate cost, liquid crystals in OCT remain an unexplored field. Perhaps in the future the interest of the community turns into these interesting correctors again.

An interesting result reported in the aforementioned work [52] was the lack of benefit when correcting aberrations through a 6-mm pupil diameter with the illuminating source of 130 nm FWHM. By the time it was published, there was no even theoretical demonstration for the need of correcting chromatic aberration for such pupils and spectral bandwidths. Actually, it was such experimental result the origin of posterior studies trying to characterize the impact of chromatic aberration in UHR OCT. An experimental outcome of this fundamental limitation is presented in Fig. 10.9.

In Fig. 10.9, two images obtained from the same subject, under similar conditions and retinal location, and both with aberration correction, show the absence of improvement when correction is accomplished through a 6-mm pupil diameter as compared with the 4-mm-pupil situation. Images are ~1,500 μm width (1,000 pixel/1.5 μm sampling), and their noisy appearance obeys to the practically lack of digital postprocessing, so that the effect of correction could be better noticed.

The possibility of rendering volumetric representation of the living retina by using frequency-domain OCT, in combination with AO, was also reported by several works [56, 57]. Using a bimorph deformable mirror, in addition to partial aberration correction of the ocular optics, accurate focusing of the sampling light was demonstrated [56]. This work reported the first AO setup exclusively designed with spherical mirrors instead of lenses. The need of conjugating different planes, two different scanning mirrors and the deformable mirror, made mandatory the implementation of several telescopes. That increased the path of the light backscattered from the retina up to about 7 m before reaching the fiberized interferometer, being necessary the subsequent compensation in the reference arm. The use of mirrors presented the advantage of avoiding back reflections from the system to the interferometer and the wavefront sensor. Particularly for the sensor is a critical issue since back reflections are usually more intense than the fraction of light back-reflected from the retina. With the mirror-based setup, it was possible using the illuminating beam as the beacon light for wavefront sensing. An evolution of the previous system [57] incorporated two deformable mirrors. One was dedicated to low-order aberrations, mainly defocus and astigmatism, while the other corrected the rest of high-order aberrations. Such approach is being used for many other applications; recent advances in control algorithms for dual mirrors can be found in Li et al. 2010 [58]. Aberration corrections were performed through a pupil size of 6.67 mm diameter, and solely normal eyes, with moderate aberrations, were compensated. Due to the large pupil diameter, the authors constrained the spectral bandwidth of the light source to 50 nm FWHM, which was within reasonable limits for avoiding the effect of the uncorrected ocular chromatic aberration. Digital postprocessing of the images produced fascinating volumetric representation of the retina, showing the three-dimensional structure of some layers.

AO was also applied to a simultaneous en-face OCT/SLO system for improving resolution of retinal images [59]. In the aforementioned work, a 37-actuator deformable mirror [Oko Technologies, Holland] was applied to 6-mm pupil diameter. Consequently, only partial correction of aberrations was accomplished. The spectral width of the light source was 17 nm, yielding an axial resolution in the retina of 13 μm. In spite of the limited stroke of the deformable mirror, the correction improved contrast of images up to the level of resolving individual photoreceptors.

Another work in the direction of combining AO with a dual instrument capable for SLO and OCT, but achieving a higher sampling speed, was reported in [60]. Effective cone separation at different eccentricities could be estimated from retinal images.

Polarization-sensitive OCT has also benefitted from AO, reporting higher resolution in the retinal tissue. Retinal pigment epithelium, and its effect of polarization of backscattered light, has been studied with unprecedented resolution in [61]. Another work used AO OCT for understanding the impact on image quality of applying different spectral sources for retinal imaging, more details in [62].

In recent times, exploring longer wavelengths for ophthalmic OCT, toward the infrared portion of the spectrum, has shown a renewed interest. AO has also been merged for such application [63]. Other setups has been reported combining the previous concept, 1 μm illumination for increased penetration and visualization of the choriocapillaris and choroid, and a dual deformable mirror configuration for monochromatic aberration correction [64]. The latter apparatus allows for SLO and OCT operation and exhibits a very promising performance for clinical use.

10.6 Aberration Pancorrection in UHR OCT: Toward Isotropic and Subcellular Resolution

10.6.1 Pancorrection: Full Aberration Correction

In previous sections of this chapter, the need for correcting chromatic aberration in addition to monochromatic aberrations when large pupils and wide spectral bandwidths are employed in OCT has been shown. The simultaneous correction of monochromatic and chromatic aberration is referred to as *pancorrection*, indicating full or complete correction [38, 65, 66]. The experimental system employed for first showing pancorrection on real eyes was the AO head of an ophthalmic UHR OCT. Figure 10.10 shows the setup.

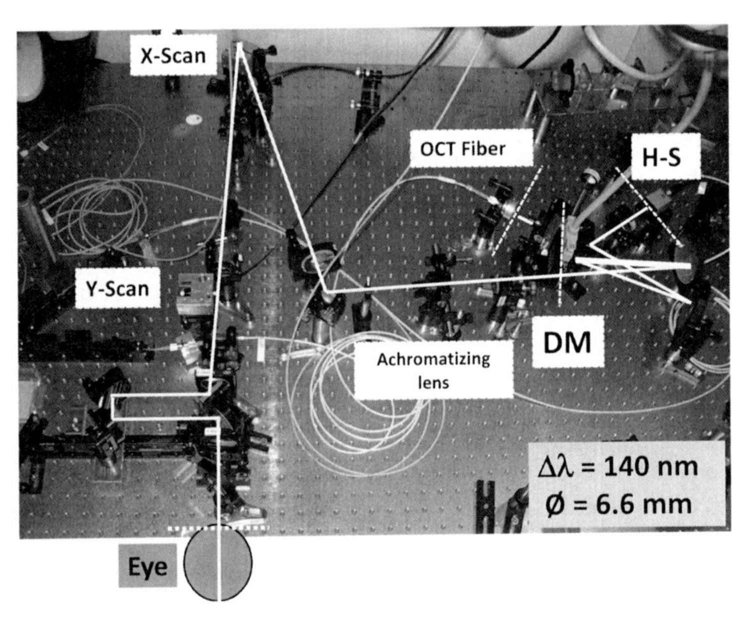

Fig. 10.10 AO system for pancorrection used in UHR OCT. The system combined a customized achromatizing lens and a magnetic deformable mirror (Mirao 52, Imagine Eyes, France) for full ocular aberration correction. Adapted from Fernández et al. [66]

The corrected pupil size was 6.6 mm, and the spectral bandwidth was 140 nm FWHM, Gaussian spectrum. The system was designed with lenses and spherical mirrors. Essentially, several telescopic relays are aligned along the light path so that optical conjugation across the different planes of interest is achieved. These planes are firstly the exit pupil of the eye, the two containing the scanning mirrors, the plane containing the correcting device, the chromatic aberration corrector, the wavefront sensor, and the entrance of the collimator connecting the AO system with the fiberized interferometer. Although correct conjugation is important for all of these planes, the case of the two scanning mirrors is critical. When using large pupil diameters, the pivot point of scanning must be at the entrance pupil of the eye, or very close, otherwise vignetting of the incoming beam will occur.

Achromatizing lenses are usually employed for correcting chromatic aberration of the eye [66–71]. Since this particular aberration is relatively constant across subjects, practically independent of age and refraction, static correction of average values is often applied. Nevertheless, one should take into account for certain applications that there are of course certain deviations among individuals. For the construction of the first *pancorrective* AO system, a triplet was designed [65] compensating average chromatic aberration found in the normal eye in the near-infrared range [38]. The design is shown on Fig. 10.11, where geometrical parameters of the lens, together with the employed materials, are also presented. Preliminary measurements reported excellent average corrections of longitudinal chromatic aberration. Those are depicted at the right plane of Fig. 10.11. Average chromatic aberration from five normal subjects is represented with and without the achromatizing lens. The smooth curve corresponds to the estimated chromatic aberration as expected from a simple water eye model in the near IR (Emsley schematic eye model). Note that the AO system might introduce additional longitudinal chromatic aberration if the implemented lenses, and in general refractive elements, are not specifically designed for minimizing such aberration. Normally, achromatic doublets or triplets of acceptable quality are available off-the-shelf.

A deformable mirror based on magnetic forces (Mirao 52, Imagine Eyes, France) was used for correcting the monochromatic aberrations in the aforementioned work.

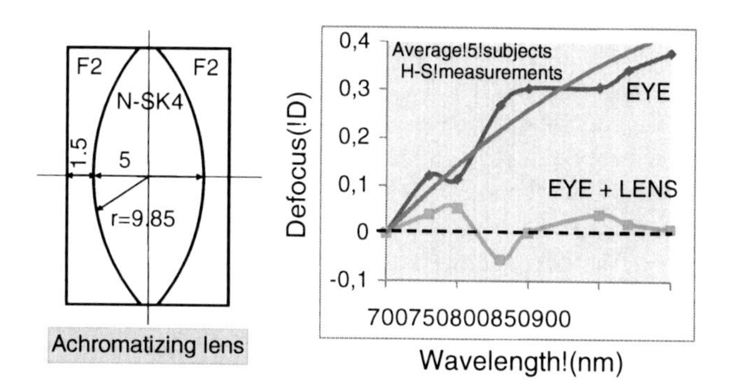

Fig. 10.11 Design of the triplet for chromatic aberration correction and performance of the compensation on real eyes measured with a Hartmann–Shack wavefront sensor

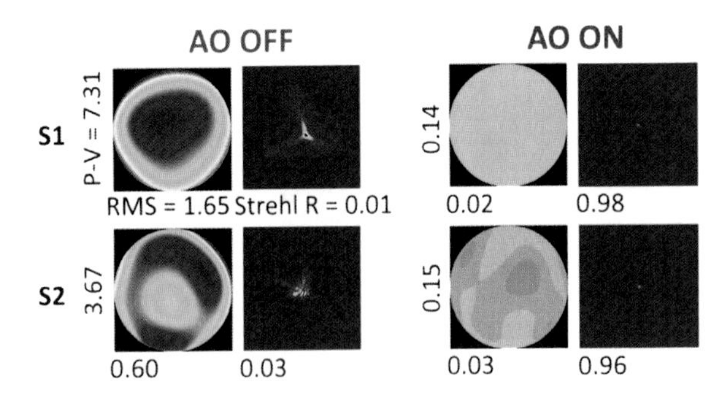

Fig. 10.12 Effect of pancorrection on real eyes through a pupil of 6.6 mm diameter. Aberration compensation reached diffraction limited performance in all cases

The mirrored surface, silver coated, is endowed with a set of 52 independent magnetic actuators. By means of a corresponding set of coils, the force exerted on the membrane can be controlled. A voltage applied to each coil creates a magnetic field pushing or pulling the magnet as a function of the sign of the voltage. The force generated between the actuators and the mirrored flexible membrane is directly proportional to the magnetic field created inside the coil. This magnetic field is proportional to the current inside the coil. The main advantage is that the exerted force depends linearly on the applied current, which is a substantial difference with electrostatic deformable mirrors, where the dependence of voltage with deformation is quadratic. Spacing across adjacent actuators is 2.5 mm, being the total mirrored surface 15 mm diameter. Optical quality of the membrane when no voltage is applied is below 0.1 μm RMS over the entire pupil. The capability of this technology for correcting ocular aberrations is very promising, due to its linear control and large stroke. Even aberrations from pathological corneas, as those affected by keratoconus, can be corrected with the magnetic mirror. Some aberration corrections, of both monochromatic and chromatic aberrations, from real eyes are displayed on Fig. 10.12.

The left columns of Fig. 10.12 show natural ocular aberrations from two subjects and their associate PSFs. In the color-coded representation of the aberrations, blue color indicates negative values while red color applies for positives values. The peak-valley parameter (P-V, given in micrometers) is given for each case, so that the reader can appreciate the maximum difference in the aberration maps. Ocular aberrations were measured by a Hartmann-Shack wavefront sensor up to the sixth order. The global value of the RMS (in micrometers, for the 6.6-mm pupil diameter) is also presented in the maps. For the generation of the associated polychromatic PSFs, average longitudinal chromatic aberration found in the normal eye was used, weighted with a Gaussian spectrum centered at 800 nm emulating the actual situation occurring in ophthalmic UHR OCT.

In the figures of the PSFs, blue color indicates zero while red corresponds to the maxima of energy. The effect of the aberration correction over the PSFs, once chromatic and monochromatic aberration have been compensated, is really eloquent, reaching Strehl ratios over 0.95. The presented AO modality, incorporating the correction of chromatic aberration, is very powerful. When combined with a capable corrector, the ocular aberrations can be fully compensated, even those generated by pathological eyes, where the benefit of AO should be more important.

10.6.2 UHR OCT and Pancorrection of Aberrations

The impact of the application of pancorrection, as compared with the use of classical AO and not correction at all, when using broad spectral bandwidth light sources was shown in [72]. In Fig. 10.13, some B-scans obtained at the same eccentricity and from the same subject are presented. Those were retrieved through a 6.6-mm pupil diameter using a 140 nm FWHM Gaussian spectrum centered at 800 nm. The correction of monochromatic aberrations, the case labeled with AO, shows a significant increment in the signal-to-noise ratio, enabling the detection of individual photoreceptors. The combination of AO with chromatic aberration correction, labeled as pancorrection in the figure, further ameliorates the images. The separation across different retinal layers shows a sharper transition in this last case. Morphological details, otherwise hidden, appear visible at the pancorrected images. At the bottom of Fig. 10.13, a portion of image labeled as 3 is zoomed for better visualization. The selected area corresponds to the photoreceptors. The outer segments are clearly resolved, along with other subtle details. The expanded image is approximately 230 μm width, and it is centered at 2.25 nasal. Note that the technique allows not only for individual photoreceptors counting but even for discerning across subcellular structures.

Figure 10.14, from the same work [72], further compares retinal sections obtained with pancorrection at ~1 (a, b) and ~2 (c, d). Several intraretinal layers are also imaged: nerve fiber layer (NFL), ganglion cell layer (GCL), IPL, outer plexiform layer (OPL), Henle fiber layer (HF), outer nuclear layer (ONL), external limiting membrane (ELM), and retinal pigment epithelium (RPE). Cones are easily resolved as separate elements with clear axial continuities. The quality of images allows for direct correlations of the individual photoreceptor signals in OCT with corresponding photoreceptor elements from histology. It must be noted that images are barely postprocessed, and solely a 2-pixel Gaussian filter for reducing speckle noise was applied. Therefore, images are essentially as they were retrieved by the interferometer.

Figure 10.14a shows that nuclear layers scatter light less than others compounded essentially by nerve fibers. As in standard OCT occurs, the outer retinal layers corresponding to the photoreceptors and retinal pigment epithelium provide the strongest signal. Panel b and d in Fig. 10.14 show an enlarged (4×) portion of the

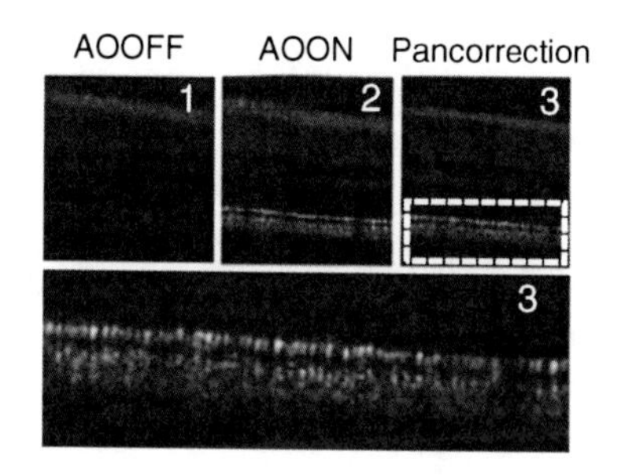

Fig. 10.13 Effect of pancorrection on retinal images obtained with UHR OCT in a normal eye. Individual photoreceptors are resolved, showing subcellular morphological details. Adapted from Fernández et al. [72]

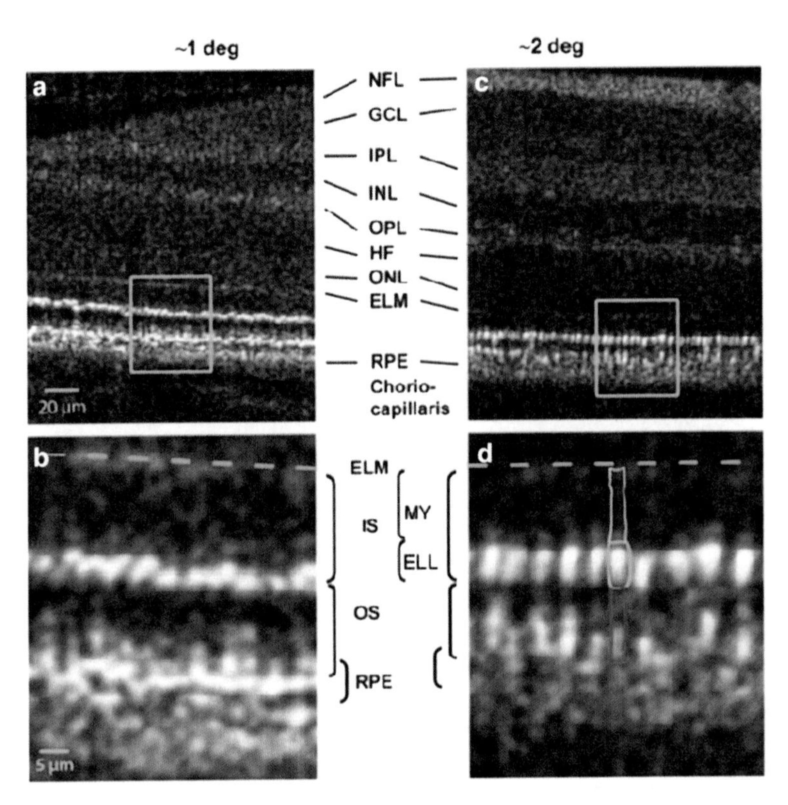

Fig. 10.14 Retinal images obtained with pancorrection and UHR OCT at different eccentricities. Adapted from Fernández et al. [72]

photoreceptor areas for better visualization of the achieved resolution and contrast. Enlarged areas present part of the retina from the external limiting membrane to the retinal pigment epithelium. Several of the axial elements of individual photoreceptors are resolved: ellipsoid (ELL), consisting of densely packed tubular mitochondria with multiple internal membranes (cristae), and myoid (MY) of the cone inner segments (IS). A second high-intensity sub-tier clearly represents the photoreceptor outer segment (OS) zone. The next evident layer is the RPE with its processes surrounding outer segments and the epithelial cellular portion.

Applying the concept of pancorrection [66] in UHR OCT, Zawadzki et al. [73] employed a triplet at the exit pupil of the AO system, before focusing the light into the fiber of the interferometer, for correcting the ocular chromatic aberration produced by a Gaussian spectrum of 112 nm FWHM centered at 836 nm. Two different deformable mirrors compensated for the rest of the aberrations. The effect of reducing speckle by multiple images averaging was also studied in his work, finding a notable benefit of such technique in the detection of some intraretinal features.

The amount of information obtained combining pancorrection and UHR OCT is enormous. Appropriate visualization of data is mandatory, moreover when new perspectives of the intraretinal tissue and cellular morphology are provided. A correct and fruitful interpretation of the data is necessary for the incorporation of novel rendering protocols. The work of Zawadzki et al. [74] provided an example of such techniques, with a clear orientation to potential use in clinics.

Parallel to the use of pancorrection, the techniques involved in ophthalmic OCT has experienced significant changes, for instance, in terms of the speed of acquisition of images, and the possibility of using longer wavelengths for deeper penetration into the outer retina. The improvement in the quality of images is evident along the last years, but sometimes, it has been not clearly stated the origin or dominating factor for such enhancement. The beautiful work of Povazay et al. [75] supplies thorough analysis across different techniques and technologies providing some light in the issue. In this direction, the combination of ultrahigh speed of acquisition, up to 120,000 A-scans/s, and pancorrection is showing amazing images of the living retina [76]. In this last work, phenotyping of the living retina is achieved. Some images even suggest certain structures which can be connected with individual ganglion cell bodies.

Several works have used correction of monochromatic and chromatic aberrations in ophthalmic UHR OCT [72–76]. The possibility of imaging the living retina with near-isotropic resolution of approximately 3 μm opens the door to novel concepts in the study of the morphology of the retina. The dramatic increase in the scanning ratio is also a major technical advance, since it is the true responsible for the astonishing three-dimensional representations of the intraretinal tissue we found in recent papers. Subcellular resolution is being achieved, with enormous potential for the better understanding of the pathogenesis of several retinal conditions. Early changes at cellular level might translate into effective diagnosis of conditions even before any functional symptom could be perceived on vision.

10.7 Summary

This chapter reviewed the impact of the eye's aberration in the resolution and contrast of OCT images. After a simplified revision of the eye's optical properties, a detailed evaluation on how eye's optics affects OCT images was presented. The subsequent sections covered the most recent literature where the eyes aberration was corrected within different OCT approaches. In particular, first was the application of adaptive optics, with either deformable mirrors or liquid crystal spatial light modulators, to correct for the monochromatic aberrations. We analyzed the special limits imposed by chromatic aberrations due to the spectral bandwidth of the OCT source. The combined correction of monochromatic and chromatic aberrations, referred as pancorrection, showed the best results in terms of improved contrast and resolution. Most of the AO systems for OCT are now research prototypes.

For all these promising advances to be translated into the everyday clinic practice, several issues must be first solved. Performing AO in the human eye is not that simple, and there are many technical constraints to be surpassed along the way. An obvious fact is that measuring correctly the ocular aberrations is fundamental prior correcting them, which is not trivial. In addition, the cost of AO, although showing a decreasing trend, is probably still borderline for its inclusion in commercial clinical devices. Appropriate aberration correctors should be carefully selected. Since AO is particularly suited for highly aberrated eyes, there are of little help those exhibiting limited stroke or amplitude in the generation of aberrations.

It is likely that in the near future, commercial systems would incorporate this technology to provide with better images and improved diagnostic capabilities.

Acknowledgments Some of the figures of the current chapter, particularly those pertaining retina images, show results obtained in close collaboration with W. Drexler and his group, who are acknowledged. This work has been financially supported by "Ministerio de Educación y Ciencia," Spain (grant FIS2007-64765), and "Fundación Séneca," Murcia, Spain (grant 04524/GERM/06).

References

1. D. Huang, E.A. Swanson, C.P. Lin, J.S. Schuman, W.G. Stinson, W. Chang, M.R. Hee, T. Flotte, K. Gregory, C.A. Puliafito, J.G. Fujimoto, Optical coherence tomography. Science **254**(5035), 1178–1181 (1991)
2. W. Drexler, U. Morgner, F.X. Kärtner, C. Pitris, S.A. Boppart, X.D. Li, E.P. Ippen, J.G. Fujimoto, In vivo ultrahigh-resolution optical coherence tomography. Opt. Lett. **24**(17), 1221–1223 (1999)
3. W. Drexler, U. Morgner, R.K. Ghanta, F.X. Kärtner, J.S. Schuman, J.G. Fujimoto, Ultrahigh resolution ophthalmic optical coherence tomography. Nat. Med. **7**(4), 502–507 (2001)
4. W. Drexler, Ultrahigh resolution optical coherence tomography. J. Biomed. Opt. **9**(1), 47–74 (2004)
5. A. Unterhuber, B. Povazay, B. Hermann, H. Sattmann, W. Drexler, V. Yakovlev, G. Tempea, C. Schubert, E.M. Anger, P.K. Ahnelt, M. Stur, J.E. Morgan, A. Cowey, G. Jung, T. Le, A. Stingl, Compact, lowcost TiAl2O3 laser for in vivo ultrahigh-resolution optical coherence tomography. Opt. Lett. **28**(11), 905–907 (2003)
6. A.F. Fercher, W. Drexler, C.K. Hitzenberger, T. Lasser, Optical coherence tomography—principles and applications. Rep. Prog. Phys. **66**, 239–303 (2003)
7. P. Artal, A. Benito, J. Tabernero, The human eye is an example of robust optical design. J. Vis. **6**(1), 1–7 (2006)
8. P. Artal, A. Guirao, E. Berrio, D.R. Williams, Compensation of corneal aberrations by the internal optics in the human eye. J. Vision **1**(1), 1–8 (2001)
9. J. Tabernero, A. Benito, E. Alcón, P. Artal, Mechanism of compensation of aberrations in the human eye. J. Opt. Soc. Am. A Opt. Image. Sci. Vis. **24**(10), 3274–3283 (2007)
10. P. Artal, J. Tabernero, The eye's aplanatic answer. Nat. Photon. **2**, 586–589 (2008)
11. P. Artal, E. Berrio, A. Guirao, P. Piers, Contribution of the cornea and internal surfaces to the change of ocular aberrations with age. J. Opt. Soc. Am. A Opt. Image. Sci. Vis. **19**(1), 137–143 (2002)
12. G.M. Pérez, S. Manzanera, P. Artal, Impact of scattering and spherical aberration in contrast sensitivity. J. Vision **9**(3), 19.1–19.10 (2009)

13. J.K. Ijspeert, P.W. de Waard, T.J. van den Berg, P.T. de Jong, The intraocular straylight function in 129 healthy volunteers; Dependence on angle, age and pigmentation. Vision Res. **30**(5), 699–707 (1990)
14. J. Liang, D.R. Williams, Aberrations and retinal image quality of the normal human eye. J. Opt. Soc. Am. A Opt. Image. Sci. Vis. **14**(11), 2873–2883 (1997)
15. F. Vargas-Martín, P.M. Prieto, P. Artal, Correction of the aberrations in the human eye with a liquid crystal spatial light modulator: limits to the performance. J. Opt. Soc. Am. A Opt. Image. Sci. Vis. **15**(9), 2552–2562 (1998)
16. E.J. Fernández, I. Iglesias, P. Artal, Closed-loop adaptive optics in the human eye. Opt. Lett. **26**(10), 746–748 (2001)
17. H. Hofer, L. Chen, G.Y. Yoon, B. Singer, Y. Yamauchi, D.R. Williams, Improvement in retinal image quality with dynamic correction of the eye's aberrations. Opt. Express. **8**(11), 631–643 (2001)
18. E.J. Fernández, P. Artal, Membrane deformable mirror for adaptive optics: performance limits in visual optics. Opt. Express. **11**(9), 1056–1069 (2003)
19. E.J. Fernández, P.M. Prieto, P. Artal, Binocular adaptive optics visual simulator. Opt. Lett. **34**(17), 2628–2630 (2009)
20. E.J. Fernández, P.M. Prieto, P. Artal, Adaptive optics binocular visual simulator to study stereopsis in the presence of aberrations. J. Opt. Soc. Am. A Opt. Image. Sci. Vis. **27**(11), A48–A55 (2010)
21. J. Liang, D.R. Williams, D.T. Miller, Supernormal vision and high-resolution retinal imaging through adaptive optics. J. Opt. Soc. Am. A Opt. Image. Sci. Vis. **14**(11), 2884–2892 (1997)
22. A. Roorda, F. Romero-Borja, W. Donnelly Iii, H. Queener, T. Hebert, M. Campbell, Adaptive optics scanning laser ophthalmoscopy. Opt. Express. **10**(9), 405–412 (2002)
23. A.G. Podoleanu, D.A. Jackson, Noise analysis of a combined optical coherence tomograph and a confocal scanning ophthalmoscope. Appl. Opt. **38**(10), 2116–2127 (1999)
24. A.G. Podoleanu, G.M. Dobre, R.G. Cucu, R.B. Rosen, Sequential optical coherence tomography and confocal imaging. Opt. Lett. **29**(4), 364–366 (2004)
25. S. Kimura, T. Wilson, Confocal scanning optical microscope using single-mode fiber for signal detection. Appl. Opt. **30**(16), 2143–2150 (1991)
26. T. Wilson (ed.), *Confocal Microscopy*. (Academic, London 1990)
27. T. Wilson, A.R. Carlini, Size of the detector in confocal imaging systems. Opt. Lett. **12**(4), 227–229 (1987)
28. E. Fernández, W. Drexler, Influence of ocular chromatic aberration and pupil size on transverse resolution in ophthalmic adaptive optics optical coherence tomography. Opt. Express. **13**(20), 8184–8197 (2005)
29. T. Wilson, C.J.R. Sheppard, *Theory and Practice of Scanning Optical Microscopy*, 1st edn (Academic, New York, 1984)
30. M. Born, E. Wolf, *Principles of Optics*, 7th ed (Pergamon, Oxford, UK, 1999)
31. R.J. Noll, Zernike polynomials and atmospheric turbulence. J. Opt. Soc. Am.**66**(3), 207–211 (1976)
32. L.N. Thibos, R.A. Applegate, J.T. Schwiegerling, R. Webb, VSIA Standards Taskforce Members, in *Standards for Reporting the Optical Aberrations of the Eyes*, ed. by V. Lakshminarayanan. OSA Trends in Optics and Photonics, vol 35 (Optical Society of America, Washington, DC, 2000), pp. 232–244
33. R.E. Bedford, G. Wyszecki, Axial chromatic aberration of the human eye. J. Opt. Soc. Am. **47**(6), 564–565 (1957)
34. W.N. Charman, J.A. Jennings, Objective measurements of the longitudinal chromatic aberration of the human eye. Vision Res. **16**(9), 999–1005 (1976)
35. P.A. Howarth, A. Bradley, The longitudinal chromatic aberration of the human eye and its correction. Vision Res. **26**(2), 361–366 (1986)
36. L.N. Thibos, M. Ye, X. Zhang, A. Bradley, The chromatic eye: a new reduce-eye model of ocular chromatic aberration in humans. Appl. Opt. **31**(19), 3594–3600 (1992)

37. H.L. Liou, N.A. Brennan, Anatomically accurate, finite model eye for optical modeling. J. Opt. Soc. Am. A Opt. Image. Sci. Vis. **14**(8), 1684–1695 (1997)
38. E. Fernández, A. Unterhuber, P. Prieto, B. Hermann, W. Drexler, P. Artal, Ocular aberrations as a function of wavelength in the near infrared measured with a femtosecond laser. Opt. Express. **13**(2), 400–409 (2005)
39. D.A. Atchison, G. Smith, Chromatic dispersions of the ocular media of human eyes. J. Opt. Soc. Am. A Opt. Image. Sci. Vis. **22**(1), 29–37 (2005)
40. S. Manzanera, C. Canovas, P.M. Prieto, P. Artal, A wavelength tunable wavefront sensor for the human eye. Opt. Express. **16**(11), 7748–7755 (2008)
41. E.J. Fernández, P. Artal, Ocular aberrations up to the infrared range: from 632.8 to 1070 nm. Opt. Express. **16**(26), 21199–21208 (2008)
42. D.T. Miller, J. Qu, R.S. Jonnal, K. Thorn, Coherence gating and adaptive optics in the eye, in *Coherence Domain Optical Methods and Optical Coherence Tomography in Biomedicine VII*, ed. by Valery V. Tuchin, Joseph A. Izatt, James G. Fujimoto, vol 4956 (Proc. SPIE, Bellingham, WA, 2003), pp. 65–72
43. B. Hermann, E.J. Fernández, A. Unterhuber, H. Sattmann, A.F. Fercher, W. Drexler, P.M. Prieto, P. Artal, Adaptive-optics ultrahigh-resolution optical coherence tomography. Opt. Lett. **29**(18), 2142–2144 (2004)
44. J. Liang, B. Grimm, S. Goelz, J.F. Bille, Objective measurement of wave aberrations of the human eye with the use of a Hartmann–Shack wavefront sensor. J. Opt. Soc. Am. A Opt. Image. Sci. Vis. **11**(7), 1949–1957 (1994)
45. P.M. Prieto, F. Vargas-Martín, S. Goelz, P. Artal, Analysis of the performance of the Hartmann–Shack sensor in the human eye. J. Opt. Soc. Am. A Opt. Image. Sci. Vis. **17**(8), 1388–1398 (2000)
46. F. Fercher, C.K. Hitzenberger, G. Kamp, S.Y. El-Zaiat, Measurement of intraocular distances by backscattering spectral interferometry. Opt. Commun. **117**(1–2), 43–48 (1995)
47. M. Choma, M. Sarunic, C. Yang, J. Izatt, Sensitivity advantage of swept source and Fourier domain optical coherence tomography. Opt. Express. **11**(18), 2183–2189 (2003)
48. M. Wojtkowski, R. Leitgeb, A. Kowalczyk, T. Bajraszewski, A.F. Fercher, In-vivo human retinal imaging by Fourier domain optical coherence tomography. J. Biomed. Opt. **7**(3), 457–463 (2002)
49. R. Leitgeb, C. Hitzenberger, A. Fercher, Performance of Fourier domain vs. time domain optical coherence tomography. Opt. Express. **11**(8), 889–894 (2003)
50. J.F. de Boer, B. Cense, B.H. Park, M.C. Pierce, G.J. Tearney, B.E. Bouma, Improved signal-to-noise ratio in spectral-domain compared with time-domain optical coherence tomography. Opt. Lett. **28**(21), 2067–2069 (2003)
51. Y. Zhang, J. Rha, R. Jonnal, D. Miller, Adaptive optics spectral optical coherence tomography for imaging the living retina. Opt. Express. **13**(12), 4792–4811 (2005)
52. E.J. Fernández, B. Povazay, B. Hermann, A. Unterhuber, H. Sattmann, P.M. Prieto, R. Leitgeb, P. Ahnelt, P. Artal, W. Drexler, Three-dimensional adaptive optics ultrahigh-resolution optical coherence tomography using a liquid crystal spatial light modulator. Vision Res. **45**(28), 3432–3444 (2005)
53. P. Prieto, E. Fernández, S. Manzanera, P. Artal, Adaptive optics with a programmable phase modulator: applications in the human eye. Opt. Express. **12**(17), 4059–4071 (2004)
54. A. Unterhuber, B. Povazay, B. Hermann, H. Sattmann, W. Drexler, V. Yakovlev, G. Tempea, C. Schubert, E.M. Anger, P.K. Ahnelt, M. Stur, J.E. Morgan, A. Cowey, G. Jung, T. Le, A. Stingl, Compact, low-cost TiAl2O3 laser for in vivo ultrahigh-resolution optical coherence tomography. Opt. Lett. **28**(11), 905–907 (2003)
55. E.J. Fernández, P.M. Prieto, P. Artal, Wave-aberration control with a liquid crystal on silicon (LCOS) spatial phase modulator. Opt. Express. **17**(13), 11013–11025 (2009)
56. R.J. Zawadzki, S.M. Jones, S.S. Olivier, M. Zhao, B.A. Bower, J.A. Izatt, S. Choi, S. Laut, J.S. Werner, Adaptive-optics optical coherence tomography for high-resolution and high-speed 3D retinal in vivo imaging. Opt. Express. **13**(21), 8532–8546 (2005)

57. R.J. Zawadzki, S.S. Choi, S.M. Jones, S.S. Oliver, J.S. Werner, Adaptive optics-optical coherence tomography: optimizing visualization of microscopic retinal structures in three dimensions. J. Opt. Soc. Am. A Opt. Image. Sci. Vis. **24**(5), 1373–1383 (2007)
58. C. Li, N. Sredar, K.M. Ivers, H. Queener, J. Porter, A correction algorithm to simultaneously control dual deformable mirrors in a woofer-tweeter adaptive optics system. Opt. Express. **18**(16), 16671–16684 (2010)
59. D. Merino, C. Dainty, A. Bradu, A.G. Podoleanu, Adaptive optics enhanced simultaneous en-face optical coherence tomography and scanning laser ophthalmoscopy. Opt. Express. **14**(8), 3345–3353 (2006)
60. M. Pircher, R.J. Zawadzki, J.W. Evans, J.S. Werner, C.K. Hitzenberger, Simultaneous imaging of human cone mosaic with adaptive optics enhanced scanning laser ophthalmoscopy and high-speed transversal scanning optical coherence tomography. Opt. Lett. **33**(1), 22–24 (2008)
61. B. Cense, W. Gao, J.M. Brown, S.M. Jones, R.S. Jonnal, M. Mujat, B.H. Park, J.F. de Boer, D.T. Miller, Retinal imaging with polarization-sensitive optical coherence tomography and adaptive optics. Opt. Express. **17**(24), 21634–21651 (2009)
62. B. Cense, E. Koperda, J.M. Brown, O.P. Kocaoglu, W. Gao, R.S. Jonnal, D.T. Miller, Volumetric retinal imaging with ultrahigh-resolution spectral-domain optical coherence tomography and adaptive optics using two broadband light sources. Opt. Express. **17**(5), 4095–4111 (2009)
63. K. Kurokawa, K. Sasaki, S. Makita, M. Yamanari, B. Cense, Y. Yasuno, Simultaneous high-resolution retinal imaging and high penetration choroidal imaging by one-micrometer adaptive optics optical coherence tomography. Opt. Express. **18**(8), 8515–8527 (2010)
64. M. Mujat, R.D. Ferguson, A.H. Patel, N. Iftimia, N. Lue, D.X. Hammer, High resolution multimodal clinical ophthalmic imaging system. Opt. Express. **18**(11), 11607–11621 (2010)
65. E.J. Fernández, A. Unterhuber, B. Povazay, B. Hermann, P. Artal, W. Drexler, Chromatic aberration correction of the human eye for retinal imaging in the near infrared. Opt. Express. **14**(13), 6213–6225 (2006)
66. E.J. Fernandez, L. Vabre, B. Hermann, A. UnterhubervB. Povazay, W. Drexler, Adaptive optics with a magnetic deformable mirror: applications in the human eye. Opt. Express. **14**(20), 8900–8917 (2006)
67. A. Ames, C.A. Proctor, Dioptrics of the eye. J. Opt. Soc. Am. **5**, 22–84(1921)
68. A.C. van Heel, Correcting the spherical and chromatic aberrations of the eye. J Opt Soc Am. **36**, 237–239 (1946 Apr)
69. I. Powell, Lenses for correcting chromatic aberration of the eye. Appl. Opt. **20**(24), 4152–4155 (1981)
70. A.L. Lewis, M. Katz, C. Oehrlein, A modified achromatizing lens. Am. J. Optom. Physiol. Opt. **59**(11), 909–911 (1982)
71. Y. Benny, S. Manzanera, P.M. Prieto, E.N. Ribak, P. Artal, Wide-angle chromatic aberration corrector for the human eye. J. Opt. Soc. Am. A Opt. Image. Sci. Vis. **24**(6), 1538–1544 (2007)
72. E.J. Fernández, B. Hermann, B. Povazay, A. Unterhuber, H. Sattmann, B. Hofer, P.K. Ahnelt, W. Drexler, Ultrahigh resolution optical coherence tomography and pancorrection for cellular imaging of the living human retina. Opt. Express. **16**(15), 11083–11094 (2008)
73. R.J. Zawadzki, B. Cense, Y. Zhang, S.S. Choi, D.T. Miller, J.S. Werner, Ultrahigh-resolution optical coherencetomography with monochromatic and chromaticaberration correction. Opt. Express. **16**(11), 8126–8143 (2008)
74. R.J. Zawadzki, S.S. Choi, A.R. Fuller, J.W. Evans, B. Hamann, J.S. Werner, Cellular resolution volumetric in vivo retinal imaging with adaptive optics–optical coherence tomography. Opt. Express. **17**(5), 4084–4094 (2009)
75. B. Povazay, B. Hofer, C. Torti, B. Hermann, A.R. Tumlinson, M. Esmaeelpour, C.A. Egan, A.C. Bird, W. Drexler, Impact of enhanced resolution, speed and penetration on three-dimensional retinal optical coherence tomography. Opt. Express. **17**(5), 4134–4150 (2009)
76. C. Torti, B. Povazay, B. Hofer, A. Unterhuber, J. Carroll, P.K. Ahnelt, W. Drexler, Adaptive optics optical coherence tomography at 120,000 depth scans/s for non-invasive cellular phenotyping of the living human retina. Opt. Express. **17**(22), 19382–19400 (2009)

Chapter 11
The SL SCAN-1: Fourier Domain Optical Coherence Tomography Integrated into a Slit Lamp

F.D. Verbraak and M. Stehouwer

Abstract The detailed cross-sectional images of OCT can be used for diagnosis and follow-up, assessing therapeutic success or failure. Recently, the OCT technology has been implemented in a small unit compatible with existing slit lamps. This increases the efficiency of the routine clinical examination of a patient, will increase the comfort of the patient, and saves time. Additionally, the posterior segment can be scanned through a handheld lens and even through a three-mirror lens.

11.1 Introduction

Optical coherence tomography (OCT) was introduced in the early 1990s. Since then, it has become an important tool in the ophthalmological clinic [1–3]. The detailed cross-sectional images can be used for diagnosis, and follow-up, assessing therapeutic success or failure. Numerous studies have demonstrated the valuable contribution of OCT in common diseases like diabetic retinopathy, age-related macular degeneration, vitreoretinal pathology, and glaucoma [4–10]. In patients with corneal pathology, like corneal dystrophies and degenerations, keratoconus, inflammations, or following surgery, OCT has proven to be useful providing objective measurements, additive to observations with the slit lamp biomicroscope alone [11–14]. With the introduction of Fourier Domain Optical Coherence Tomography (FD-OCT), an exceptional high resolution and a very fast acquisition time can be realized, providing even more detailed images of the layered structure of the retina, the peripapillary retinal nerve fiber layer, the cornea, iris, and anterior chamber. The cross-sectional images obtained with the FD-OCT systems add the third dimension to clinical observation and have considerably improved the quality of the diagnostic process.

F.D. Verbraak (✉)
Department of Ophthalmology, Academisch Medisch Centrum, Postbus 22660, 1100 DD, Amsterdam
e-mail: f.d.verbraak@amc.nl

R. Bernardes and J. Cunha-Vaz (eds.), *Optical Coherence Tomography*,
Biological and Medical Physics, Biomedical Engineering,
DOI 10.1007/978-3-642-27410-7_11,

Recently, the OCT technology has been implemented in a small unit compatible with existing slit lamps, the basic device in all ophthalmic practices [15]. This add-on FD-OCT imaging device (SL SCAN-1, Topcon) is developed through a collaboration of the department of Biomedical Engineering and Photonics in the AMC and Topcon Europe Medical Research. With the slit lamp-integrated OCT, one can obtain OCT images of the posterior or anterior segment during a normal standard examination of the eye. This increases the efficiency of the routine clinical examination of a patient, will increase the comfort of the patient seated behind just one device, and saves time. Scans can be made directly of the anterior segment. The posterior segment can be scanned through a handheld lens and even through a three-mirror lens. This proved to extend the area of the retina that can be imaged to the far periphery [16].

11.2 The SL SCAN-1 Device

The slit lamp with integrated OCT used in this study consists of a slit lamp (Topcon SL-D7), with a slit lamp-adapted FD-OCT named SL SCAN-1, and a camera (Topcon camera DV3, resolution = 3.24 Mp). The complete OCT setup is placed between the oculars and objective lenses of the slit lamp (Fig. 11.1).

The SL SCAN-1 is a FD-OCT with a broadband superluminescent diode (SLD) light source ($\Delta\lambda$ 30 nm, central wavelength 840 nm). The axial scan resolution is 8–9 μm in tissue, and with a scan speed of 5,000 a-scans per second, a b-scan of up to 1,028 a-scans can be composed. Dependent on the slit lamp magnification, a single b-scan has a length of up to 12 mm. The OCT scanner is switched on by a click on the joystick button. A b-scan can be made without interrupting the standard slit lamp examination. Positioning of the b-scan is accomplished with the help of an aiming beam (650 nm), visible through the right ocular. When the OCT scanner is switched on, continuously b-scans are made and shown in a live window on a PC screen. The screen is positioned in a way to allow inspection of the b-scan with a short sideway glance while keeping the observed area of interest in position. A complete b-scan, or set of b-scans, is made by a second click on the joystick.

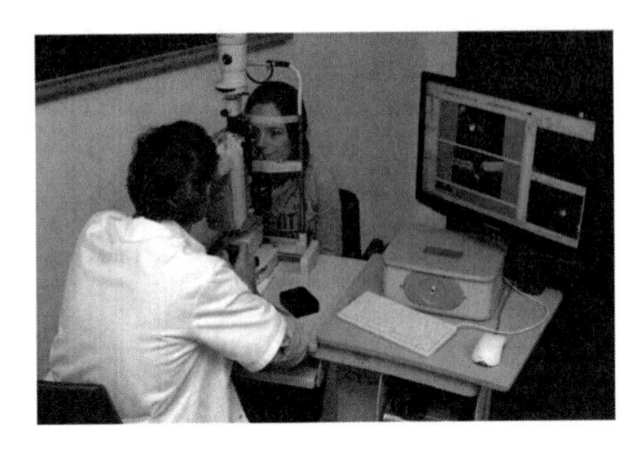

Fig. 11.1 Complete set-up of slit lamp integrated FD-OCT (SLSCAN, Topcon)

A color fundus photograph of the observed area can be made simultaneously with the scan and shows the exact position of the scan(s) with a green line (Topcon camera DV3, resolution = 3.24 Mp).

Different scanning patterns can be obtained (horizontal single line, raster, cross, grid, radical, and circle pattern), and up to 10 single b-scans (consisting of 512 or 1,028 a-scans) can be averaged.

11.3 Scanning the Anterior and the Posterior Segment

The SL SCAN-1 can be used in two modes: the posterior mode, for scanning the posterior pole, and the anterior mode, for scans of the anterior segment. The focus of the OCT in posterior mode is adjusted with the aid of a fast z-alignment in the reference arm of the device. Through this z-alignment, one can make OCT scans of the posterior segment through any common handheld lens while simultaneously performing slit lamp biomicroscopy of the fundus. In the anterior mode, the fast z-tracking system is switched off. The focus of the slit lamp and the OCT are then fixed and coincide, enabling to obtain directly OCT scans of parts of the anterior segment observed with the slit lamp. With the help of the aiming line, the b-scan can be positioned at exactly the area of interest (anterior or posterior segment) and makes it less dependent of an internal fixation point.

11.4 Scanning the Periphery with the SL SCAN

In posterior mode, scanning through a handheld lens, the SL SCAN-1 expands the range of the scanned area of the retina compared to the range of the common commercially available OCT devices. Not only the posterior pole but even the periphery becomes within reach. Basically, all areas that can be observed through the handheld lens can be scanned.

Till now, cross-sectional images of the peripheral retina could only be obtained with ultrasonographic biomicroscopy (UBM) or magnetic resonance imaging (MRI), with a rather low resolution. The SL SCAN-1 can provide cross-sectional images with much higher resolution, showing details of the different layers of the peripheral retina. This could provide new insights and additional information of peripheral retinal abnormalities and diseases.

11.5 Scanning Through a Three-Mirror Contact Lens with the SL SCAN

Although OCT scans of peripheral lesions in the retina can be obtained through a handheld lens, the quality of the scans seemed to be inversely related to the distance from the posterior pole. This inverse relationship could be the result of an increase in optical aberrations, introduced by the lens and the optics of the examined eye, when observing the more peripheral retina. The use of a three-mirror contact lens improves the optics observing the peripheral retina. With the SL SCAN-1 in the posterior mode (z-tracking system switched on), OCT scans can also be made through a three-mirror contact lens, with the gonio-mirror as well as the peripheral retina mirrors.

The length of the sample arm will differ considerably between a handheld lens and a three-mirror lens. However, independent of the used lens and its corresponding sample arm length, the fast z-tracking system is able to line out the correct length of the reference arm. Once the retinal signal is found, the system will lock on to this signal and will continuously compensate for the dynamic character of the examination. The range of the standard z-tracking system has to be slightly enlarged to enable the SL SCAN-1 to line out the correct length of the reference arm while using a three-mirror contact lens. This range adjustment can be made with any SL SCAN-1 device.

The SL SCAN-1 proved to be capable to scan the more peripheral retina through a three-mirror contact lens. In addition, the anterior chamber angle can be observed with the gonio-mirror of the three-mirror contact lens, and OCT scans can be made of the structures in the anterior chamber angle like the trabecular meshwork.

11.6 Scanning with the Fundus Viewer, an Add-On Lens

Single cross-sectional b-scans of the retina provide detailed information of the morphology of the retina and enable to detect and measure small changes in local retinal thickness. With a certain number of scans made in succession, a topographic retinal thickness map can be provided of the complete posterior pole, including the standard nine ETDRS-defined regions (Early Treatment of Diabetic Retinopathy Study). With the SL SCAN-1, it is possible to cover this whole area by obtaining six or twelve successive b-scans, with 6-mm length, in a radial scanning pattern. From these radial scans, a macular topographic thickness map can be constructed with the help of specific software (FastMap, Topcon).

With a handheld lens, it proved to be difficult to make six successive scans precisely through the foveal center. The small movements of the handheld lens and movements of the patient's eye during scanning (lack of an internal fixation) result mostly in a decentralized fovea on the thickness map. To overcome this problem, Topcon has developed the fundus viewer system (lens holder with a 60D Volk lens). This system provides a rigid setup which allows obtaining a more stable set of OCT images of the central retina. One may use the smallest dimensions of the slit lamp illumination as a surrogate point light source providing an internal fixation point. Reliable topographic maps of the central retina can be made with the fundus viewer lens, including the ETDRS regions.

In mydriasis, the fundus viewer system also provides a minimum of disturbing reflections during funduscopy and provides a steady image of the retina with a field of view of 30°, which is ideal to obtain detailed fundus photographs with a digital slit lamp camera.

11.7 Examples of Subjects Studied

As part of an ongoing study to evaluate the slit lamp-integrated OCT, consecutive patients seen in the outpatient clinic of the ophthalmic department of the Academic Medical Center, Amsterdam, the Netherlands, between December 2009 and September 2010, were asked to participate. The study followed the tenets of the declaration of Helsinki. Patients were informed about the nature of the SL SCAN-1 used in the study and gave their written informed consent.

To explore the first impressions of the OCT scans made with SL SCAN-1 and the setup of a slit lamp with an integrated OCT, patients were included with representative corneal or retinal pathologies. Patients with posterior pole pathology with unclear ocular media (the cornea, lens, and vitreous) or a pupil of insufficient size were excluded.

A number of patients were also studied with the help of a three-mirror contact lens. These patients were selected to provide the first impression of OCT scans obtained through the three-mirror contact lens. They were diagnosed with diseases affecting the anterior segment like pigment dispersion syndrome, primary open-angle glaucoma, iris cysts, or peripheral retinal abnormalities, like peripheral neovascularization, peripheral local retinal detachment, and peripheral retinoschisis.

Following standard ophthalmic examination, all patients were reexamined with the slit lamp with integrated OCT. All patients with posterior segment pathology had received mydriatic eye drops in the affected eye(s) (tropicamide and phenylephrine). All patients received two drops of oxybuprocaïne to anesthetize the eye before applying the three-mirror contact lens with methocel 2%.

11.8 Results

Using the slit lamp with integrated OCT was almost identical to standard slit lamp examination either with a handheld lens or a three-mirror contact lens. No limitations were observed in the optics of the slit lamp with integrated OCT that interfered with visualization of the anterior or posterior segment. The slight increase in the anterior-posterior dimension of the slit lamp caused by the add-on FD-OCT, placed between the oculars and objective lenses, did not hamper operating the device. The area of interest could be brought into focus, the aiming line was clearly visible, and simultaneous observing and scanning could be performed in posterior mode through a handheld lens (78D Volk lens) or through a three-mirror lens (HAAG-STREIT), or without a lens in the anterior mode observing the anterior segment.

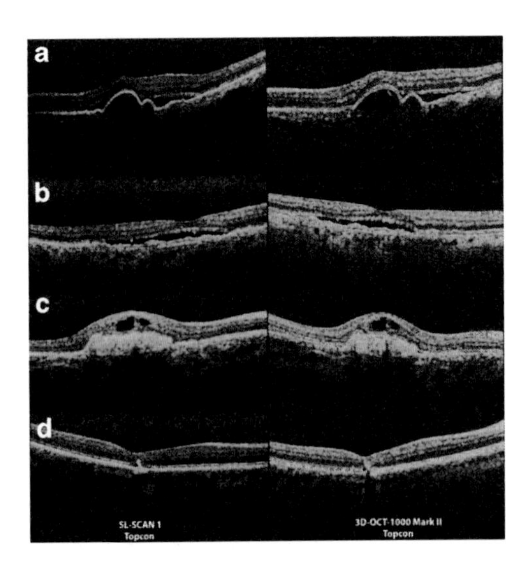

Fig. 11.2 A to D four examples of patients with different macular pathology (for explanation see text)

11.8.1 Scans of the Posterior Pole Obtained Through a Handheld Lens (Posterior Mode)

Figure 11.2 shows images of patients with pathology of the posterior pole. The shown scans obtained through a handheld lens with the SL SCAN-1 are made at almost identical location as the scans made with the 3D-OCT-1000 Mark II, allowing comparison between both systems.

Figure 11.2a shows the presence of pigment epithelial detachments and flanking neurosensory detachment in a patient known with exudative age-related macular degeneration (AMD), following treatment with ranibizumab (and PDT). The scans shown in Fig. 11.2b are obtained in a patient known with AMD, an occult choroidal neovascularization and a neurosensory detachment. Figure 11.2c shows a patient with a central fibrotic lesion due to an idiopathic neovascularization and treated with Avastin injections and PDT, who had a stable OCT image over two years without leakage on the fluorescein angiography. Figure 11.2d is a patient with a recurrent uveitis, most likely based on toxoplasmosis, known with a small old chorioretinal scar in her right eye.

11.8.2 Scans of the Peripheral Retina Obtained Through a Handheld Lens or Three-Mirror Lens (Posterior Mode)

The SL SCAN-1 is capable of scanning outside the standard range of the common commercially available (posterior segment) OCT devices, as illustrated in Fig. 11.3,

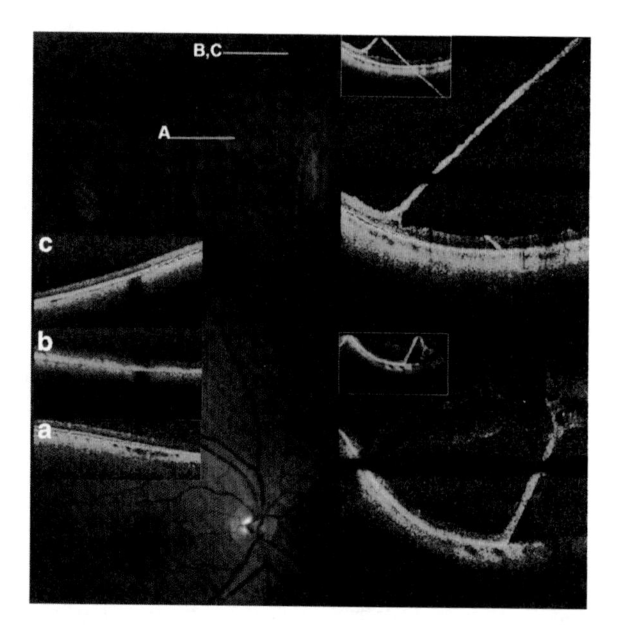

Fig. 11.3 Patient with peripheral naevus, A, B, and C shows b-scans and corresponding location on the fundus photograph (*left side*), Patient with retinal detachment (*lower half, right side*), and patient with retinal schisis (*upper half, right side*)

left side. The naevus shown in this figure is scanned with the SL SCAN-1 (scan B and C) but was out of reach for the 3D-OCT-1000 Mark II (most peripheral scan = A). The scan made through the handheld lens with the SL SCAN-1 shows the expected shadowing below the naevus but lacks fine details of the retina (scan B). In contrast, the scan made through a three-mirror lens did not only show the characteristic shadowing but also details of the different layers of the retina in this location (scan C).

The right side of Fig. 11.3 shows the differences in an OCT scan between a peripheral retinoschisis and a localized peripheral detachment. The upper scan, the peripheral retinoschisis, shows a clear splitting of the retina into an inner and outer retinal layer. Whereas the lower scan, the localized peripheral detachment, shows vitreoretinal traction and a detached neurosensory retina. The shown images are an interpretation of the scan showing the symmetric overlapping image artifact referred to in the literature as the complex conjugate or mirror image artifact.

11.8.3 Scans of the Anterior Segment Obtained Directly (Anterior Mode) and Through a Three-Mirror Lens (Posterior Mode)

The scans of the anterior segment showed a good image quality. The image shown in Fig. 11.4a is obtained in a patient who received Descemet's stripping endothelial keratoplasty. Figure 11.4b shows a patient with a Baerveldt implant, a glaucoma

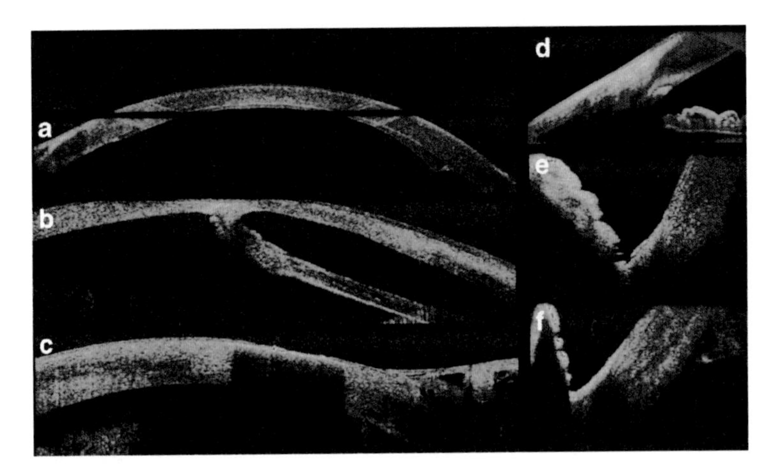

Fig. 11.4 OCT scans made of the anterior segment, direct mode (*left*), and indirect through three-mirror contact lens (*right*)

drainage device with a drainage tube in the anterior chamber. In this patient, the drainage tube is too long and wrongly positioned, and the tube can touch the endothelium of the cornea. Figure 11.4c is a vertical scan obtained in a patient who suffers from band keratopathy.

Besides the cornea, the anterior chamber angle is important to scan. Figure 11.4d shows an image of a patient with a normal anterior chamber angle. This image is obtained with the SL SCAN-1 in anterior mode and shows an open anterior chamber angle, but without details of the structures composing the angle itself.

Scanning with the SL SCAN-1 through the gonio-mirror of a three-mirror contact lens provides images which visualize the entire structure of the angle. With the slit lamp magnification at 40×, iris processes and the triangular trabecular meshwork become visible (Fig. 11.4e, f). Figure 11.4f is obtained in a patient with pigment dispersion. Due to the accumulation of pigment in the trabecular meshwork, this triangular trabecular structure demonstrated a higher reflectivity on the obtained scan.

11.8.4 Thickness Measurements of the Posterior Segment with the Fundus Viewer and FastMap (Posterior Mode)

Figure 11.5 shows two thickness maps made with the SL SCAN-1, six radial scans of 6 mm, through the fundus viewer, a lens holder with a 60D Volk lens. The lower thickness map is from the same patient shown in Fig. 11.2c. A photograph of the central retina made simultaneously with the set of six radial scans shows the central scar. It was not possible in all patients to make scans with the foveal dip in the center. The upper thickness map is made in a healthy person. To make the six radial

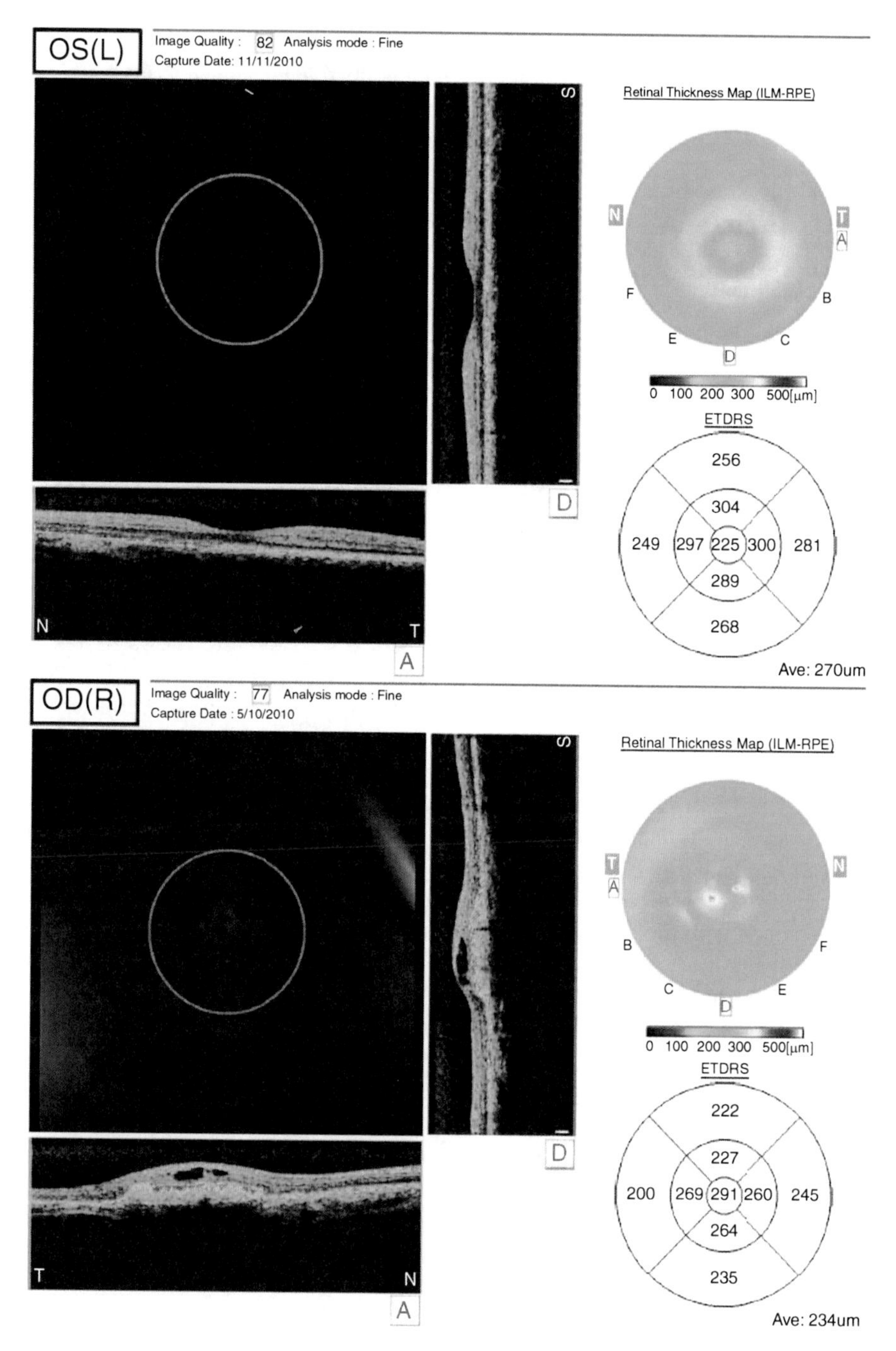

Fig. 11.5 Topographic map reconstructed from six radial scans, with retinal thickness of the nine ETDRS regions

scans, the slit lamp illumination was turned to its smallest dimension and used as a surrogate internal fixation point. With this technique, reliable sets of scans could be made with the foveal dip in the center of the b-scans.

11.9 Discussion

The first results of images proved that the SL SCAN-1 is able to produce OCT images of sufficient quality and to correctly identify common retinal pathologies. Scans can also be made of the anterior segment which is a unique combination of the slit lamp with integrated FD-OCT. The add-on OCT did not interfere with the functionality of the slit lamp during examination of the retina or the anterior chamber neither through a handheld lens or through a three-mirror contact lens. Essentially, OCT scans could be obtained of all structures which were observed during slit lamp biomicroscopy through any lens.

The SL SCAN-1 incorporates both the higher resolution and the faster scanning speed of a Fourier domain OCT. The fast scanning speed of 5,000 scans per second, with 512 a-scans per b-scan, allows a full b-scan to be made in approximately 0.1 s. In the posterior mode, the automatic z-alignment adjusts the reference arm length and corrects for small movements in the sample. These features make the OCT system fast enough to make a full and high-quality scan, even in an eye observed through a handheld lens introducing involuntary movements due to both the handheld lens and the observed retina. The high resolution of 8–9 μm in tissue is sufficient to show all relevant layers of the retina in detail, as shown in Fig. 11.2.

Patients with exudative AMD treated with anti-VEGF intraocular injections need regular examinations with OCT scans during follow-up to detect the presence or absence of signs of still active leakage [7, 17]. Over 50 patients with exudative AMD were scanned with the SL SCAN-1 and analyzed for the presence of leakage. The conclusions in these patients based on the slit lamp-integrated OCT were identical to the conclusions based on a stand-alone FD system (the 3D-OCT-1000, Mark II, Topcon). To screen for local signs of leakage in patients with AMD, a number of successive scans can be made with the SL SCAN-1 to cover the whole macular area.

With a handheld lens, lesions in the far periphery could be scanned with the SL SCAN-1, which could not be scanned with commercial OCT devices. Although the quality of the scans diminishes with the distance from the fovea. This opens new regions of the eye to be imaged with a high resolution and could be of value in patients with peripheral retinal lesions.

The scans of the anterior segment with the SL SCAN in anterior mode are not completely comparable to the scans made with commercially available systems. On the one hand, there is a difference in tissue penetration caused by the central wavelength of the SLD light sources. The SL SCAN-1 has a central wavelength of 830 nm versus 1310 nm in the commercially available systems. A longer wavelength allows for a slightly better penetration of the highly scattering structures in the anterior segment. On the other hand, the SL SCAN-1 is a FD-OCT in contrast to most of the commercially available systems, which are time domain (TD) with exception of the Casia SS-1000 (Tomey Corporation) [12, 18, 19]. The FD-OCT has a faster scanning speed and an increased resolution, but a reduced scan depth (2 mm in the SL SCAN-1 versus > 5 mm in TD-OCT). As a consequence of the limited depth range, it is not possible to show in a single scan the complete anterior segment without the symmetric overlapping image artifact, referred to in the literature as the complex conjugate or mirror image artifact [20, 21]. Nevertheless, the images of the anterior segment made with the SL SCAN-1 seemed to be of acceptable quality (Fig. 11.4).

In several patients, it was demonstrated that with the SL SCAN-1, OCT scans of the peripheral retina and the anterior chamber angle through a three-mirror contact lens can be made. To examine lesions in the peripheral retinal, the ophthalmologist can use indirect funduscopy, biomicroscopy with a contact lens, and/or photography (with or without angiography), and ultrasonography (UBM), or MRI to come to a final diagnosis. None of these examinations provide detailed cross-sectional images of the peripheral retina. The OCT scans of the peripheral retina made with the SL SCAN-1 (through a three-mirror contact lens) do show retinal structures with a resolution of 15–20 μm. These detailed images of the peripheral retina will provide additional information in retinal diseases like peripheral degenerations, retinoschisis, Coats' disease, peripheral angiomatous proliferations, and peripheral neovascularizations.

OCT scans of a peripheral retina would also add in the diagnosis and follow-up of peripheral retinal defects. The presence of vitreous traction on a peripheral retinal defect is of importance in the choice of treatment and timing of follow-up. It will help to answer the question whether a laser treatment for a peripheral defect is effective and sufficient, looking for the presence or absence of (progressive) leakage beyond the laser coagulates. In peripheral tumors, the presence of leakage on top of the tumor is important for the differential diagnostics.

Examinations of the anterior chamber angle can be made with gonioscopy, UBM, adjusted posterior segment OCT devices, and dedicated anterior segment OCT devices. The best resolution is obtained with the dedicated anterior segment OCT devices. However, the anterior chamber angle structures are for a large part obscured by the high scattering properties of the limbus, the peripheral sclerocorneal transition. With the SL SCAN-1, OCT scans of the anterior chamber angle can be obtained through the goniolens of a three-mirror contact lens. This could provide images of peripheral anterior synechiae, the structural abnormalities in goniodysgenesis (Rieger anomaly, Peters anomaly, congenital glaucoma), and tumors in the angle, like extensions of iris, or ciliary body melanomas. Figure 11.3f shows an increased reflectivity of the triangular trabecular structure, due to accumulation of pigment in this meshwork in a patient with pigment dispersion. Visualizing the changes in the trabecular meshwork could give more information about the pathophysiology in pigment dispersion, pseudoexfoliation, and perhaps even show trabecular changes due to corticosteroid in so-called high responders [22, 23].

In combination with add-on lens of the fundus viewer system, the quality and stability of scans made with the SL SCAN improve, and pattern scans of the central macular region can be made with sufficient precision to construct topographic maps of the macular retinal thickness. These topographic maps are accurate and reproducible, allowing for reliable follow-up measurements. The fundus viewer system is also designed to reduce the light reflexes during fundus biomicroscopy. Photographs with a field of view of 30°, covering the macula and the optic nerve head, can now be made with a digital slit lamp camera.

The SL SCAN-1 is a unique FD-OCT system, able to scan the posterior pole and anterior segment, but also the anterior chamber angle and the more peripheral retina. These four modalities combined into one device make the SL SCAN-1 a very powerful aid in daily practice. Reliable topographic maps of the macula can be made with the add-on fundus viewer lens, increasing the clinical potential of the system considerably. The possibility to make photographs of all lesions observed and scanned is another important asset of the system. All OCT images and photographs can be made during the normal standard examinations with the slit lamp, including observations through a three-mirror contact lens.

OCT has become increasingly important in ophthalmology for both diagnosis and follow-up of patients. With the slit lamp with integrated OCT, one can make OCT images during the normal standard examination with the slit lamp, without interfering with the functionality of the slit lamp. This increases the efficiency of the routine clinical examination of a patient, will increase the comfort of the patient seated behind just one device, and will save time.

In many clinics, OCT examinations are performed by technicians, not directly familiar with the pathology at hand. This could lead to inadequate scans missing the relevant pathology. With the slit lamp with integrated OCT, the ophthalmologist himself will make the OCT scans, directing the scan to the area of interest, observed with slit lamp biomicroscopy, avoiding this problem.

It can be concluded that the SL SCAN, an FD-OCT system integrated into a slit lamp, can make excellent images of both the anterior and posterior segment. The build in FD-OCT did not interfere with the standard examinations performed with the slit lamp, including the examination with a three-mirror contact lens. With the possibility to obtain OCT images directly during routine examination with the slit lamp, an improvement has been realized in the efficiency of patient examination. With the SL SCAN, also the peripheral retina and the anterior chamber angle can be scanned.

References

1. W. Drexler, J.G. Fujimoto, State-of-the-art retinal optical coherence tomography. Prog. Retin. Eye Res. **27**(1), 45–88 (2008)
2. M.E. van Velthoven, D.J. Faber, F.D. Verbraak, T.G. van Leeuwen, M.D. de Smet, Recent developments in optical coherence tomography for imaging the retina. Prog. Retin. Eye Res. **26**(1), 57–77 (2007)
3. D. Huang, E.A. Swanson, C.P. Lin, J.S. Schuman, W.G. Stinson, W. Chang, M.R. Hee, T. Flotte, K. Gregory, C.A. Puliafito, Optical coherence tomography. Science **254**(5035), 1178–1181 (1991)

4. D.J. Browning, A.R. Glassman, L.P. Aiello, N.M. Bressler, S.B. Bressler, R.P. Danis, M.D. Davis, F.L. Ferris, S.S. Huang, P.K. Kaiser, C. Kollman, S. Sadda, I.U. Scott, H. Qin, Diabetic Retinopathy Clinical Research Network, Optical coherence tomography measurements and analysis methods in optical coherence tomography studies of diabetic macular edema. Ophthalmology **115**(8), 1366.e1–1371.e1 (2008)
5. D.L. Budenz, M.J. Fredette, W.J. Feuer, D.R. Anderson, Reproducibility of peripapillary retinal nerve fiber thickness measurements with stratus OCT in glaucomatous eyes. Ophthalmology **115**(4), 661.e4–666.e4 (2008)
6. L.K. Chang, H.F. Fine, R.F. Spaide, H. Koizumi, H.E. Grossniklaus, Ultrastructural correlation of spectral-domain optical coherence tomographic findings in vitreomacular traction syndrome. Am. J. Ophthalmol. **146**(1), 121–127 (2008)
7. M. Fleckenstein, P. Charbel Issa, H.M. Helb, S. Schmitz-Valckenberg, R.P. Finger, H.P. Scholl, K.U. Loeffler, F.G. Holz, High-resolution spectral domain-OCT imaging in geographic atrophy associated with age-related macular degeneration. Invest. Ophthalmol. Vis. Sci. **49**(9), 4137–4144 (2008)
8. I. Krebs, S. Ansari-Shahrezaei, A. Goll, S. Binder, Activity of neovascular lesions treated with bevacizumab: comparison between optical coherence tomography and fluorescein angiography. Graefes Arch. Clin. Exp. Ophthalmol. **246**(6), 811–815 (2008)
9. S.Y. Lee, S.G. Joe, J.G. Kim, H. Chung, Y.H. Yoon, Optical coherence tomography evaluation of detached macula from rhegmatogenous retinal detachment and central serous chorioretinopathy. Am. J. Ophthalmol. **145**(6), 1071–1076 (2008)
10. O. Tan, G. Li, A.T. Lu, R. Varma, D. Huang, Advanced Imaging for Glaucoma Study Group, Mapping of macular substructures with optical coherence tomography for glaucoma diagnosis. Ophthalmology **115**(6), 949–956 (2008)
11. R. Chang, D.L. Budenz, New developments in optical coherence tomography for glaucoma. Curr. Opin. Ophthalmol. **19**(2), 127–135 (2008)
12. S. Dorairaj, J.M. Liebmann, R. Ritch, Quantitative evaluation of anterior segment parameters in the era of imaging. Trans. Am. Ophthalmol. Soc. **105**, 99–108 (2007)
13. A. Konstantopoulos, J. Kuo, D. Anderson, P. Hossain, Assessment of the use of anterior segment optical coherence tomography in microbial keratitis. Am. J. Ophthalmol. **146**(4), 534–542 (2008)
14. L.M. Sakata, R. Lavanya, D.S. Friedman, H.T. Aung, H. Gao, R.S. Kumar, P.J. Foster, T. Aung, Comparison of gonioscopy and anterior segment ocular coherence tomography in detecting angle closure in different quadrants of the anterior chamber angle. Ophthalmology **115**(5), 769–774 (2008)
15. M. Stehouwer, F.D. Verbraak, H. de Vries, P.H. Kok, T.G. van Leeuwen, Fourier Domain Optical Coherence Tomography integrated into a slit lamp; a novel technique combining anterior and posterior segment. Eye (Lond.) **24**(6), 980–984 (2010)
16. M. Stehouwer, F.D. Verbraak, H. de Vries, P.H. Kok, T.G. van Leeuwen, Scanning beyond the limits of standard OCT with a Fourier domain optical coherence tomography integrated into a slit lamp: the SL SCAN-1. Eye (Lond.) **25**(1), 97–104 (2011)
17. H. Dadgostar, N. Waheed, The evolving role of vascular endothelial growth factor inhibitors in the treatment of neovascular age-related macular degeneration. Eye (Lond.) **22**(6), 761–767 (2008)
18. C.K. Leung, H. Li, R.N. Weinreb, J. Liu, C.Y. Cheung, R.Y. Lai, C.P. Pang, D.S. Lam, Anterior chamber angle measurement with anterior segment optical coherence tomography: a comparison between slit lamp OCT and Visante OCT. Invest. Ophthalmol. Vis. Sci. **49**(8), 3469–3474 (2008)
19. M. Gora, K. Karnowski, M. Szkulmowski, B.J. Kaluzny, R. Huber, A. Kowalczyk, M. Wojtkowski, Ultra high-speed swept source OCT imaging of the anterior segment of human eye at 200 kHz with adjustable imaging range. Opt. Express. **17**(17), 14880–14894 (2009)
20. M.V. Sarunic, S. Asrani, J.A. Izatt, Imaging the ocular anterior segment with real-time, full-range Fourier-domain optical coherence tomography. Arch. Ophthalmol. **126**(4), 537–542 (2008)

21. A.B. Vakhtin, K.A. Peterson, D.J. Kane, Demonstration of complex-conjugate-resolved harmonic Fourier-domain optical coherence tomography imaging of biological samples. Appl. Opt. **46**(18), 3870–3877 (2007)
22. N. Niyadurupola & D.C. Broadway, Pigment dispersion syndrome and pigmentary glaucoma—a major review. Clin. Exp. Ophthalmol. **36**(9), 868–882 (2008)
23. O.Y. Tektas, E. Lütjen-Drecoll, Structural changes of the trabecular meshwork in different kinds of glaucoma. Exp. Eye. Res. **88**(4), 769–775 (2009)

Index

Achromatizing lenses, 227
Acute retinal ischemia, 28
Adaptive optics (AO), 102, 119, 210
Age-related macular degeneration (AMD), 54, 65, 87, 157, 163, 187
Albumin, 159
Angle blockade, 132
Angle-opening distance, 133
Angle recess area, 134
Annular light reflex, 30, 42
Anterior chamber angle, 133
Anterior segment, 186
Anti-VEGF, 58, 62
Apoptosis, 71
AREDS, 66
Arterioles, 42
Artifact, 243
A-scan, 159, 165
Astigmatism, 225
Atrophy, 64
Autofluorescence imaging, 189
Automatic classification, 168, 170
Axial resolution, 145, 213, 214

Backscattering, 111
Bayesian, 170
Bessel function, 214
Best corrected VA (BCVA), 68
Biomicroscopy, 55, 247, 248
Birefringence, 110, 179, 194
Birefringent axis orientation, 176, 178
Blood–aqueous barrier, 158
Blood–brain barrier, 158
Blood–retinal barrier (BRB), 2, 158, 159, 161
  breakdown of, 3
Blood–retinal barrier permeability, 162
Borders of GA, 94
Border zone of GA, 93
Branch retinal artery occlusion (BRAO), 33, 62
Branch retinal vein occlusion (BRVO), 59, 61
Broadband spectra, 210
Bruch's membrane, 65, 187

Calcification, 189
Canaloplasty, 136
Central retinal artery (CRA), 24
Central retinal artery occlusion (CRAO), 29, 62
Central retinal vein occlusion, 54, 59
Central retinal vein occlusion (CRVO), 37, 40, 59
Central serous chorioretinopathy (CSC), 71, 81
Cherry-red spot, 29, 37
Choriocapillary network, 159
Choroid, 24
Choroidal effusion, 133
Choroidal ischemia, 43
Choroidal neovascularization (CNV), 66, 68, 187
Choroidal nevi, 179
Chromatic aberration, 212, 214, 216, 218, 225–228, 230
Ciliary artery, 33
Cilioretinal artery, 32
Cilioretinal artery occlusion, 33
Circumpapillary RNFL, 111
Classification, 171
Clinical evaluation of macular edema, 4
Clinically significant macular edema (CSME), 4, 54
Closed macula hole (CMH), 77
CNV. *See* Choroidal neovascularization (CNV)

R. Bernardes and J. Cunha-Vaz (eds.), *Optical Coherence Tomography*,
Biological and Medical Physics, Biomedical Engineering,
DOI 10.1007/978-3-642-27410-7, © Springer-Verlag Berlin Heidelberg 2012

Coherence length, 143
Coherence light, 209
Coherent Fluorotron Master, 160
Collateral circulations, 24
Color fundus photography, 158, 197, 239
Coma, 211
Complex conjugate, 243, 247
Cones, 54
Confocal microscopy, 137, 213
Confocal plane, 161
Confocal scanning laser ophthalmoscopy (CSLO), 94, 110, 160, 161
Conjunctiva, 186
Contrast, 215, 216
  functions, 216
  perception, 129
Contrast sensitivity function (CSF), 212
Cornea, 127, 179, 186
Corneal curvature, 129
Corneal grafts, 127
Corneal pachymetry, 127
Corneal pathology, 237
Cotton wool spots, 35
Cross-interference, 145
Cross-sectional eye fundus imaging, 141
Cross-sectional images, 6
C-scan, 210
CSLO. *See* Confocal scanning laser ophthalmoscopy (CSLO)
Cup-disc area ratio, 112
Cup volume, 113
Cystic retinal edema, 76
Cystoid macular edema (CME), 4, 71, 163, 170
Cystoid spaces, 11

Decision trees, 170
Deconvolution, 161
Deformable mirror, 212, 219, 222, 225, 230
Degree of polarization uniformity (DOPU), 179
Depolarization, 176, 179
Descemet stripping automated endothelial keratoplasty (DSAEK), 128
Detector, 160
Diabetes, 1
Diabetic ischemic maculopathy, 41
Diabetic macular edema (DME), 54, 157, 162, 171, 196
  incidence and prevalence of, 3
Diabetic maculopathy, 196
Diabetic retinopathy (DR), 157, 159, 161, 163, 170, 237
Diabetic retinopathy progression, 2
Diagnosis, 237
Diagnosis and classification, 4
Diattenuation, 176, 179
Diffraction, 211
Diffuse macular edema, 4
3D imaging, 112
Disc area, 113
Disrupted blood-retinal barrier, 165
Drusen, 87, 88, 183, 187
Drusen volume, 91
DSAEK. *See* Descemet stripping automated endothelial keratoplasty (DSAEK)

Early treatment diabetic retinopathy study group (ETDRS), 4, 54, 240
Effective resolution, 216, 217
Electroretinogram (ERG), 71
Emmetropia, 211
Endothelial cells, 158
Endothelial graft, 127
En face, 115
Enhanced depth imaging (EDI) SD-OCT, 44
Epiretinal membranes (ERMs), 75, 77, 81
ETDRS. *See* Early treatment diabetic retinopathy study group (ETDRS)
External fovea, 27
External limiting membrane, 7, 51
Extracellular edema, 2
Eye fundus, 139
Eye tracking, 115

FAF. *See* Fundus autofluorescence (FAF) imaging
False color-coded map, 8
Features, 170
Fibrosis, 191, 192
Fibrous tissues, 179
Fiducial markers, 165
Fisher linear discriminant, 170
Fixation point, 246
Flood illumination cameras, 219
Fluorescein angiography (FA), 4, 24, 55, 158–160, 162, 163
Fluorescein concentration, 159
Fluorescein leakage, 161
Fluorescence, 161, 162
Fluorescence light, 160
Focal macular edema, 4
Fourier domain, 148, 221, 237
Fourier-domain (FD)-OCT, 112
Foveal avascular zone (FAZ), 24, 41
Foveoschisis, 80

Frequency domain, 221
Full-field, 150
Full-thickness macular hole (FTMH), 77
Full width at half maximum (FWHM), 161, 210
Functional information, 152
Fundus autofluorescence (FAF) imaging, 96, 99
  patterns, 98
Fundus photography, 189
Fundus references, 165
Funduscopy, 241, 247
FWHM. *See* Full width at half maximum (FWHM)

Gamma function, 169
Ganglion cell layer (GCL), 7, 24, 43
Ganglion cells, 110, 195
Ganglion cells complex (GCC), 115
Gaussian function, 169, 215
Geographic atrophy (GA), 66, 68, 87, 93, 183, 187, 189
  lateral spread of, 101
Glaucoma, 110, 186, 193
Glaucoma progression, 112
Goniopuncture, 136
Gonioscopy, 132, 247
Grid laser, 198

Hard exudates, 14, 196
Hartmann-Shack, 228
Heidelberg Retina Angiograph, 161
Henle's fiber layer, 179
High-penetration, 153
Human retina, 139
Hyperpigmentation, 88
Hyperreflective foci, 196
Hyperreflectivity, 10
Hypertension, 43
Hyporeflectivity, 10
Hypoxia, 64

Image, 115
Image coregistration, 161, 165
Imaging in glaucoma, 110
Infarction, 28
Infraclinical foveolar detachments, 11
INL. *See* Inner nuclear layer (INL)
Inner blood–retinal barrier, 158, 167
Inner limiting membrane (ILM), 163, 167
Inner nuclear layer (INL), 7, 24, 43
Inner plexiform layer, 115
Inner retinal layer (ganglion cell complex), 118
Inner segment, 51
Intact, 165
Interference, 126
Interferometer, 219
Interferometric, 209
Internal limiting membrane, 6
Inverse problem (deconvolution), 162
Iridocorneal angle, 131, 132
Iris, 179
Ischemia, 27, 37, 58, 64, 159

Jones matrix, 176

Keratoconus, 186

Lamellar keratoplasty, 127
Laser, 160
Laser photocoagulation, 162
LASIK, 127
Leakage, 58, 162, 246, 247
Leave-one-out, 171
Length of coherence, 145
Linear space, 169, 170
Logarithmic space, 169, 170
Lorentzian, 161
Low-coherence interferometry, 52, 141

Müller, 176
Macular edema (MO), 38, 61
Macular hole (MH), 54, 75, 81
Magnetic resonance imaging (MRI), 239
Melanin, 179, 186, 189
Melanoma, 181
Metamorphopsia, 71, 75
Microexudates, 197
Microfolds, 130
Microperimetric, 102
Micropsia, 71
Microtubuli, 194
Middle limiting membrane (MLM), 28
Monochromatic aberrations, 216, 218, 219, 227, 229
Monochromatic ocular aberrations, 216
Morbus Best, 199
Morbus Stargardt, 199
Multiple sclerosis, 199
Mydriasis, 241
Myopic foveoschisis, 80, 81

Near-infrared autofluorescence image, 189
Necrosis, 64
Nerve fiber layer, 52
Neural networks, 170
Neuropathy, 194
Noncystoid, 4
Nonperfusion area, 43
Numerical aperture, 210
Numerical map, 8
Nyctalopia, 71

Ocular aberrations, 210, 211, 222, 228
Ocular ischemic syndrome (OIS), 37
Ocular muscles, 179
Open-loop, 222
Ophthalmic artery occlusion, 33
Ophthalmoscope, 161
Optic axis orientation, 179
Optic disc, 52, 110
Optic nerve, 110
Optic nerve head (ONH), 111, 112
Optic nerve head scanning, 116
Optical aberration, 211, 240
Optical biopsy, 51
Optical coherence tomography, 4, 5, 139, 163, 165, 167–172
  instrumentation, 5
Optical properties, 163
Optical spectrum, 52
Outer blood–retinal barrier, 158, 159
Outer nuclear layer, 7, 198
Outer retinal thickening, 14
Outer segment, 51

Pancorrection, 226, 229–231
Pan-corrective AO, 227
Pathogenesis, 231
Pathophysiology of retinal edema, 2
Patterns of progression, 162
Peak-valley parameter, 228
Perfusion, 159
Peripapillary RNFL, 111, 113, 116
Peripheral tumors, 247
Phakic intraocular lens (IOL), 131
  implantation, 132
Phase wrapping, 222, 223
Phenotyping, 231
Photocoagulation, 198
Photodynamic therapy (PDT), 68
Photoreceptor atrophy, 75
Photoreceptor's outer segments, 7
Pigmentary alterations, 188
Pigment epithelial detachments, 192
Pinhole, 160, 161, 213
Plaques, 197
Plexiform layers, 7
Point spread function (PSF), 161, 213
Polarization scrambling, 179, 186, 188, 196, 198
Polarization-sensitive OCT, 100, 119, 149, 225
Polarization state, 176
Polypoidal choroidal vasculopathy (PCV), 68
Posterior hyaloidal traction, 15
Posterior segment, 54
Preeclampsia, 43
Progression of atrophy, 94
Prominent middle limiting membrane (MLM), 28, 30
Protein, 159
Pupil size, 118, 215

Quantification of atrophy, 100
Quantum OCT (Q-OCT), 151

Radial basis function (RBF), 170
Refractive index, 211
Resolution, 161, 216, 238, 246, 247
Retardation, 178, 194
Reticular pseudodrusen, 91
Retina, 186
Retinal artery occlusion, 62
Retinal detachment (RD), 54, 71, 75, 81
Retinal leakage analyzer (RLA), 161–165
Retinal nerve fiber layer (RNFL), 7, 110, 179, 194
  thickness, 113, 117
Retinal pigment epithelium (RPE), 51, 113, 115, 158, 163, 179, 182, 186, 188, 189, 198
  migration, 88
Retinal thickness, 8, 53
Retinal thickness measurements, 9
Retinitis pigmentosa (RP), 71, 81, 199
Rhegmatogenous retinal detachment (RRD), 75
Rim area, 113
RNFL. *See* Retinal nerve fiber layer (RNFL)
Rods, 54
RPE. *See* Retinal pigment epithelium (RPE)
RPE/choriocapillaris complex, 7

Saccades, 161
Scanning laser ophthalmoscopy, 194, 219

Scanning laser polarimetry (SLP), 110, 194
Scanning speed, 246
Scar tissue, 179, 193
Scattered light, 212
Scattering, 211
Sclera, 179, 186
Sclerectomy, 136
Scotoma, 75
SD-OCT. *See* Spectral-domain OCT
Segmentation, 111, 182, 191
Serous retinal detachment, 11
Shadowing effects, 10
Shannon information, 152
Signal-to-noise, 219, 221
Signal-to-noise ratio, 229
Silicone oil tamponade, 80
Simultaneous, 94
Simultaneous fundus autofluorescence, 98
Single-mode fibers, 213
Slit lamp, 240, 241, 246, 248
Slit lamp biomicroscopy, 237, 239
Sodium fluorescein, 159, 161
Soft drusen, 68
Speckle, 230
Spectral-domain OCT, 5, 52, 94, 110, 118, 147
Speed, 238
Spherical aberration, 211
Sponge-like appearance, 11
Starling's law, 3
Statistics, 172
Stereophotography, 55, 110, 112
Stereoscopic fundus photographs, 4
Stokes vector, 176, 178, 182
Strehl ratio, 221, 222, 228
Stretched exponential distribution, 169
Structure–function correlation, 102
Subcellular resolution, 226, 231
Subretinal fluid, 58, 191
Subretinal fluid-3D, 192
Subretinal hemorrhage, 190
Sum of the squared differences (SSD), 167
Superficial capillary plexus, 24
Superluminescent diode, 238
Supervised classification, 170
Support vector machines (SVM), 170, 171
Swept laser source (SS)-SD-OCT, 119
Swept-source, 148

TD-OCT, 52, 118
Tendons, 179
Thickness, 194
Thickness map, 183, 185, 190
Three-mirror contact lens, 239
Tight junctions, 158
Time domain, 5, 52, 146, 219, 247
Trabecular meshwork, 179, 186
Trabecular-iris space area, 134
Trabeculectomy, 136
Trabeculoplasty, 137
Transverse resolution, 212, 217

Ultrahigh-resolution OCT, 210
Ultrahigh-speed, 152
Ultrasonographic biomicroscopy (UBM), 239
Ultrasonography, 247
Uvea, 186

Vascular endothelial growth factor (VEGF), 58
Vascular network, 165, 167
Visual acuity (VA), 212
Vitreomacular interface, 15
Voxel, 160

Z-alignment, 239, 246
Zernike polynomials Z_i, 214
Z-tracking, 240

Printed by Publishers' Graphics LLC